The human biology of
circumpolar populations

THE INTERNATIONAL BIOLOGICAL PROGRAMME

The International Biological Programme was established by the International Council of Scientific Unions in 1964 as a counterpart of the International Geophysical Year. The subject of the IBP was defined as 'The Biological Basis of Productivity and Human Welfare', and the reason for its establishment was recognition that the rapidly increasing human population called for a better understanding of the environment as a basis for the rational management of natural resources. This could be achieved only on the basis of scientific knowledge, which in many fields of biology and in many parts of the world was felt to be inadequate. At the same time it was recognized that human activities were creating rapid and comprehensive changes in the environment. Thus, in terms of human welfare, the reason for the IBP lay in its promotion of basic knowledge relevant to the needs of man.

The IBP provided the first occasion on which biologists throughout the world were challenged to work together for a common cause. It involved an integrated and concerted examination of a wide range of problems. The Programme was co-ordinated through a series of seven sections representing the major subject areas of research. Four of these sections were concerned with the study of biological productivity on land, in freshwater, and in the seas, together with the processes of photosynthesis and nitrogen-fixation. Three sections were concerned with adaptability of human populations, conservation of ecosystems and the use of biological resources.

After a decade of work, the Programme terminated in June 1974 and this series of volumes brings together, in the form of syntheses, the results of national and international activities.

INTERNATIONAL BIOLOGICAL PROGRAMME 21

The human biology of circumpolar populations

EDITED BY

F. A. Milan

Professor of Anthropology and Human Ecology
Institute of Arctic Biology
University of Alaska
Fairbanks, Alaska

CAMBRIDGE UNIVERSITY PRESS
CAMBRIDGE
LONDON NEW YORK NEW ROCHELLE
MELBOURNE SYDNEY

CAMBRIDGE UNIVERSITY PRESS
Cambridge, New York, Melbourne, Madrid, Cape Town, Singapore, São Paulo, Delhi

Cambridge University Press
The Edinburgh Building, Cambridge CB2 8RU, UK

Published in the United States of America by Cambridge University Press, New York

www.cambridge.org
Information on this title: www.cambridge.org/9780521112666

First published 1980
This digitally printed version 2009

A catalogue record for this publication is available from the British Library

Library of Congress Cataloguing in Publication data
Main entry under title:

The Human biology of circumpolar populations.

(International Biological Programme; 21)
Includes index.

1. Anthropometry–Arctic regions. 2. Man–Influence
of environment–Arctic regions. 3. Human physiology.
4. Adaptation (Physiology) 5. Human biology.
I. Milan, Frederick A.   II. Series.
GN58.A7H85   573'.6   79–322

ISBN 978-0-521-22213-6 hardback
ISBN 978-0-521-11266-6 paperback

This volume is dedicated
to the memory of
J. 'Sandy' Hart, PhD (1916–1973)

Chairman, Canadian Sub-Committee
Human Adaptability, IBP
and
original Editor

Contents

Table des Matières

Содержание

Contenido

List of contributors

F. Auger, Department of Anthropology, University of Montreal, CP 6128, Montreal H3C3J7 PQ, Canada

J. Balslev-Jørgensen, Laboratory of Anthropology, University of Copenhagen, Copenhagen 2100, Denmark

A. A. Dahlberg, Zoller Memorial Dental Clinic and Department of Anthropology, University of Chicago, USA

H. H. Draper, Nutrition Department, University of Guelph, Guelph, Ontario, Canada N1G 2W1

A. W. Eriksson, Free University of Amsterdam, Institute of Human Genetics, Amsterdam, The Netherlands

Harriet Forsius, Department of Paediatrics, University of Oulu, Kajaanintie 50, 90220 Oulu 22, Finland

Henrik Forsius, Department of Ophthalmology, University of Oulu, Kajaanintie 50, 90220 Oulu 22, Finland

D. R. Hughes, Department of Anthropology, University of Toronto, Canada

S. Itoh, Emeritus Professor of the Hokkaido University, 6–31, Nishikori 2-chome, Otsu 520, Japan

P. L. Jamison, Department of Anthropology, Indiana University, Bloomington, Indiana, USA

W. Lehmann, Institut für Humangenetik der Universität Kiel, 330 Kiel, W. Germany

T. Levin, Department of Anatomy, University of Gothenburg, Medicinaregatan 3, 400 33 Göteborg 33, Sweden

F. A. Milan, Institute of Arctic Biology, University of Alaska, Fairbanks, Alaska, USA 99701

J. F. de Peña, Department of Anthropology, University of Manitoba, Winnipeg, MB R3T 2N2, Canada

Yu. G. Rychkov, Department of Anthropology, Leninskie Gory, MSU, Moscow 117 234, Russia

List of contributors

R. J. Shephard, Department of Environmental Health, School of Hygiene, University of Toronto, Canada

V. A. Sheremet'eva, Department of Anthropology, Leninskie Gory, MSU, Moscow 117234, Russia

N. E. Simpson, Department of Paediatrics, Queen's University, Kingston, Ontario, Canada

J. Skrobak-Kaczynski, Laboratorium for Miljøfysiologi, Oslo, Norway

Foreword

Admirably edited by Dr Fred Milan this is the fifth of the six volumes in the International Biological Programme synthesis series dealing with the investigations carried out in the Human Adaptability Section. Like that edited by P. T. Baker (*The Biology of the High-Altitude Peoples*, 1978, IBP vol. 14) this volume provides a conspectus of the whole range of peoples inhabiting a distinctive and demanding habitat – the circumpolar zone. Complementary to these two ecosystem syntheses is the collection of studies brought together and edited by G. A. Harrison (*Population Structure and Human Variation*, 1977, IBP vol. 11) which covers such diverse habitats as the Amazonian forest, the African savanna and the African tropics, Islands in Polynesia and Melanesia, and the ecological transitions of migration and urbanization.

In addition to the multidisciplinary studies of the communities described in these three ecosystem volumes, many other groups were the subject of more restricted investigations during IBP. Taking a comprehensive world view of all the populations sampled, two highly important parameters were selected for synthesis presentation in this series. These comprise *Worldwide Variation in Human Growth* (edited by Phyllis B. Eveleth and J. M. Tanner, 1976, IBP vol. 8), and *Human Physiological Work Capacity* (edited by R. J. Shephard, 1978, IBP vol. 15). A third volume *Work and Life in Hot Climates* (by J. S. Weiner) will centre on the human heat tolerance studies carried out within the IBP.

The present volume testifies again to the remarkable level of multinational co-operation achieved within the HA section, and particularly so in the circumpolar sector. Over the ten-year period, and in every phase of planning, field work, review meetings and writing-up, a high measure of close co-ordination was achieved by teams from Canada, USA, Denmark, Norway, Finland, France and the USSR.

The present synthesis, as a review of a wealth of both new and earlier data, ranks both as an indispensable source book and as a 'base-line' of the circumpolar peoples – Eskimos from Greenland to Siberia, Lapps and Ainu – at a period in their history when the traditional forms of life were beginning to give way to new life and work-styles influenced by modern technology of transport, housing, hygiene, and their attendant benefits and stresses.

Thus the material assembled here displays at the same time the biological properties characteristic of self-sufficient small isolates in close interaction with a demanding environment as well as those induced by the newer conditions of cultural and technological change. These two states, of tradition and transition, are closely analysed in respect of such biological attributes as genetic constitution, demographic and reproductive performance, nutritional status, growth trends, cold tolerance, work capacity, vision, dental development and bodily physique. A valuable chapter on behavioural patterns in the Arctic has been included.

It is not possible of course in this bald outline to do justice to the scope and depth and the uniformly high quality of the contributions or to mention in detail the many original observations brought to light by the HA workers. The successes achieved in this many sided project are well indicated in the summary chapter by the editor. Dr Milan and his contributors are to be congratulated on providing an outstanding work of biological synthesis. It stands as a worthy tribute to the late Sandy Hart, to whom the book is dedicated, and who did so much as theme co-ordinator to ensure the success of this major endeavour within the International Biological Programme.

J. S. WEINER

IBP Human Adaptability section publications

J. S. Weiner (1969).
A guide to human adaptability proposals
Blackwell Scientific Publications, Oxford. 88 pp.

J. S. Weiner & J. A. Lourie (1969).
A guide to field methods
Blackwell Scientific Publications, Oxford. 652 pp.

S. Biesheuvel (ed.) (1969).
Methods for the measurements of psychological performance
Blackwell Scientific Publications, Oxford. 110 pp.

P. B. Eveleth & J. M. Tanner (1976).
Worldwide variation in human growth
Cambridge University Press, London. 498 pp.

G. A. Harrison (ed.) (1977).
Population structure and human variation
Cambridge University Press, London. 342 pp.

P. Baker (ed.) (1978).
The biology of high-altitude peoples
Cambridge University Press, London.

R. J. Shephard (1978).
Human physiological work capacity
Cambridge University Press, London. 303 pp.

F. Milan (ed.) (1979).
The human biology of circumpolar peoples
Cambridge University Press, London.

J. S. Weiner.
Components of human physiological function: thermal responses and respiratory function
Cambridge University Press, London. (In preparation.)

A detailed guide to all the IBP projects contributing to the theme of this volume is given in *Human adaptability: a history and compendium of research within the IBP* (1977) by K. J. Collins & J. S. Weiner, published by Taylor & Francis Ltd, 10–14 Macklin Street, London WC2B 5NF. A collection of HA reports, reprints and archival material is held in the Library of the British Museum (Natural History), Cromwell Road, London SW7 5BD.

1. Introduction

D . R . H U G H E S and F . A . M I L A N

The purpose of this volume is to present an international synthesis of the research findings obtained from studies of circumpolar human populations carried out in one of the human adaptability projects of the International Biological Programme (IBP). A synthesis of the information about the demography, genetics, odontology, growth and development, nutrition, physiology and psychology of selected Arctic populations is presented in this volume in chapters prepared by the scientific workers who participated in the field research.

From 1967 through 1974, IBP scientists from the USSR, Japan, USA, Canada, Denmark, Norway, Finland, West Germany and France conducted research on indigenous human populations in the circumpolar zone. US scientists studied an Alaskan Eskimo population on the north coast of Alaska in the villages of Wainwright, Point Hope, Point Barrow and an inland population at Anaktuvuk Pass. Canadian investigators worked at Igloolik and Hall Beach in the central Arctic. Scandinavians worked jointly on the Greenlandic populations at Aupilagtoq and Kraulshavn in West Greenland and on the Lappish people at Svettijärvi and Nellim in Finland. The French continued research that had been started in the 1930s on the East Greenlandic population at Ammassalik, and Scoresbysund. Soviet scientists investigated a number of populations in the USSR, and Japanese scientists studied the Ainu on the islands of Hokkaido.

This human adaptability project focussed on populations living in an environment that is characterized by low temperatures, seasonal extremes in light and darkness and relatively meagre ecological resources. Despite those environmental constraints, the populations studied have occupied the Arctic for many generations. Their long-term occupancy in small kin-based groups has provided human biologists with a marvelous opportunity to study aspects of human adaptability in considerable detail. The objective of the studies presented in this volume was to elucidate the biological and behavioral processes responsible for the successful adaptation of aboriginal circumpolar populations to their environment and its resources.

Development of circular studies

Professor Weiner has described the evolution of the Human Adaptability (HA) component of the IBP in Volume 1 of the IBP series (Worthington,

1975). An important event for the development of the HA studies was the symposium held at Burg Wartenstein, Austria, in 1964. The consensus of that symposium appeared in a WHO Technical Report *Research in Population Genetics of Primitive Groups* (1964), and the papers presented at that symposium appeared in a volume edited by P. T. Baker & J. S. Weiner entitled *The Biology of Human Adaptability* (1966). Chapters in this volume written by W. S. Laughlin and J. A. Hildes summarized the state of knowledge available at that time about the genetics, anthropology, health and physiological adaptations in Arctic populations.

The first meeting of the US National Committee for the IBP occurred on 6–7 March 1965. The HA sub-committee then consisted of F. Sargent *chairman*, S. Robinson, D. B. Shimkin, J. V. Neel and W. S. Laughlin. The IBP Eskimo program developed shortly afterwards. The late J. S. Hart *chairman for HA in Canada* suggested to F. Sargent that the US and Canada jointly pursue cooperative inter-disciplinary studies on Arctic populations and he organized a meeting at the National Research Council in Ottawa on 25 November 1966 to discuss the matter. Present were W. S. Laughlin from the USA and D. R. Hughes, G. Beaton, J. S. Hart and J. A. Hildes from Canada. An Air Canada strike disrupted the plans of F. Sargent and F. A. Milan to attend.

At this meeting W. S. Laughlin outlined a study plan which essentially followed the outline presented in the WHO Technical Report, for a human biological investigation which he proposed for the community of Wainwright, Alaska. The committee decided that other Eskimo populations in the central Arctic at Igloolik and in West Greenland should also be studied to compare them with their Alaskan kinsmen. It was agreed that these studies could best be carried out by a number of integrated international teams of scientists for each discipline working across selected study sites in the Arctic.

A second Canadian meeting was held on 30 March–1 April 1967 at the University of Manitoba Medical School in Winnipeg. Present were F. Sargent, W. S. Laughlin, L. Irving and F. A. Milan from the USA and J. S. Hart and J. A. Hildes from Canada. It was decided then that a working party conference should be scheduled for November and should be convened by J. A. Hildes and chaired by F. A. Milan. Milan was appointed scientific coordinator of the joint US–Canadian project by the two HA Chairmen.

The IBP–HA sponsored Point Barrow Working Party Conference held at the Naval Arctic Research Laboratory in November 1967, which was attended by 41 scientists representing all of the northern countries except the USSR, provided the format and methods for the study of northern indigenous populations under the auspices of the IBP. This was previous to the publication of the IBP instructional manuals. Published papers by Milan (1968) and Laughlin (1970) have discussed those plans for the international

Eskimo study. A report of the Working Party Conference by Milan was prepared in mimeograph form and disseminated to attendees and other interested persons.

At that Conference it was decided that each national group would organize its own study team rather than use the method of joint international teams suggested at Winnipeg. Attendees agreed to the establishment of a committee to oversee an 'International Study of Eskimos'. The appointed project directors from each country were members of this committee: US, F. A. Milan; Canada, D. R. Hughes; Denmark, J. B. Jørgensen; and France, R. Gessain. The chairmanship, first held by F. A. Milan, was rotated among the other directors. The function of this committee was to coordinate and review scientific proposals, expedite the movement of scientists between countries and promote the exchange of data for cross-national comparisons. One of the more important aims of the IBP-sponsored studies of northern indigenous populations has always been the comparison of human biological data obtained by similar methods across national boundaries.

Before the Barrow Conference, the HA sub-committees from the four Scandinavian countries had met at Sandefjørd in Norway in December 1966 to decide upon a general course of action for collaborative activities in the IBP. Earlier, in the summer of 1966, H. Forsius, A. W. Eriksson and W. Lehmann, representing Finland and West Germany, had conducted WHO-sponsored biological research on Skolt Lapps at Svettijärvi in Finland. Under IBP sponsorship, these studies were continued the following year at Nellim in Finland and they became a major part of the Scandinavian HA projects. A second part of the HA program was the joint Scandinavian studies in West Greenland organized and led by J. B. Jørgensen of Denmark.

In May 1967, the first in a series of Nordic symposia on human biology and medicine was held at Uleåborg in Finland to discuss the results of the Lappish studies. Papers from this symposium were published in *Nordisk Medicin* (1968). Since that time, HA investigators conducting research on northern populations (Ainu, Lapps, Eskimos) have met annually to exchange data and ideas at HA-sponsored meetings in Copenhagen, Denmark (May 1967); Hurdal, Norway (June 1968); Kiel, Germany (June 1970); Oulu, Finland (June 1971); Husavik and Reykjavik, Iceland (June 1973); and Yellowknife, Yukon Territory, Canada (June 1974). Canadians, Danes and Americans met at the annual meeting of the American Association for the Advancement of Science in Boston in December 1969 to discuss preliminary results of the joint study in a symposium organized by D. Hughes.

At the Kiel meeting in 1970, J. Lange Andersen *Chairman HA for the Nordic countries* organized an 'International Steering Committee for Cross-National Comparisons of Circumpolar Populations'. Members of this committee were J. S. Hart *Chairman* (Canada); D. Hughes *secretary* (Canada); F. A. Milan (USA); T. Lewin (Sweden); J. B. Jørgensen (Denmark); K. L.

3

Andersen (Norway); A. Eriksson (Finland); R. Gessain (France); S. Itoh (Japan).

At the program review of the Canadian Igloolik Project, Trinity College, Toronto (12–14 March 1971), where the review committee of R. J. Harrison, Oxford; L. D. Carlson, University of California, Davis; F. A. Milan, Alaska; G. Beaton, Toronto; P. Larkin, University of British Columbia; W. H. Cook, Canadian Committee for the IBP; and J. S. Hart, Canadian Research Council, met with D. Hughes and all HA Canadian investigators, international data synthesis was again discussed. J. S. Hart then formulated an agenda for the meeting of the Steering Committee for the next meeting in Oulu, Finland.

It was decided that the Steering Committee should arrive at Oulu the following year prepared to discuss (1) objectives, (2) common elements in all studies with a check list of all measurements, (3) sub-groupings of investigators for international comparisons, (4) plans for the first international workshop and (5) national commitments for support.

The Oulu meeting was held on 20 June 1971. Members of the Steering Committee, who also were attending the Second International Symposium on Circumpolar Health, met at the Medical Technology Institute to discuss the agenda items formulated at Toronto. Present at this meeting were J. S. Hart, D. Hughes, S. Itoh, O. Wilson, R. Gessain and F. A. Milan. J. S. Hart was elected Editor of this volume. The plans for international data comparisons were discussed again at Husavik in June 1972 and at the Vth General Assembly of the IBP in Seattle in September 1972.

Unfortunately, and sadly, J. 'Sandy' Hart died on 6 May 1973 before he was able to conclude the task which he had assumed and pursued vigorously. At the Reykjavik Meeting in June 1973, it was decided by those present that this volume should be written and F. A. Milan was elected to assume the editorship.

IBP representatives from the Soviet Union were present at the meetings in Kiel, Oulu and Reykjavik, and Z. I. Barbashova *Chairperson for Human Adaptability Studies in the USSR*, attended the IBP Seattle Assembly and visited the Institute of Arctic Biology in Fairbanks in 1972.

This particular IBP program was characterized by a certain amount of exchange of scientific personnel between countries, for example, Henrik Forsius, a Finnish ophthalmologist, T. Lewin, a Swedish anatomist and S. Haraldson, a Swedish public health specialist, worked in Alaska as did P. Di Prampero, an Italian physiologist and Jens Brøsted, a Danish student. J. Krog and M. Wicka, physiologists from Norway, Harriet Forsius, Finland, B. Chiarelli, a geneticist from Italy, and W. Mather, US, worked at Igloolik. In addition, D. R. Hughes, Canada, worked at Inari in Finland and R. S. MacArthur, Canada, a psychologist, worked in Greenland.

Theoretical and Practical Justifications for Human Adaptability Research in Circumpolar Populations

At the initial planning meeting at Point Barrow in 1967, W. S. Laughlin made a very succinct statement justifying the study of Arctic populations. He was speaking mainly of Eskimos, but many of his remarks are equally applicable to the Lapps and the Ainu. His statement is quoted here:

'The world's Eskimos are a remarkably successful group of mankind that has demonstrated its ability to adapt to difficult physical circumstances and, furthermore, to expand around a large sector of the northern circumpolar world. If their close relatives the Aleuts of Alaska are included, then they occupy the longest linear distance of any single human population group. This expansion demonstrates their ability to do more than simply survive harsh conditions despite limited natural resources. Archaeological evidence suggests that this expansion began some ten thousand years ago, on the coasts of the now-submerged Bering land bridge with peoples considered ancestral to the Eskimo and Aleut of today, culminating in the occupation of parts of what is now Siberia, Alaska, Canada and Greenland. Circumpolar adaptation, particularly cultural adaptation, may therefore be shown to have considerable time depth by means of this archaeological documentation.

The time involved obviously allows a sufficient number of human generations for the evolutionary agencies of natural selection, genetic drift and hybridization to operate, in conjunction with such environmental stresses as climate, disease and nutrition, to bring about many of the characteristics associated with Eskimos of today.

No one trait or character can be singled out as having played the most important part in this evolutionary process, or to account for this notable evolutionary success. At both the individual and population level of biological organization, therefore, the data of several scientific disciplines must be integrated in any inquiry. This is the main justification for a multidisciplinary approach.

One consequence of the geographical distribution of Eskimos is their citizenship in four different countries viz., the USSR, the USA, Canada, and Denmark. The Lapps, too, are citizens of Norway, Sweden, Finland and the USSR. We shall see that when definitions of circumpolar peoples are considered both the Lapps and the Ainu can fall within the scope of these studies. These circumstances, then, justify the international aspects of the human adaptability inquiry. There are clearly some common environmental factors, particularly during the winter months, that affect these populations, and have affected them for many, many generations. Equally clearly, there are also many significant variations. What factor of human adaptability, then, may depend on these circumstances? Are there common explanations for any of these?' (Laughlin, 1967, 1970).

The human biology of circumpolar populations

The Arctic environment

As seen by an astronaut, planet earth is an oblate-shaped spheroid floating in deep space. The $23\frac{1}{2}°$ tilt of its axis in relation to its orbital plane and its annual march around the sun gives rise to the seasons. The sun, viewed from the earth, is directly overhead at the Tropic of Capricorn at latitude 23° 50′ S on the winter solstice and at the Tropic of Cancer at latitude 23° 50′ N on the summer solstice. The Arctic Circle, a cartographer's line encircling the earth at latitude 66° 30′ 03′ N, delimits a northern area from which the sun is not visible on the winter solstice and does not set on the summer solstice. The preciseness of the timing of these events is affected by the fact that the sun has an appreciable diameter, and that the sun's rays are refracted when passing through the atmosphere. Thus, the sun may appear on the horizon at the Arctic Circle after the date of the winter solstice (Haggett, 1972).

Precisely delimiting the Arctic environmentally is a difficult task (Ives & Barry, 1974). In a general way, the Arctic is bounded by the northern limit of the tree line, where, according to Köppen (1936), no summer month attains a mean temperature of 10 °C. The Arctic delimited this way includes the tundra and polar deserts of northern Alaska, a large part of Canada north of Hudson's Bay in the east and all of the Arctic archipelago in the north, all of Greenland, Spitzbergen and other eastern Arctic islands, the northern fringe of Norway and the Soviet Union north of latitude 70° N.

The greater part of the Arctic is taken up by the Polar Basin. This covers almost three million square miles, or 80% of the total area. In winter, 80% of the Polar Basin is ice-covered and in summer 60% (Budyko, 1966). The central portion of the Basin is a lifeless desert of floating ice, moving with the wind and ocean currents from the American and Asiatic continents north across the Pole and then to the south along the coasts of Greenland.

Solar energy powers the atmospheric circulation, ocean currents and food chains. Latitude, continentality and altitude account for the largest share of the systematic environmental variation on the earth. Whereas solar radiation is approximately 500 cal/cm²/day year round at the equator, at latitude 85° N, solar radiation varies between 650 cal/cm²/day in June and zero in January (Gates, 1962). Low temperatures at high latitudes in the summer, despite a higher radiation flux than at the equator at that time of year, are due to the high albedo of the snow and ice. An albedo of 0.8, compared with an albedo of 0.4 for the earth in general, reflects much of the incoming radiation back to outer space. Although the Arctic receives higher amounts of solar radiation in summer than the equator, only a small proportion is absorbed (Budyko, 1966).

According to Sater, Ronhovde & Van Allen (1971), the climate of the Arctic may be characterized as follows:

1. A distinctive regime of daylight and darkness with a low solar elevation resulting in a pronounced radiation heat loss.
2. Surface weather systems associated with the large scale, cold-cored circumpolar vortex present in the free atmosphere over the area of the Arctic. This is a function of the differential heating between the equator and the North Pole and the earth's rotation, the so-called Coriolis Effect.
3. Snow and ice cover which reflects some 80% of the total solar radiation back to the sky.
4. A strong temperature inversion above the snow or ice surface resulting from radiational cooling. The development of an inversion is strongest under calm anticyclonic conditions and most pronounced in Arctic regions with 'continental' type climates.

In ecological terms, the Arctic has limited ecological resources and a limited 'carrying capacity' for a human population owing to its climatic characteristics.

Climatic data, collected by long-term weather stations in the Arctic are presented in Table 1.1 as representative of the IBP study areas. Barrow is 90 miles north-east of Wainwright, Coral Harbor is 380 miles south of Igloolik and Karesuando is approximately 220 miles west of Svettijärvi and Nellim.

Of interest to us are the low mean daily temperatures, the low minimum temperatures and the low amounts of precipitation, most of which falls as snow. The high barometric pressures and the direction of the prevailing winds, easterly or north-easterly, are due to the presence of the cold-cored circumpolar vortex discussed by Sater *et al.* (1971). Upernavik's wind directions are influenced also by a katabatic, or gravity flow, of cold air from the Greenland Ice Cap. NE Atlantic Ocean currents and the Gulf Stream have an ameliorating effect on the climate of Fenno-Scandia which is reflected in the winter temperatures at Karesuando.

By way of contrast, Boston, Massachusetts, USA (42° 13' N, 71° 07' W), a reasonably habitable place, at least since the 1620s, has a mean daily temperature of 10.5 °C, with a minimum of – 29 °C and a maximum of 38 °C. The annual total precipitation in Boston is 1206 mm and the mean barometric pressure is 992.1 mbar.

The main characteristics of the climate of northern Alaska, the location of the IBP study sites at Wainwright, Point Hope and Barrow are: year-round aridity, low temperatures and frequent high winds with drifting snow in winter and cool temperatures and a high incidence of fog in the summer. The Arctic Ocean is frozen for about eight months of the year and the sea ice attains a thickness of between two and three meters. During the summer, the ocean is covered with progressively disintegrating floe-ice of the present and past years. This floe ice is, as a rule, not far from shore all summer (Milan, 1964). Igloolik is similar to the Arctic coast of Alaska in its climate but has twice as much precipitation. Upernavik, on the west coast of Greenland, is slightly warmer than Wainwright or Igloolik, and Karesuando is warmer than all of the other Arctic locations.

Table 1.1. *Climatic data for IBP study areas (data from Landsberg, 1970)*

Location	Barrow, Alaska (71° 18′ N, 156° 47′ W)	Coral Harbor, Canadian Northwest Territories (64° 12′ N, 83° 22′ W)	Upernavik, Greenland (72° 47′ N, 56° 10′ W)	Karesuando, Finland (68° 27′ N, 22° 30′ E)
Temperature (°C)				
Daily mean	— 12.4	— 11.3	— 6.4	— 1.5
Extremes:				
Maximum	25.6	25.0	19.0	32.2
Minimum	— 48.9	— 52.0	— 40.0	— 34.0
Precipitation (mm)	110	249	186	380
Mean wind direction	Northeast	North	East, Northeast	Southwest
Mean wind speed (m/s)	5.4	5.7	1.8	1.5
Mean pressure (mbar)	1016.5	1005.3	1010.7	1011.0
No. of days with fog	65	32.5	35.2	17

By looking at the climatological data, it can be concluded that the Arctic regions of the world are not a hospitable place for man at any state of cultural development. Man is required to make a number of biological, behavioral, technological and cultural adjustments to survive. Yet, according to the archaeological evidence, the indigenous populations have survived and thrived in the Arctic for many generations.

Man and the northern circumpolar regions – a perspective

Finally we come to man himself in the Arctic. The purpose of this volume is to show how man has adapted in a biological and behavioral way to his surroundings. We need some perspective in which to put the information collected.

The precise antiquity of man in the Old World Arctic is still undetermined. Archaeological remains of his presence have been found between the Ob River and Lake Baikal in the USSR, dating back some 10 000 to 12 000 years. This is the time of the Third Würm Glaciation in Europe, and of the Wisconsin Glaciation in North America, i.e. conditions of extreme cold. Man's movement into and through the Arctic regions is therefore a fairly recent event in human prehistory, and there is no doubt, of course, that these early inhabitants of the north were of the same species as ourselves, viz., *Homo sapiens*. With the exception of a few islands in the Canadian Arctic archipelago, there are no large sections of the American Arctic and sub-Arctic, and even of northernmost Greenland, that have not, at one time or another, been inhabited by Eskimos in the north or Indians in the sub-Arctic. Most of the Eurasian Arctic and sub-Arctic has also been occupied for thousands of years, with the exception of some of the off-shore islands and most of the Taimyr Peninsula. It appears, too, that man was not driven into these Arctic regions, or forced to live there – he settled there by his own choice.

Originally, man evolved in the tropics, perhaps five million years ago, spreading northwards and southwards from those latitudes probably some half-a-million years ago. The palaeoclimatology of the Pleistocene period suggests that early bands of hunters in western and central Europe were living in the periglacial zone possibly a quarter of a million years ago, obviously under conditions of great cold.

Culturally speaking, however, it was towards the end of the Paleolithic or Old Stone Age, perhaps 70 000 years ago, that we have the first actual evidence of man having learnt the art of living in a cold environment. From the Aurignacian, Solutrean, and particularly the Magdalenian cultural periods, in the Old World, numerous archaeological finds demonstrate his ingenuity and his adaptability. By this time, he had developed specialized tools and weapons of chipped stone. The presence of scrapers and other tools for preparing skins, and of needles and buttons of bone and ivory, indicate the manufacture and use of protective clothing of some degree of sophistication. The magnificent cave art of those times shows the numerous animals that he hunted and are characteristic of cold climates. There are pictures of seal, muskox, mammoth, cave bear and the woolly rhinoceros. As the glaciers receded northwards, so these hunters of the Upper Palaeolithic followed them, ranging across the Eurasiatic land mass, until, in the closing phases of the Pleistocene period, bands of northern hunters crossed to the New World. By about 2000 B.C., man had conquered the circumpolar Arctic from Norway round to Greenland, with the few exceptions previously noted.

If we focus for a moment upon Eskimos, we find that to identify their earliest appearance on the Arctic scene we must rely again on cultural evidence. Eskimo culture is distinctive in that it is an adaptation to living in a treeless region. It reflects a mixed hunting economy, with emphasis upon sea mammals such as seal, walrus and whale. The earliest archaeological evidence of this kind of activity, practiced by what we can term Proto-Eskimos, dates back about 5000 years, and is known as the Cape Denbigh Flint complex of northwestern Alaska. At this time, some of the Denbigh people hunted seal in the summer months, probably with boats, whilst others hunted caribou in the interior. The types and styles of tools found show much resemblance to finds at older Asian sites, so that many archaeologists believe that the Denbigh culture originated in the Palaeolithic and Mesolithic of the Far East, and in the early Neolithic of Siberia (4000 B.C.). Obviously these Denbigh people were well adapted, culturally, to survive in the Arctic. With impressive speed they moved eastwards across Arctic Canada to northeast Greenland, which they reached by 2000 B.C. In Canada, this first migratory wave of the descendants of the Denbigh people is represented by the Pre-Dorset culture. By about 800 B.C., the Pre-Dorset stage had evolved into a Dorset stage, with some minor differences in tool-making and in the range of tools made. The oldest skeletal remains of Eskimos in Canada come from a Dorset site on the south side of Hudson's Bay, and are dated to about 500 B.C. The Dorset culture lasted from about 8000 B.C. to A.D. 1300 in certain Arctic areas, but in others it was pushed aside, about A.D. 900, by a new culture, known as the Thule, that was moving eastwards from northern Alaska and was eventually to reach Greenland and Labrador. There may well have been cultural exchanges between Dorset and Thule Eskimos, the latter learning, for example, to make snow houses from the former. The Thule people were even more effectively adapted to life on the Arctic shores, being expert whale hunters. The Thule Eskimo, then, are the direct ancestors of the modern Canadian Eskimo, including the Caribou Eskimo who were to turn inland west of Hudson's Bay.

From the archaeology, therefore, we see evidence of at least 5000 years of Eskimo survival in an extremely harsh environment, accomplished by means of the evolution of a culture highly responsive to environmental change and variation, yet little affected by foreign influences. Until the eighteenth century, any change in this was very gradual. From that time onwards, the Arctic was to become inadequate to secure survival for the Eskimo in his homeland, at least in his traditional way of life (Hughes, 1968; Dumond, 1977).

The biological and behavioral information obtained on the Arctic populations studied by the IBP teams should be regarded in the light of the historical perspective described above.

References

Baker, P. T. & Weiner, J. S. (ed.) (1966). *The Biology of Human Adaptability*. Oxford: The Clarendon Press.

Budyko, M. I. (1966). Polar Ice and Climate. In *Proc. Symp. Arctic Heat Budget and Atmospheric Circulation*, ed. J. Fletcher. Santa Monica: RAND Corporation.

Dumond, D. E. (1977). *The Eskimos and Aleuts*. London: Thames & Hudson.

Gates, D. M. (1962). *Energy Exchange in the Biosphere*. New York: Harper and Row.

Haggett, P. (1972). *Geography: A Modern Synthesis*. New York: Harper & Row.

Hughes, D. (1968). An eclectic view of the physical anthropology of the Eskimo. In *Eskimo of the Canadian Arctic*, ed. V. F. Valentine & F. G. Vallee. Toronto: Van Nostrand.

Ives, J. D. & Barry, R. G. (1974). *Arctic and Alpine Environments*. London: Methuen.

Köppen, W. (1936). Das Geographische System der Klima. In *Handbuch der Klimatologie*, ed. W. Köppen & R. Geiger. Berlin: Borntraeger.

Landsberg, H. E. (ed.) (1970). *World Survey of Climatology*. Paris: Elsevier.

Laughlin, W. S. (1967). The point of studying Eskimos for the IBP. In *Report of the Working Party Conference at Point Barrow*, ed. F. A. Milan. Madison: University of Wisconsin. (Mimeograph.)

Laughlin, W. S. (1970). Study of Eskimos in the IBP. *Arctic*, **23**, 3.

Milan, F. A. (1964). The acculturation of the contemporary Eskimo of Wainwright, Alaska. *Anth. Papers Univ. Alaska*, **11**, 1–95.

Milan, F. A. (1968). The international study of Eskimos. *Arctic*, **21**, 123.

Sater, J. E., Ronhovde A. G. & L. C. Van Allen. (1971). *Arctic Environment and Resources*. Washington, D.C.: Arctic Institute of North America.

Worthington, E. B. (ed.) (1975). *The Evolution of IBP, International Biological Programme, vol. 1*. Cambridge University Press.

2. The demography of selected circumpolar populations

F. A. MILAN

This chapter summarizes information on the demography of selected populations that were studied in the circumpolar zone under the auspices of the IBP. These population groups include Eskimos of the northwest coast of Alaska, Eskimos presently living at Igloolik and Hall Beach in the Canadian Arctic, Eskimos in Northwest and East Greenland and the Skolt and Inari Lapps of Finland. The demography of the Ainu has been well described by Kodama (1970). Information from published or prepared papers from the IBP studies are reviewed here. Other information on demography is discussed where pertinent.

Except for the East Greenlanders at latitude 65° N, and the Ainu in much more southerly locations, the populations discussed here all live north of the Arctic Circle. As was emphasized in the first chapter, these populations have occupied the Arctic areas of the world for many generations despite the environmental constraints of the Arctic.

Census data show that Arctic human populations in the areas studied have increased their numbers at rapid rates. It is generally considered that this increase is in response to improvements in public health and in nutrition since the 1950s. In 1970 the world's Eskimo population was approximately 91 000 (Table 2.1). In the period 1750–1800, it has been estimated that Eskimos numbered approximately 48 000 across the Arctic. According to this reckoning, there were approximately 26 000 in Alaska, 2000 on St Lawrence

Table 2.1. *Eskimo population*

Country	Population	Source of information
Alaska	32 311	US census, 1970
Canada	17 550	Canadian census, 1971
Greenland	40 089[a]	Ministry for Greenland, 1972
USSR	1 064	Chapter 3, this volume
Total	91 014	

[a] For census purposes the distinctions are between those born in Greenland (40 089) and those born outside Greenland (8492).

Special gratitude is expressed to J. B. Jørgensen, Copenhagen, for unpublished material on the population of northwest Greenland, and to Graham Rowley, Ottawa, for census information on Canada's Eskimo population.

Island and the Siberian mainland and 20 000 in Canada, Greenland and Labrador (Oswalt, 1967).

In order to understand the present demographic situation in the small, isolated Arctic settlements studied by the IBP teams, it is necessary to be aware of the historical demography of the regions in which these communities are located.

From an analysis of the historical demography of these areas and the summaries of the present demographic situations, it is apparent that the Arctic communities are more similar to each other than they are to communities south of them in their own national units.

The population numbers of Arctic peoples today reflect their past experience with communicable diseases and the relatively recent introduction of public health. According to Rausch (1974), the state of health of the indigenous human populations of Arctic North America previous to European contact was similar to that of any natural mammalian population in a favorable habitat. Indigenous diseases were generally the natural focal zoonoses. The human population supported parasites, which included head lice, pinworms and helminths from fish, and suffered from cystic and alveolar hydatid disease, trichinosis, and amebiasis. Rabies and brucellosis were present. These natural diseases plus starvation and accidents, according to Rausch (1974), were the only factors controlling population numbers in the Arctic at the time of contact with Europeans.

Europeans introduced communicable diseases, such as tuberculosis, syphilis and smallpox. The effects of these communicable diseases are still noticeable today. For example, during the IBP medical examinations at Wainwright, over 40% of the 300 people seen at the health clinic stated that they had had pulmonary tuberculosis. Over 82% of them were adults, showing the effects of the tuberculosis control programs in reducing the incidence of this disease in the younger members of the population (Robinhold & Rice, 1970).

The historical demography of Alaska's native population

The historical demographic reckoning from Alaska shows that the native population enumerated by Petrof in Alaska's first census in 1880 declined rapidly, owing to mortality from communicable diseases introduced by Siberian Russians and Europeans. The 1880 population figure totaled 32 977 persons; this figure was not again attained until 1947, some 67 years later. The period of Russian Administration of Alaska, which began in 1747 with the arrival of Vitus Bering and lasted until the Cession of Alaska to the United States in 1867, saw a marked decrease in the indigenous populations. For example, a decrease of some 11 000 Aleuts occurred in 140 years (Hrdlicka, 1945). Delarof, in Petrof (1884), stated that the population of Kodiak Island and the

Alaska Peninsula decreased by 2389 persons over a 27 year period. Petrof (1884) wrote that smallpox epidemics over the period 1838–9 had killed about one-half of the population of Alaska. Syphilis, tuberculosis, influenza, pneumonia, measles, smallpox and scurvy all made their appearance during this period. Veniaminov (1840) described the first recorded case of tuberculosis in two Aleuts in Siberia in 1770.

At the time of purchase of Alaska from Russia in 1867 Public Health Service physicians were assigned to revenue cutters. One of these physicians reported the following (Woolfe, 1893).

The native people inhabiting northwestern Alaska are in a fair way to become exterminated even within the present century. The primary causes that will operate to bring about this end are the increasing prevalence of syphilitic, bronchial and pulmonary diseases. The whalemen, who in bygone years, during their sperm whale cruises amid the South Sea Islands, did much evil in this direction, are now effecting the spread of the most loathsome type of venereal disease among these natives.... Pulmonary and bronchial diseases are very prevalent.... Rheumatism, swelling from fractures, and a numerous train of boils, tumors, and suppurating sores that follow an unhealthy condition of the blood, are types of bodily ailments that are common.... The surgeon on the United States revenue steamer *Bear*, on her cruise to the Arctic during 1890, examined and prescribed for several hundred natives. The results of his examination proved that 85% of those treated were afflicted with either secondary or tertiary syphilis.

A Federal Act of 1905 declared that the Federal Bureau of Education was to be responsible for the education and general welfare, including health, of the native population of the Territory of Alaska. For the next 25 years this Bureau was the only government agency directly concerned with native health. The Territorial Legislature paseed a Vital Statistics Act in 1913. It was reported that the influenza epidemic of 1918 accounted for over 2000 deaths in the Territory. The death rate for the Territory in 1925, based on a population of 60 000 and admittedly inaccurate vital statistics, was estimated at 12.5/1000 population (Parks, 1927). In 1931 the responsibility for health care for the native population was delegated to the Bureau of Indian Affairs along with other major responsibilities of education, economic development and welfare. Health care was a secondary component of the mission in an agency whose primary responsibilities were unrelated to health. The Bureau of Indian Affairs, like the Bureau of Education, was never adequately funded to provide good health care. In 1936 when funds became available through the passage of the federal Social Security Act, health services in the Territory expanded. This marked the real beginning of public health activities in Alaska.

For a long time in Alaska's history there were few improvements in the health status of the native population because of their geographical isolation and the lack of an adequate public health program. In 1943, according to Morrison (1944), who was employed to evaluate the vital statistics work in the Territory, crude birth rates were 18.8/1000 population for non-natives and 29.8/1000 population for natives. Death rates were reported at 13.3/1000 for non-natives and 24.0/1000 for natives. Infant mortality rates were reported as 37.9/1000 live births for non-natives and 180.3/1000 live births for native Alaskans. Of a total of 1322 deaths reported in the Territory in 1943, in natives 21% were due to tuberculosis as compared to only 1% of all deaths in non-natives.

In 1945, the legislature created a Territorial Department of Health, under a full-time Commissioner, which began sharing with the Federal Government the responsibilities for native health. In 1946 the legislature passed a tuberculosis control program proposed by the Health Department. In 1948, Alaska's Territorial Governor, Ernest Gruening, and his Commissioner of Health, C. Earl Albrecht, MD, while testifying before a Congressional Hearing in Washington, DC stated that the tuberculosis death rate for native Alaskans was 359/100 000 population, or 14 to 16 times the overall rate in the US at that time. Due to the personal and diligent efforts of Governor Gruening, C. Earl Albrecht and US Senator E. L. Bartlett in calling the attention of the US Congress to the poor state of native health in the Territory, a generous appropriation of $1 115 000 was received from the Federal Government in 1949. This so-called 'Alaska Grant' was administered by the Alaska Department of Health and the US Public Health Service. It was used for instituting control programs for venereal diseases and tuberculosis, inoculating with Bacillus Calmette-Guérin vaccine, establishing a tuberculosis case register, delivering general health services, conducting epidemiological surveys, improving nutrition, establishing a public health program in veterinary medicine, improving sanitation and insect control. An additional grant was used for constructing a tuberculosis facility at Mt Edgecumbe, a 400-bed hospital at Anchorage and 25-bed Quonset Hut hospital facilities at Barrow, Kotzebue, Bethel and Kanakanak.

A study team, appointed by the Secretary of the Department of the Interior and headed by Dr Thomas Parran, former Surgeon General of the US Public Health Service, conducted a year and one-half study of Alaska native health in 1953–4. As regards tuberculosis, they reported '...the morbidity and mortality-related tuberculosis...exceeded in severity anything so far recorded in the annals of public health, even in the poorest areas of China where tuberculosis was rampant' (Wherritt, 1964). Tuberculosis was described as the 'Alaska scourge'. Deaths from infectious diseases other than tuberculosis and for diseases of pregnancy and childbirth were found to be ten times that of the US white population. Sanitary conditions generally throughout Alaska were described as poor. In 'native Alaska', they were described as deplorable. Water supplies were derived from every available source and of the 34 public systems for communities larger than 100 persons only six were considered satisfactory. In the field of sanitation there were

16

problems associated with a safe water supply, those associated with a lack of proper sewage disposal and housing conditions that fostered the spread of disease. Dental health status was found to be extremely poor. Hospital facilities were inadequate. Health personnel in all categories were judged to be inadequate in numbers (Parran *et al.*, 1954).

In 1955, the US Public Health Service assumed all health functions previously performed by the Alaska Native Service of the Bureau of Indian Affairs. The US Public Health Service and the Alaska Department of Health launched a determined attack against tuberculosis, the acknowledged primary source of mortality and morbidity. Technological advances and great medical progress during World War II improved air transportation and an expanded budget contributed to this attack on tuberculosis (Milan, 1974).

Although the crash attack on tuberculosis commanded most of the resources, maternal and child health were also improved through the efforts of teams of itinerant public-health nurses and physicians who operated prenatal and well-baby clinics throughout rural Alaska. Village women were trained in midwifery and supplied with medical equipment and aseptic supplies. An attempt was made to clean up villages, to improve the state of drinking water and solid waste disposal by trained village sanitarians who were supervised by public health personnel.

The rapid increase in the Alaska native populations which occurred after 1950 was in direct response to the tuberculosis control programs, especially effective after streptomycin became available after 1945 for specific therapy, improvements in health care and in nutrition. These improvements resulted in a general decrease in mortality, especially that of infants, in a population with a high fertility potential. Improved nutrition, i.e. higher caloric intake, could have had an effect on birth weights and on increasing the survival rates of newborns. Gonorrhea, when chronic, may produce sterility (Davis *et al.*, 1973). The recent work of Berman, Hanson & Hellman (1972) has also shown the importance of breast-feeding as a child-spacing mechanism among Eskimo women at Barrow. The eight to ten months of lactation amenorrhea seen in nursing mothers, as compared to the average post-partum period of 51 days for non-nursing mothers, delayed ovulation and the possibilities of pregnancies in the nursing mothers. In the Eskimo area, in general, the shift from breast-feeding to bottle feeding has had an important influence on increasing pregnancies.

In Fig. 2.1 are shown the crude death rates for Alaska natives from 1950 to 1966. The dramatic decline in the death rates between 1950 and 1955 is illustrated in this figure. The increase in life expectancy for the native population in Alaska in the period 1950–60 exceeded any such change in the US population during any period of similar length. By 1969, native life expectancy at birth was calculated to be 60.4 yr, or only 9.3 yr less than the life expectancy for the US population at the same time (Gurunanjappa, 1969).

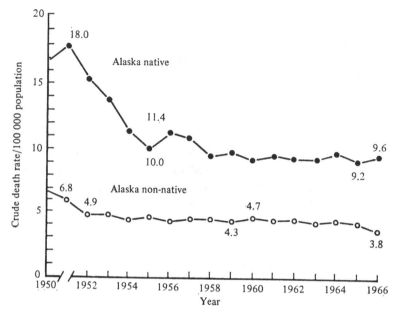

Fig. 2.1. Crude death rates for the Alaskan population for 1950–66. Source: Alaskan Department of Health.

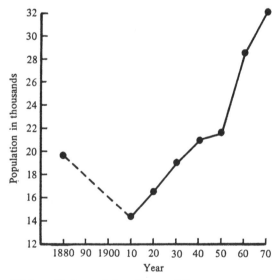

Fig. 2.2. Alaskan Eskimo–Aleut population. Source: US census.

The growth of the Alaskan Eskimo–Aleut population over a 90-yr period is shown in Fig. 2.2. This figure shows the initial decline, the increase after

18

Table 2.2. *Causes of death of Alaska natives compared with US citizens of all races for 1960–8*

Rank order Natives	Causes of death	1968 Alaska Natives	Rate/100 000 1960 US (all races)	Rank order US
1	Accidents	188.2	57.5	4
2	Heart disease	89.4	372.6	1
3	Malignant neoplasms	70.3	159.4	2
4	Diseases of early infancy	60.8	21.9	6
5	Vascular lesions of the central nervous system	41.8	105.8	3
6	Influenza and pneumonia	36.1	36.8	5
7	Suicide	26.6	10.7	10
8	Homicide	26.6	7.3	13

Data derived from Alaska Native Health Service, US Public Health Service, Anchorage, Alaska.

1910 and the rapid population increase following 1950. The growth rates of the 1950s decreased in the following decade in response to family planning programs. The number of natives and non-natives in Alaska were essentially equal in the 1920–30 decade. Since 1940, the non-native segment of the population in Alaska has increased almost six-fold from 40 000 to 234 000 in 1970. Alaska natives represented about one-fifth of the total population in 1970 and they have gone from equal to minority status in a generation's time.

As mentioned earlier, health care of Alaskan natives was transferred from the Bureau of Indian Affairs to a newly created Division of Indian Health within the Department of Health, Education and Welfare on 1 July 1955. During the period of IBP studies in Alaska, the Alaska Division of Indian Health administered a 276-bed hospital, the Alaska Native Medical Center in Anchorage, and six Service Units with hospitals at Barrow, Bethel, Kanakanak, Kotzebue, Mt Edgecumbe and Tanana. The Medical Center serves as a supporting and consulting center to the six Service Units. The Service Unit hospitals were in daily radio contact with Village Health Aides in each native community (Lee, 1971).

As compared to the situation earlier, there was a significant change in the attributed cause of death for Alaska natives due to improvements in their health care. This is illustrated in the Table 2.2. Tuberculosis was rigidly controlled, there were no deaths reported for 1971, and the leading cause for hospitalization in 1971 was *otitis media* (10%) followed by psychotic, psychoneurotic and personality disorders (8%).

Alaska's northern Eskimo population

During the IBP (1968–71) studies were conducted on Eskimos living in Point Hope, Wainwright, Barrow and Anaktuvuk Pass. The same historical events affected these communities as has been discussed for the entire native population of Alaska.

In 1882 Lieutenant Ray (Ray, 1885), officer in charge of the meteorologic and magnetic station at Pt Barrow, counted only 410 people in 72 family groups living on the Arctic coast between Wainwright Inlet and Pt Barrow. The settled coastal villages in this area reported 276 people in 1880 and 295 people in 1890 (Porter, 1893). At the turn of the century, there were approximately 750 persons living in coastal settlements from Pt Hope to Pt Barrow. Gubser (1965) estimated that the Inland Eskimo population was not over 1000 in the 1880s. The total population of north-west Alaska can be estimated at less than 2000 in 1900.

This population has nearly doubled since 1900, for the Arctic Slope Regional Corporation, a local business-for-profit organization incorporating all Eskimos of the region which was formed after the Alaska Land Claims Settlement of 1972, reported 3678 Eskimos on the north coast of Alaska in 1973.

Diamond Jenness (1928) stated that the original coastal populations at Pt Barrow and the mouth of the MacKenzie River had earlier died of disease introduced at the end of the nineteenth century by commercial whalers. The majority of the Eskimo inhabitants by 1913 to 1916, when he travelled along the north coast, were immigrants, or their descendants, from inland areas.

Wainwright, the main village studied in the IBP, is named after Wainwright Inlet, so named by the Royal Navy Captain Beechey of HMS *Blossom* in honor of his navigator as they sailed past in 1826 while searching for Sir John Franklin. The construction of a school house at Wainwright during the summer of 1904 caused the resettlement in that location of remnants of the original coastal population and Inland Eskimo then living in that area of north-west Alaska. A similar event, the construction of a hospital and school house at Barrow, 1896, served to settle the people in that area. Pt Hope, however, had been a settled area for several thousand years according to the archaeologists Larsen & Rainey (1948).

About 1900, northern Alaska was populated by Eskimos in a small number of inland bands and Eskimos in coastal settlements. The contemporary population in that area is descended from Eskimos from the entire region of north-west Alaska.

Barrow, with a population of over 2000 in 1970, of which only some 5% are non-Eskimo, is the ninth largest city in Alaska. In 1939 the Wainwright population was about 341, only 22 less than Barrow; in 1970 the Wainwright population numbered 315 while the Barrow population had increased by a

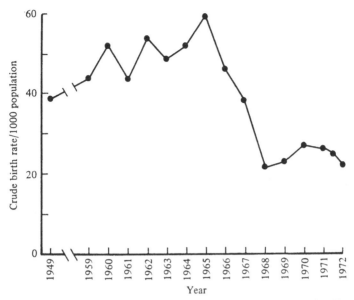

Fig. 2.3. Crude birth rate for the Barrow Eskimo population. Source: Alaskan Department of Health.

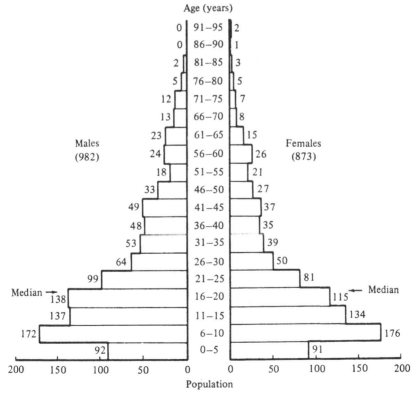

Fig. 2.4. Population profile for the Barrow Eskimo population. Source: Bureau of Indian Affairs.

The human biology of circumpolar populations

factor of six due to natural increase and the in-migration of Eskimos from other villages.

The crude birth rate for Barrow is shown in Fig. 2.3. The decline in birth rate after 1965 was due to the availability and success of a family planning program. The population profile for Barrow is shown in Fig. 2.4. Note the enlarged bottom of this profile due to a recent reduction in the birth rate. It also should be noted that there are fewer women than men over age 16 yr. This is due to a differential out-migration of the women to other parts of Alaska. Vital statistics data for the 1960–70 decade show births at Barrow to total 1012, deaths to total 250, but the net increase was only 550 persons. 312 persons had migrated out of the census district.

Wainwright had a population similar to that of Barrow with a low median age, a shortage of women in marriageable ages due to out-migration and a recent reduction in birth rate (Fig. 2.5). Wainwright's population was 75 in 1890, increased to 388 in 1940, decreased to 225 in 1950 due to out-migration to Barrow and other parts of Alaska during the war years, and has increased slowly since.

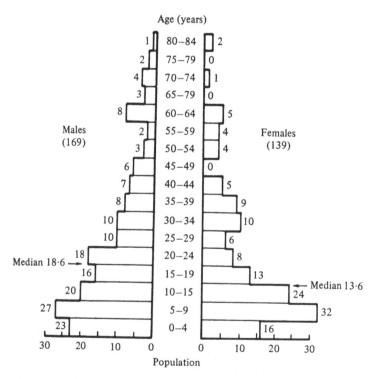

Fig. 2.5. Population profile for the Eskimo population of Wainwright, July 1968.

A registration list of births and deaths kept at the Wainwright school house covering that village, Point Lay, Icy Cape and Atanik (a small village to the north of Wainwright) was examined for the 20-yr period 1917–37 to obtain early vital statistics. During this period, according to the registration list, 114 males and 101 females were born while 108 males and 84 females died. The net increase over 20 yr was 23 persons. Assuming a mean population size of 206 persons for this period (based on 137 persons in 1917 and 275 in 1937), the crude birth rate was calculated at 52.18/1000 population and the crude death rate at 46.60/1000 population. 40% of all deaths listed were children under five years of age.

Canada's Eskimo population

According to census data, the Eskimo population of Canada has increased recently in a similar fashion to the Alaskan Eskimo. Although the British North America Act of 1871 provided for the first census of Canada, Eskimos were only enumerated for the first time in 1921 at 3269, a number probably not anywhere near the real total. In 1941, exclusive of Newfoundland and Labrador, there were 7178 Eskimos. By 1951 this number had increased to 8646 plus 1468 counted in Newfoundland and 500 in Labrador. Less than 44% were over age 21 yr as compared to 60% for the whole of Canada.

By 1971 the Canadian Eskimo population totaled 17 550, as shown in Table 2.3, with over 64% in the Canadian Northwest Territories, the location of the IBP study sites of Igloolik and Hall Beach. The population profile for the Eskimos of the Northwest Territories for 1971 is shown in Fig. 2.6. 51% of the population was under age 14 yr. In comparison, 43% of the indigenous

Table 2.3. *Canadian Eskimo population: 1 June 1971 census*

Province	Number
Newfoundland	1055
Prince Edward Island	–
Nova Scotia	20
New Brunswick	5
Quebec	3755
Ontario	760
Manitoba	130
Saskatchewan	75
Alberta	135
British Columbia	210
Yukon Territory	10
Northwest Territories	11 400
Total for Canada	17 555

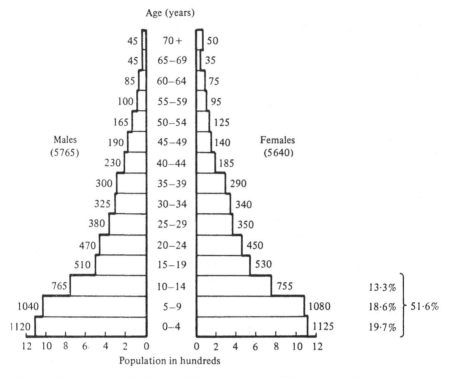

Fig. 2.6. Population profile for the Eskimo population of the Canadian Northwest Territories. Source: Canadian census.

Indian population and only 36% of the non-native segment of the population were in the same category.

Crude death rates for the Eskimo population in the Northwest Territories for the period 1962 through 1969 are graphed in Fig. 2.7. from data presented in an earlier demographic study by Lu & Mathurin (1973). As in Alaska, there is a difference in these rates between the Eskimos and non-natives, with Eskimo death rates falling to lower levels over time. Death rates for infants under one year old averaged over the periods 1962–71 for Eskimos, Indians and others were 98/1000, 55/1000 and 36/1000 respectively. After four years old, death rates were similar for the three groups until age 50 yr, and by age 65 yr they were 48/1000, 19/1000 and 16/1000 for Eskimos, Indians and others.

Freeman (1971) has written that the Northwest Territories are regions of low population density and rapid population growth, demographic characteristics of less developed nations in Africa, the Near East and Latin America. He reported the Eskimo population of the Northwest Territories to be in-

24

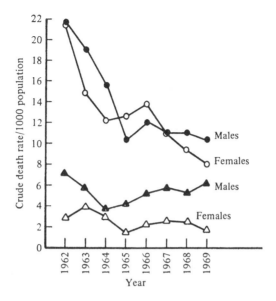

Fig. 2.7. Crude death rates for Eskimos and others in the Canadian Northwest Territories for the period 1962–9. —●—, —○— Eskimos; —▲—, —△—, others, including Indians. Source: Lu & Mathurin (1973).

creasing at 5.2%/year in 1964. This means that the population doubling time is only 13½ yr.

Schaefer (1973) has described the recently improved health conditions in the Canadian North. In the 1930s and 1940s the death rate among Eskimos from tuberculosis was higher than the birth rate. In 1947 this death rate was 70/1000 population. The construction of an elaborate system of nursing stations and small modern hospitals was responsible for controlling tuberculosis. Schaefer (1973) has pointed out that the foremost causes of death in the Northwest Territories in recent years have been accidents, violence and poisoning, usually alcohol related.

Igloolik and Hall Beach

According to McAlpine & Simpson (1976), the Eskimos presently living at Igloolik and Hall Beach originally lived in isolated small camps scattered throughout Foxe Basin. Starting in the 1940s the Eskimos were encouraged by the government to move into centrally located communities. This move was virtually complete by the time the population was studied by the IBP team. However, all persons over 40 yr of age were born outside of Igloolik in

25

The human biology of circumpolar populations

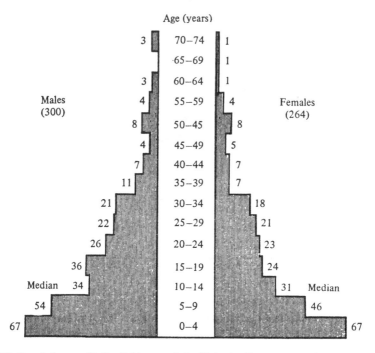

Fig. 2.8. Population profile for Eskimos at Igloolik in the Canadian Northwest Territories.

small camps illustrating the recency of their location. Igloolik's population profile is presented in Fig. 2.8.

Greenland's population

In the census of 1972, Greenland's total population was enumerated at 48 581 persons. Of this number 40 089 were born in Greenland, that is are presumed to be Greenlanders, and 8492 were born outside of Greenland, or are presumed to be Danes. These categories are those used by the census takers. Of those enumerated in the 1972 census, 36 392 lived in towns, 10 907 in outlying communities and 1282 in weather stations. On 1 January 1975, Greenland's total population was enumerated at 49 502 of whom 39 979 were born in Greenland.

Between 1950 and 1972, the Greenlanders nearly doubled in numbers while the Danes living in Greenland increased by a factor of eight. In the 1972 census, 45.3% of the Greenlanders, as compared to 21.7% of the Danes, were under age 15 yr. Some 66.1% of the Danes were males as compared to 50.3% of the Greenlanders. Vital statistics for Greenlanders are shown in Fig. 2.9. The average life expectancies for Greenlanders were 32 and 37 yr for males and

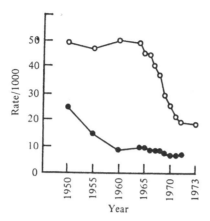

Fig. 2.9. Vital statistics for Greenland 1950–73 for persons born in Greenland. —●—, crude death rates; —○—, crude birth rates. Source: Arsberetning 1973, Ministry for Greenland.

females respectively in the period 1946–51. In the period 1966–70, life expectancies increased and were calculated at 57 yr for male Greenlanders and 65 yr for female Greenlanders.

A similar population increase has occurred in the relatively isolated settlements of Ammassalik and Scoresby Sound in East Greenland. At Ammassalik, the population doubled in a 20-yr period, numbering 1247 persons in 1950 and 2517 persons in 1971 (Robert-Lamblin, 1975). The Scoresby Sound Administrative District was established in 1925 and the total District population increased by 500% over the period 1925 to 1971 'due to the extraordinarily efficient medical care provided by the Danes' (Langaney, Gessain & Robert, 1974). In a novel attempt to study 'dependency ratios', Gessain (1968) related the number of inhabitants in the Amassalimiut to the number of kayaks, hypothesizing that in this sea-mammal hunting society, it was impossible to be a hunter without a kayak. The proportion of numbers of inhabitants to kayaks was initially 3.4 in 1884, 5.3 in 1930 and 20 in 1968. In 1971, 49.7% of the population of Ammassalik was under 15 yr of age (Robert-Lamblin, 1975). Gessain (1969) described the strong 'founder effect' seen in one kindred of 259 persons who were descended from two East Greenlanders alive in 1800. Ducros (1976) has pointed out the importance of this 'founder effect' in studying the blood-group gene frequencies and the anthropometric measurements in this East Greenland population.

Thus we see the Greenlandic population undergoing a rapid population increase after 1950 and a migration into centrally located towns. This was essentially the result of the work of a Royal Commission established by King Frederik in 1948 to consider Greenland's social, economic, medical and housing problems. Their recommendations included relocation and the

27

concentration of the population in fewer but larger communities, and the depopulation of some of the smaller settlements and districts (Schurman, 1970).

IBP Greenland communities

Studies were conducted in Aupilagtoq and Kraulshavn in the Upernavik District in West Greenland and in Ammassalik and Scoresby Sound in East Greenland.

The Soviet Arctic

Population numbers for the minority peoples in the Soviet Arctic have been provided by Rychkov & Sheremet'eva in Chapter 3. Elsewhere Armstrong (1965) has described the Russian settlement of the Soviet Arctic and has analyzed the available census material for that region. Census figures are available for 1897, 1926, 1939, 1959 and 1970. Armstrong states that the 1939 census is generally considered unreliable. He also states that between 1926 and 1959 the Russian population of the North increased by a factor of 15 or 16 from a position of a 1:3 minority to a position of a 4:1 majority in relation to the native population. The decrease of the Asiatic Eskimo population from 4000 to 1064 enumerated in Chapter 3, according to Armstrong, can be partly explained by their assimilation into the neighboring Chukchi.

Common observations in the Eskimo IBP communities

Similar observations and calculations on the demographic structure, median age, age at menarche, age at menopause, number of consanguinous marriages, number of pregnancies, twinning rates and sex ratios were made in each community under study. This information has been presented in a series of papers. Milan (1970, 1971) has written on Wainwright; McAlpine & Simpson (1976) have described Igloolik and Hall Beach; Hansen & Jørgensen (1970) have described Kraulshavn; Gessain (1968, 1969), Robert-Lamblin (1975) and Langaney et al. (1974) have described the demographic situation in East Greenland.

Eskimo community size varied from 564 (1971) at Igloolik to 149 (1969) in Kraulshavn. Wainwright, at 308 (1968), Hall Beach at 348 (1971) and Aupilagtoq, at 161 (1969), were intermediate in size. Ratios of males to females were 300/264 at Igloolik, 169/139 at Wainwright and essentially equal in the other communities. Disproportionate sex ratios reflected the effects of migration, usually of young women, out of the community. The median age for males and females was 18.6 and 13.6 yr in Wainwright and 10–14 yr for both sexes in Igloolik and Kraulshavn. In Ammassalik 49.7% of the population was under 15 yr.

28

The communities differed as to percent of non-Eskimo genes in the population. In Wainwright some 27% of the population, as compared to 6% of Igloolik, carried non-Eskimo genes according to pedigree analysis.

In Wainwright three persons were one-half non-Eskimo, 19 were one-fourth non-Eskimo and 45 were one-eighth non-Eskimo. The population had been hybridized through admixture with crews of whaling vessels which sailed along that coast after 1850. The whalers who contributed genes to Wainwright were 'Old Americans', Portuguese and sailors from the South Pacific.

Age at menarche was determined at 13.78 (± 0.78) yr at Wainwright, 14.3 (± 0.2) yr at Igloolik and 13.4 (± 0.2) yr at Hall Beach. There was a period of 5.4, 3.6 and 3.5 yr respectively for each of the communities between menarchial age and the birth of the first child. Accordingly, mothers were 19.9 (± 0.6) yr, 17.9 (± 0.3) yr and 16.9 (± 0.3) yr of age at the birth of their first child. Women at Wainwright reported an average of 7.9 (± 4.0) pregnancies each. These numbers were 7.5 and 7.8 in Aupilagtoq and Kraulshavn and 6 each at Igloolik and Hall Beach. The seventeen post-menopausal women interviewed at Wainwright reported a total of 159 pregnancies or an average of 9.9 (± 4.08) pregnancies each.

The twinning rate was high in Wainwright at 27.3/1000 births, a figure twice that at Igloolik and Hall Beach (12.1/1000 and 10.9/1000). The twinning rate for Aupilagtoq was 2.1/1000 births. Sex ratios at birth were 108.2 males for every 100 females at Wainwright and 111 males per 100 females at Igloolik. Aborted fetuses averaged about 8 to 13% of all births in the Alaskan and Canadian communities.

Approximately 25% of all children born alive in the Alaskan, Canadian and West Greenlandic communities under study had died by 15 yr. Up to four years at Wainwright, hazards to life appear to have been 'flu, pneumonia, high fever of unknown cause and diarrhoea; after this age environmental hazards took their toll: freezing, drowning, and plane crashes. A similar situation was observed at Igloolik.

Birth control practices are now common in Arctic communities. For example, there were 47 women in Wainwright of reproductive age. Seventeen were post-menopausal, one was infertile, three had their reproductive capabilities terminated by surgery and 14 of the remaining 26 were practicing birth control. Twelve used intrauterine devices and two used pills. In Aupilagtoq seven out of 21 (33%) and two out of 15 in Kraulshavn (13%) were using intrauterine devices. It has been estimated that 1/3 of Greenlandic women in the ages of 15–49 yr were using intrauterine devices in 1970.

Family planning, which started in Alaska in 1965 and was encouraged earlier in Greenland, had started by 1970 in the Northwest Territories. In 1973 there were 93 women of child-bearing age at Igloolik. Thirty-one were using contraceptive pills and one had an IUD (Haraldson, 1974). According to Otto Schaefer (personal communication, 1976) almost all Igloolik women are now using birth-control methods.

Lapps

The prehistory and present status of Lapps has been described by Lewin (1971) who reported a total number of 35 000 according to the census of 1969. In 1972 (von Bonsdorff, Fellman & Lewin, 1974) there were 3799 Lapps in Finland and they represented 18% of the total population of northern Finland. Immigration of Finns into the Lappish areas started some 200 yr ago. Lapps had become a minority 50 yr ago. About 2000 subjects of all ages were studied by the IBP teams. This included 600 Skolt Lapps and 1400 Inari Lapps. The Inari Lapp gene pool was estimated to consist of 46.4% Fisher Lapp genes, 25.0% Mountain Lapp genes, 0.6% Skolt Lapp genes and 28.0% Finnish genes.

The Skolt Lapps, members of the Greek Orthodox Church, and earlier called the East Sea Finns or Russian Finns, were living in Petsamo as early as the fifteenth century. After 1944 when the Soviet Union acquired Petsamo, the Skolts moved voluntarily into Finland. The Skolts of Suenjel settled at Svettijärvi northeast of Lake Inari while the Pasvik Skolts settled at Nellim south of Lake Inari.

The inbreeding effect in Arctic populations

Although Laughlin (1966), in discussing the genetic and anthropological, characteristics of Arctic populations, wrote that inbreeding was not common IBP workers found evidence of a strong 'inbreeding effect' due to the small population size and consequent limited mate choice. A coefficient of inbreeding F, which is the probability that two allelic genes are identical by descent from a common ancestor (Cotterman, 1951), or put another way, is the frequency of homozygotes in a specific population as compared with that of a panmictic population, was calculated for Eskimos and Lapps. F, a measure of probability, can vary between 0.250 and 0. Thus, $F = 0$ signifies the absence of inbreeding and $F = 0.250$ signifies complete inbreeding for the children of an incestuous brother-and-sister marriage.

Milan (1971) calculated the inbreeding coefficient F for Wainwright using the genealogical method devised by Wright (1922) from the expression:

$$F_W = (\tfrac{1}{2})^{n+1} (1 + F_A)$$

where n equals the number of links connecting each individual in each path between one of W's parents and the other, and F_A equals the inbreeding coefficient of the ancestral population. Pedigrees were collected for analysis.

In Wainwright four out of 32 marriages in the parental generation were consanguinous being with cousins of various degrees. Any two persons of 65 yr in the parental generation taken at random were related slightly less than second cousin once removed, but not as distantly as third cousins. The mean *F* for this generation was 0.007. The mean *F* for 210 in the children generation was 0.004, or equivalent to that between third cousins.

At Igloolik eight out of 98 marriages (8.2%) were consanguinous. *F* was calculated to be at least equal to 0.0010 or that of fourth cousins.

Genealogical trees for 513 pure Skolts, 155 in Pasvik–Nellim and 358 in Svenjel–Svettijärvi and 204 half-Skolts were prepared by Lewin (1971) using parish church registers. Coefficients of inbreeding were then calculated by Vollenbruck, Lewin & Lehman (1974) using the genealogical method of Wright (1921), the hierarchial method of Wright (1922), and the indirect method of Spuhler & Kluckhohn (1953) using M and N blood group gene frequencies.

According to the genealogical method, the average *F* value for the 245 Suengel–Skolt children was 0.0055, a value between half second cousin (0.0078) and full third cousin (0.0039). This was 27 times higher than the *F*-value for Pasvik–Nellim Skolts and due, according to those authors, to the greater isolation of the first group. The average value for both populations together was 0.0045, quite close to the mean *F* of 0.004 found in Wainwright.

The average *F* for the children of the grand-parental generation was 0.0023. The average *F* for children today is 0.0045 showing the increase due to genetic isolation.

The coefficients of inbreeding seen in these three Arctic populations are low compared to populations in which endogamy is culturally sanctioned. In Southern India (Sanghvi, 1966) the mean *F* is 0.015–0.048. A value of 0.0216 has been reported for the S-leut colony of Hutterites in the US (Mange, 1964). A value of 0.0434 has been found for the Samaritans in Israel (Bonné, 1963).

Inbreeding in the Arctic communities is the result of small population size and limited mate choice rather than any approved endogamy.

Inbreeding tends to bring out recessive alleles present in heterozygote carriers. However, the evidence for deleterious effects of consanguinity in these populations is not clear probably due to the high infant mortality rates which would mask these effects.

Migration

The US census of 1970 showed that the native population of Alaska has increased by some 10 000 since 1960. The average annual growth rate for that population increased substantially only in the urban areas of the state. As pointed out long ago by Ravenstein (1885), there is usually a predominance of females among migrants. The results of this are still to be seen in Wainwright's population. Because of a shortage of marriage partners, 41% of the men are unmarried; in contrast, only 10% of the women are unmarried (Milan, 1970). The demography of the native population of an Alaskan city studied by Milan & Pawson (1975) showed that approximately one-half of all marriages were inter-racial with native women married to non-native men, pointing to the tendency of native women to leave their own communities and to marry non-native men.

In a recent study it was pointed out by Barfod, Nielsen & Nielsen (1973) that the central population registry in Copenhagen listed 2853 Greenlanders in Denmark over 14 yr in November of 1971. Greenlanders resided all over Denmark and were found in 249 of Denmark's 277 communes. Only 6% of this population was over 45 yr. Under 25 yr there were the same number of men as women; over 25 yr there were three times as many women as men. 60% of the women and 25% of the men over 20 yr were married. Of a total of 968 marriages, 98% of the women and 90% of the men were married to non-Greenlanders.

The census data and the examples just discussed for Fairbanks and Copenhagen show evidence for a strong trend to migrate among contemporary Alaskan and Greenlandic Eskimos. Additionally, the migrants are marrying outside their own ethnic groups into the larger population.

Summary

Eskimo populations spread over a wide geographic range from Siberia to Greenland, originally isolated and in balance with ecological resources, experienced a reduction in numbers due to the effects of communicable diseases which were introduced by Europeans about the turn of the century. A rapid population increase has occurred in the last 20 yr in response to public-health programs and an improved nutritional situation. These populations are now in a period of 'demographic transition' with control of their increase rates since the introduction of family planning programs in the 1960s.

Diamond Jenness (1968) has pointed out that this population increase has placed great stress on the delicate balance that existed for thousands of years between Arctic man and the limited resources of the environment.

Although the populations studied still have some consanguineous marriages due to their relatively small size and the consequent limited mate choice, their long-time genetic isolation has been broken. Eskimos in Arctic Canada remain the least racially mixed due to their geographical isolation. Village out-migration is occurring, at least in Alaska and in Greenland, and is reducing the number of available women for marriage in Alaska in the small communities.

Everywhere in the Arctic there has been an appreciable increase in the number of non-natives. This migration of 'southerners' into the Arctic has already had profound social and genetic effects on the indigenous populations.

References

Armstrong, T. (1965). *Russian settlement in the north.* Cambridge University Press.
Barfod, P., Nielsen, L. & Nielsen, J. (1973). *Grønlaendere i Danmark 1971–72.* Copenhagen: Nyt Fra Samfundsvidenskaberne.
Berman, M. L., Hanson, K. & Hellman, I. L. (1972). Effect of breast feeding on post-partum menstruation, ovulation, and pregnancy in Alaskan Eskimos. *Am. J. Obstet. & Gyn.* **114**, 424–524.
Bonné, B. (1963). The Samaritans: A demographic study. *Human Biology*, **35**, 61–89.
Cotterman, C. (1941). Relatives and human genetic analysis. *Scientific American*, **53**, 227–34.
Davis, B. D., Dubelco, R., Eisen, H. N., Ginsberg, H. S. & Wood, W. B. (1973). *Microbiology.* New York: Harper & Row.
Ducros, A. (1976). L'isolat Eskimo du Scoresbysund. Une illustration de l'effet du fondateur en Anthropologie. In *L'étude des Isolates*, pp. 307–16. Paris: INED.
Freeman, M. R. (1971). The significance of demographic changes occurring in the Canadian East Arctic. *Anthropologica*, N.S. xiii, 1–2, 215–36.
Gessain, R. (1968). Le Kayak des Ammassalimiut: évolution démographique. *Objet et Mondes*, VIII: 4. Paris: Musée de l'homme.
Gessain, R. (1969). *Ammassalik ou la civilisation obligatoire.* Paris: Flammarion.
Gubser, N. J. (1965). *The Nunamiut Eskimos: hunters of caribou.* New Haven & London: Yale University Press.
Gurunanjappa, B. S. (1969). Life tables for Alaska Natives. *Public health Reports.* **84**, 65–9.
Hansen, K. & Jørgensen, J. B. (1970). Befolkningstilvaeksten i et fangersamfund: Kraulshavn, Upernavik Distrikt. *Tidskriftet Grønland*, 173–9.
Haraldson, S. (1974). Evaluation of the Alaska Native Health Service. *Alaska Medicine*, **16**, 51–60.
Hrdlicka, A. (1945). *The Aleutian and Commander Islands and their inhabitants.* Philadelphia: Wistar.
Jenness, D. (1928). Comparative vocabulary of the western Eskimo dialects. *Report of the Canadian Arctic Expedition, 1913–18.* **15**. Ottawa: Kings Printer.
Jenness, D. (1968). Eskimo Administration. *Arctic Institute of North America technical Paper* 21.
Kodama, S. (1970). *Ainu: historical and anthropological studies.* Sapporo, Japan: Hokkaido University School of Medicine.
Langaney, A., Gessain, R. & Robert, J. (1974). Migration and genetic kinship in eastern Greenland. *Human Biology*, **21**, 272–8.
Larsen, H. & Rainey, F. (1948). Ipiutak and the Arctic Whaling Culture. *Anthropological Papers American Museum of Natural History*, **42**.
Laughlin, W. S. (1966). Genetical and anthropological characteristics of arctic populations. In *The biology of human adaptability*, ed. P. T. Baker & J. S. Weiner, pp. 469–95. Oxford: Clarendon Press.
Lee, J. F. (1971). Alaska Area Native Health Service. *Alaska Medicine*, **13**, 47–50.

The human biology of circumpolar populations

Lewin, T. (1971). History of the Skolt Lapps. In *Introduction to the biological characteristics of the Skolt Lapps.* Finnish Dental Society, **67**, Supplementum 1.

Lu, C. M. & Mathurin, D. C. E. (1973). *Population projection of the Northwest Territories to 1981.* Ottawa: Indian and Northern Affairs.

Mange, A. P. (1964). Growth and inbreeding of a human isolate. *Human Biology,* **36**, 104–31.

McAlpine, P. & Simpson, N. E. (1976). Fertility and other demographic aspects of the Canadian Eskimo communities of Igloolik and Hall Beach. *Human Biology,* **48**, 113–38.

Milan, F. A. (1970). A demographic study of an Eskimo village on the North Slope of Alaska. *Arctic,* **23**, 82–99.

Milan, F. A. (1971). Über Blutverwandschaft und Inzucht in einer Eskimo-Gemeinde. *Anthrop. Anz.* **33**, 126.

Milan, F. A. & Pawson, S. (1975). The demography of the native population of an Alaskan city. *Arctic,* **28**, 275–83.

Milan, L. J. (1974). Ethnohistory of disease and medical care among the Aleut. *Anth. Papers of the University of Alaska,* **16**, 15–40.

Morrisson, F. S. (1944). *Evaluation of Vital Statistics Work in Alaska.* Federal Security Agency. US Public Health Service. Mimeo. pp. 1–95.

Oswalt, W. H. (1967). *Alaskan Eskimos.* San Francisco: Chandler Press.

Parks, Governor George C. (1927). Letter to Governor Parks from J. H. Edwards, Asst Secretary of the Interior, Washington, DC in Archives of the Office of the Governor, Federal Record Center, Seattle.

Parran, T., Ciocco, A., Crabtree, J. A., McNerney, W. J., McGibony, J. R. & Wishik, S. M. (1954). *Alaska's Health.* Pittsburgh: Graduate School of Public Health, University of Pittsburgh.

Petrof, I. (1884). Report on the Population, Industries and Resources of Alaska. *Tenth Census of the USA 1880.* Washington, D.C

Porter, R. P. (1893). Report on the Population and Resources of Alaska. *Eleventh Census of the USA 1890.* Washington, DC.

Rausch, R. L. (1974). Tropical problems in the Arctic. Infectious and parasitic diseases, a common denominator. *Industry and Tropical Health. VIII,* pp. 63–70. Boston: Harvard School of Public Health.

Ravenstein, E. G. (1885). The laws of migration. *J. roy. stat. Soc.,* **48**, 167–277.

Ray, P. H. (1885). Report of the international polar expedition to Point Barrow, Alaska. *House of Representatives Executive Document, 48th Congress, 2nd Session,* **23**, 44.

Robert-Lamblin, J. (1975). Mortalité et expansion démographique à Ammassalik au XXe siècle. *Objets et Monde,* **15**. Paris: Musée de l'Homme.

Robinhold, D. & Rice, D. (1970). Cardiovascular health of Wainwright Eskimos. *Arctic Anthropology,* **7**, 83–5.

Sanghvi, L. D. (1966). Inbreeding in India. *Eugenics Quarterly,* **16**, 291–300.

Schaefer, O. (1973). The changing health picture in the Canadian north. *Canad. J. Ophthalmol.* **8**, 196–204.

Schurman, H. (1970). Apartment houses for Greenlanders. *North,* **17**, 16–25.

Spuhler, J. N. & Kluckhohn, C. (1953). Inbreeding coefficients of the Ramah Navaho Population. *Human Biology,* **25**, 295–317.

Veniaminov, I. (1840). *Notes on the Islands of the Unalaska District.* St. Petersburg. Ann Arbor: University Microfilms Inc.

Vollenbruck, S., Lewin, T. & Lehman, W. (1974). On the inbreeding of Skolts. *Arctic Medical Research Report,* **9**. Finland: Oulu.

von Bonsdorff, C., Fellman, J. & Lewin, T. (1974). Demographic studies on the Inari Lapps in Finland with special reference to their genealogy. *Arctic Medical Research Report*, **6**, 4.

Wherritt, H. (1964). US Public Health Service News. *Alaska Medicine*, **6**, 1–4.

Woolfe, H. (1893). *Eleventh Census of the USA 1890*. Washington, DC.

Wright, S. (1921). Systems of mating. II. *Genetics*, **6**, 124–43.

Wright, S. (1922). Coefficients of inbreeding and relationships. *American Naturalist*, **56**, 330–8.

3. The genetics of circumpolar populations of Eurasia related to the problem of human adaptation

YU. G. RYCHKOV and V. A. SHEREMET'EVA

There are 16 indigenous peoples (not counting Indo-Europeans) in the circumpolar belt of the northern hemisphere, from the Lapps of Scandinavia to the Eskimos of Greenland. All these peoples, including the Lapps and the Eskimos, also appear within the Soviet part of Eurasia's circumpolar zone. Thus, the extremely non-uniform distribution of circumpolar peoples to the west and east of Greenwich offers opportunities for research into the populational aspects of the process of adaptation at different levels of population organization within a single geographical–climatic belt. To the west of Greenwich the process of adaptation of the Eskimo part of the population proceeds in a relatively homogeneous ethnic and economic-cultural environment, whereas to the east of Greenwich the same process proceeds in an extremely heterogeneous social environment. Most amenable to research in the western part of the circumpolar area is the phenomenon of population differentiation in the process of adaptation in a comparatively homogeneous social environment in both its general biological and its particular genetic aspects. Conversely, in the eastern part of the area, with its heterogeneous social environment, more attention should be given to integrative phenomena in the biological characteristics of populations. Just as different are the possible bases for data processing. Dealing, as we are, with the eastern half of the circumpolar belt, we adhere to a generalized plan in reviewing population genetics data about the indigenous population of this Eurasian zone.

Ethnolinguistic and economic-cultural structure

The structure of the indigenous population of the circumpolar part of Eurasia is marked by numerous social characteristics, including those of language, culture and economy.

The linguistic differentiation of the population has gone very far indeed. The peoples of this zone belong to several language families, between which relations can in some cases be easily established, yet in others only tentatively. The Finno–Ugric family of languages is represented by the Lapps (1687), Komi (17 054) and Khants (5519); the Samodian family by the Nenets

(26 132), the Enets (439), Nganasans (682), and Selkups (1245); the Tungusic–Manchu family by the Evens (2109), the Turkic family by the Yakuts (11 639) and Dolgans (4083). The populations of the people within the circumpolar belt are given in parentheses. Opposed to all these languages which may be united into a still larger Uralic–Altai linguistic community, is the Paleo–Asiatic language group, conditionally regarded as a whole. It includes: the Yukaghirs (440), Itel'men (1096), Koryaks (6168) and Chukchi (10 267). The Asian Eskimos (1064) and the Aleuts of the Commander Islands (399) may be included here too, but they are usually considered in the framework of the linguistic system of New World peoples. The genealogical tree of the circumpolar peoples, depicted in Fig. 3.1, is an attempt to reflect roughly the development of their modern ethnolinguistic structure.

In so far as one may compare the ethnolinguistic structure with the ethnocultural one, archaeological investigations point to the existence of several cultural regions in the circumpolar zone by the Neolithic Age.

These regions are the following: Kola, Dvina–Pechora, Shigirian (Uralian), Ob, North-East (Yakutia and Chukotka), Kamchatka (Okladnikov *et al.*, 1968; Okladnikov & Beregovaya, 1971; Mochanov & Fedoseeva, 1973). Archaeologists assume the existence of genetic continuity in the relationships between neolithic and contemporary ethnocultural structures.

It follows, therefore, that the initial stages in the formation of this structure should be placed in the Mesolithic and possibly in the late Upper Paleolithic epoch, i.e. in the early Holocene. That was the time when the modern horobioclimatic situation in northern Eurasia took shape, with its characteristic latitudinal zonality and great stability of the zonal borderlines of the tundra and taiga throughout the entire subsequent period. Archaeological findings indicate that in this period, from the tenth to the fifth millennium B.C., northeastern Asia to the east of the Yenisey was populated by pre-Neolithic tribes of nomadic reindeer and other big-game hunters (Mochanov, 1973).

That the development of the peoples of northern Asia has never been interrupted since the final stages of the Paleolithic Age is also confirmed by genetic data. It has been established, from the distribution of many independent gene frequencies of blood groups, serum proteins and phenyl thiocarbamide (PTC) sensitivity, that the genetic distance between the total population of northern Asia and the total population of American Indians is no greater than that between two adjacent populations of some single modern people of Siberia (Rychkov, 1973*b*; Sheremet'eva, 1973). It follows that the genetic continuity of the extant peoples of northern Asia can be traced back right to the time of ancient man's penetration from Siberia across the Bering land bridge to America. It has also been established that all the peoples of northern Asia, including the circumpolar peoples regarded as a whole, completely preserve their identity with the Neolithic population of Siberia (Baikal area) with regard to the concentration of more than a score

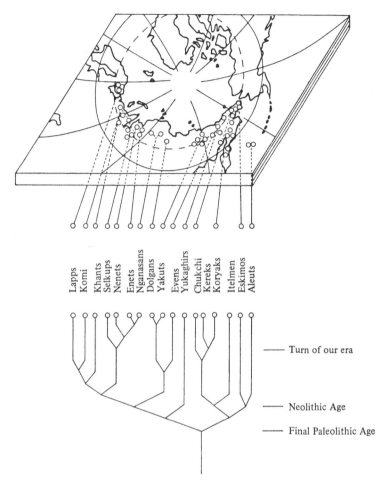

Fig. 3.1. Approximate ethnolinguistic tree of the indigenous population of the circumpolar zone of Eurasia, projected onto the populations covered by the present study.

of independent variants of cranial anatomy (Rychkov & Movsesyan, 1972; Movsesyan, 1973).

However, within the Eurasian circumpolar zone *per se*, not all the extant peoples are equally aged. From archaeological, linguistic and ethnographic evidence, it appears that this zone, throughout the entire Holocene and right down to our time was under the constant pressure and diverse influences of the peoples that populated the continental hinterland of northern Asia. Significantly later than others, it came to be settled by the Yakuts (second half of our era). Still earlier there migrated here Samodian and Ugro–Finnic peoples whose legends preserve the memory of a preceding, more ancient aboriginal population of the circumpolar zone. These legends find confirmation in archaeological and ethnographic data (Lashuk, 1968; Simchenko, 1968; Chernetsov, 1973). A living example of the changing ethnocultural strata in the circumpolar zone can be seen in the rapid decrease of the area formerly occupied by Yukaghir tribes, with the settling there, witnessed by the Russians, of the Nganasans, Dolgans, Evens, Yakuts and Chukchis.

Such migration of peoples from the continental parts of Eurasia to the circumpolar zone is significant for the biology of human populations, for it involves essentially a migration of genes, which must and will be taken into account when studying the genetic peculiarities of sub-Arctic human populations. Nothing of the kind is known in the western part of the circumpolar zone.

The most archaic type of economic-cultural structure in the circumpolar belt is that of reindeer hunters and/or fishermen, surviving among the Yukaghirs and in some measure among the Evens, Nganasans, the Kola Lapps, Khants and Itel'men. Of the same or slightly less antiquity is the type of fishermen and sea-mammal hunters of the circumpolar coast, still surviving among the Koryaks, coastal Chukchi, Asian Eskimos, and in some measure among the Lapps of the Kola peninsula; according to archaeological findings this type is known from the north coast of western Siberia and the north of eastern Europe. Both of these types, known by the archaeological monuments of the metallogenic epoch, already existed in the Neolithic and earlier epochs and may be regarded as being directly descended from a preceding type of Upper Paleolithic big-game hunter of the Pleistocene (Okladnikov, 1968). No less characteristic is the historically later (after the turn of our era) cultural-economic type of tundra reindeer breeders, represented quite strikingly among the Chukchi, Nenets, Nganasans, and to a certain extent among the Koryaks, Evens, Komis and Lapps. Perhaps, with the exception of the Yukaghirs and Eskimos, all circumpolar peoples manifest an overlapping of several economic-cultural types of varying antiquity. This led to the further differentiation of the population and an increase of the ecological niches it developed. Thus, the Chukchi and Koryaks divided into 'Coastal' and 'Reindeer' and the Nenets and Yukaghirs into 'Forest' and 'Tundra'. A similar division into 'Mountain' and 'Fisher' is known among the Lapps. The latest type, brought to the circumpolar zone by the Yakuts, is that of cattle herdsmen, which, in the historical perspective of the circumpolar zone's development, may be regarded as continuous to the reindeer breeding system. However, outside this zone one presupposes a reversed continuity relationship between them.

Thus, the process of evolution in the social organization of populations in the eastern part of the circumpolar zone was continuous from the most ancient stages to our time. The changing ethnic traditions and, particularly, the changing economic–cultural structures, as types of social adaptation by

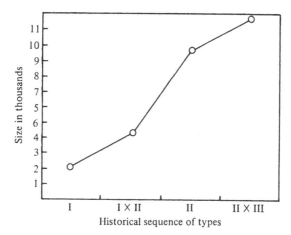

Fig. 3.2. Average size of ethnos (in thousands) during the traditional economic and cultural structure, as a type of social adaptation in the circumpolar zone of Eurasia. I, hunters and fishermen in inland and littoral areas of the circumpolar zone; II, reindeer breeders; III, cattle herders. I × II, II × III, overlapping of types.

populations to the environment, influenced those parameters of population organization, on which the biological process, and in particular genetic adaptation of the population is dependent. This is manifested particularly vividly in the relationship between the historical age of the given economic-cultural type and the average size of the peoples representing this particular type of social adaptation (Fig. 3.2). This one example is sufficient to show how great is the developmental dynamics of the circumpolar population, even if we dismiss for the moment the still more profound modern changes due to the complete and active involvement of these peoples in the general process of the socio-economic development of the peoples of the USSR.

The authors' approach to reviewing materials on circumpolar peoples with a description of the materials used

Is there still any ground, in the light of what has been said above, for attempting to draw up an integral picture of the cirumpolar population in Eurasia?

Let us proceed from the following premises. (1). On a general plane, all the traditional economic–cultural types which have just been described, are not only very similar to ancient types of natural economy, but, most significantly, are all based on the exploitation of the natural environment, on the single principle of the preservation and maintenance of a population–environment equilibrium. The limited natural resources available for utilization within the framework of the economic–cultural types, as well as the necessity of observing the equilibrium principle, are reflected in the small

41

Table 3.1. *A description of materials on circumpolar population genetics used in the survey*

Ethnic group	Number of populations	Genetic markers	Year of study and publication
Material collected during expeditions organized by the authors			
Aleuts, the Commander Islands	2	ABO, MNSs, Rhesus: D, C, C^w, c, e, E, P, Lewis, Duffy, Kell, Diego, Tf, Hp, Gc, PTC	1970; Rychkov & Sheremet'eva, 1972*a*, 1973
Eskimos, the Chukotka peninsula	3	ABO, MNSs, Rhesus: D, C, C^w, c, E, e, P, Lewis, Duffy, Kell, Diego, Tf, Hp, Gc, PTC	1970; Rychkov & Sheremet'eva, 1972*b*
Chukchi, the Chukotka peninsula	2	ABO, MNSs, Rhesus: D, C, C^w, c, E, e, P. Lewis, Duffy, Kell, Diego, Tf, Hp, Gc, PTC	1970; Rychkov & Sheremet'eva, 1972*b*
Koryaks, the Kamchatka peninsula	11	ABO, MNSs, Rhesus: D, C, C^w, c, E, e, P, Lewis, Duffy, Kell, Hp, PTC	1965; Spitsin, 1967; Sheremet'eva, 1973
Yakuts, the Yana river basin	9	ABO, Rhesus: D, PTC	1972; Gubenko, unpublished data
Nenets, the Yamal peninsula	1	ABO, MN, Rhesus: D, C, E	1963; Rychkov, 1965; Sheremet'eva, 1971
Nenets, the Ob, the Pur, the Taz river basin	1	ABO, MNS, Rhesus: D, C, E	1963; Rychkov, 1965; Sheremet'eva, 1971
Khants, the Ob river lower reaches basin	1	ABO, MNS, Rhesus: D, C, E	1963; Rychkov, 1965; Sheremet'eva, 1971
Selkups, the Taz river basin	1	ABO, MNS, Rhesus: D, C, E	1963; Rychkov, 1965; Sheremet'eva, 1971
Komi, the Arctic Ural and the Kola peninsula	2	ABO, MNS, Rhesus: D, C, E, PTC	1963, 1967; Rychkov, 1965; Khazanova & Shamlyan, 1970; Sheremet'eva, 1971; Khazanova, Sheremet'eva & Spitsin, 1972
Lapps, the Kola peninsula	3	ABO, MNSs, Rhesus: D, C, c, E, e, P. Lewis, Hp, Tf, PTC	1967, 1969; Khazanova & Shamlyan, 1970; Khazanova, Sheremet'eva & Spitsin, 1972

Table 3.1—contd.

Ethnic group	Number of populations	Genetic markers	Year of study and publication
		Material cited in the literature	
Eskimos, the Chukotka peninsula	4	ABO, MN, Hp, Tf	1958, 1970; Levin, 1959; Spitsin, 1973
Chukchi, the Chukotka peninsula	6	ABO, MN, Hp, Tf	1957, 1970; Levin, 1959; Spitsin, 1973
Chukchi, lower reaches of the Kolyma river	1	ABO, MN	1959; Zolotareva, 1968
Yukaghirs, the Kolyma tundra	1	ABO, MN	1959; Zolotareva, 1968
Evens, the Kolyma tundra	1	ABO, MN	1959; Zolotareva, 1968
Yakuts, the Indighirka river	4	Hp, Tf, Gc	1970; Spitsin, 1973
Dolgans, the Taimyr peninsula	1	ABO, MN, Rhesus: D	1959; Zolotareva, 1968
Nganasans, the Taimyr peninsula	2	ABO, MN, Rhesus: D, P, Lewis, Hp, Tf, Gc, PTC	1959, 1972; Zolotareva, 1968; Spitsin, 1973
Nenets, the Yenisci and the Ob river mouth	2	ABO, MN, Rhesus: D, P, Lewis, Hp, Tf, PTC	1959, 1972; Zolotareva, 1968; Spitsin, 1973; Shluger, 1940

absolute sizes of all the above-listed peoples. (2). There is not a single people in the circumpolar zone of Eurasia typical enough for a characterization of the entire zone. (3). Only all the indigenous peoples together may reflect the properties of the circumpolar human population, so that research must be based on sampling data from all the peoples and the populations making them up. (4). In the limited period of the IBP such a study could not be expected with adequate detail for such a vast multi-disciplinary programme, suitable for taking up only one individual population, or small groups of such populations, and must therefore be based on a limited number of characteristics of each of the populations included in a sample.

For this study, therefore, the authors, with the assistance of postgraduate students, carried out research on population genetics among the populations listed in Table 3.1. A preliminary account of some of the work has already been published (IBP News, **19**, 1969). Also used were the published works of the following authors: Shluger (1940), Levin (1958, 1959), Zolotareva (1968), Spitsin (1973); in addition a number of preliminary data on the Nganasans, Yakuts and Nenets were kindly put at our disposal by V. A. Spitsin and I. V. Perevozchikov.

The review includes material pertaining only to those populations that are localized in the immediate vicinity of the Arctic Circle or still higher latitudes. Thus, the authors sought to avoid uncertainties regarding the climato-geographic border of the circumpolar sub-Arctic zone which, especially in the far eastern part, shifts considerably southward, down to the areas of Kamchatka and the Sea of Okhotsk populated by the Koryak, at which latitude in more westward areas lies the zone of the taiga, developed by other peoples. By taking the Arctic Circle as the southern border we exclude some of the influence of the peoples of the continental part of Eurasia on the characteristics of the circumpolar population. This persists in the form of outside gene migration pressure, which will be taken into account during further analysis.

Size of population

In the USSR the indigenous population in the circumpolar zone of Eurasia north of the Arctic Circle comes to about 82 500, which is about 8000 less than the total Eskimo population of the western hemisphere. However, the Eurasian population is divided into ethnic sub-populations whose sizes within the boundaries of the USSR were indicated in a preceding section.

The total area occupied by the circumpolar population within these borders is 3750 km². Taking into account the indigenous population alone, the area is populated at an average density of 0.02 persons per km². This is an average density, but distribution is not uniform: the discrete character of population distribution over this territory is due to the scarcity of sites along river banks and sea and lake shores situated conveniently close to the migration paths of animals. The same holds true for the distribution of reindeer grazing lands in the tundra. Therefore, the basic modern centers of population are also the most ancient habitats, and this has been confirmed archaeologically (Dikov, 1960; Lashuk, 1968).

Below the ethnic level, the simplest structural element of the total circumpolar population is an elementary population. The number of populations at this elementary level, estimated according to data at our disposal, is 320–30.

As was noted earlier, all the traditional sub-Arctic cultural–economic types that evolved in the process of social adaptation were founded and function on the principle of equilibrium between the human population and the

Table 3.2. *An estimate of the total size Nt of the elementary population of the Eurasian circumpolar zone*

Nt	Number of populations	Percentage from total sample
34–136	16	25.81
136–238	16	25.81
238–340	15	24.14
340–442	8	12.90
442–544	5	8.06
544–646	1	1.61
646–748	0	0
748–850	1	1.61
Totals	62	100.00

$Nt \pm s_{\bar{N}t} = 253 \pm 20$
$V_{Nt} = 25098$
$\text{H}_{Nt} = 159$

environment. Hallowed by tradition, the principle of equilibrium strictly regulates the economic activities of the collective and, particularly, the time, place, sequence and scope of man's action on objects of nature, be it a wild deer, a marine animal or reindeer lichen on the pastures. This inevitably leads to a reverse regulation of the properties and parameters of the human population itself, which establishes the existence of rather narrow limits of variability for the size of an elementary population in the sub-Arctic zone.

This size, designated as Nt has been estimated by sampling data making up 19% of the sum total of the populations. Appropriate size distributions are presented in Table 3.2. As one can see, the size actually proves rather stable in the strip from the Kola peninsula to Chukotka, changing within three orders of population size at an average value of $\bar{N}t = 253 \pm 20$ and a standard deviation of $\sigma_{Nt} = 158$. This estimate is very close to the average for the Eskimo population of the western part of the circumpolar zone $\bar{N}t = 267 \pm 31$, $\sigma_{Nt} = 97.0$ (Rychkov & Sheremet'eva, 1972). However, the essential difference in the level of economic–cultural differentiation of the populations west and east of Greenwich which has already been described is reflected in the different magnitudes of the standard deviation, which in the eastern part of the circumpolar realm is 1.5 times greater than in the west.

The average value found for the total size of an elementary circumpolar population is not only statistically significant. It accords with the practice of contemporary construction in the Soviet North (Vasiliev, Gurvich & Simchenko, 1967). The new settlements that are being built for the indigenous population are designed to house 200 people. This apparently reflects man's centuries-old experience in developing the circumpolar zone of Eurasia, in the course of which an optimal and rather stable population size has evolved characterized by the above cited estimates.

From these estimates, it is possible to estimate the average area developed by a population. This area comes to 11 520 km² for the circumpolar zone which on the whole agrees well with actual data on the areas occupied by individual populations in the north of Yakutia (12 000 km²) (Chersky, Kondakov & Morov, 1968). From such an estimate of a population's area, placing the population in its center, it can be said that the average distance between adjacent populations is in the range of 100–120 km. Taking into account the ribbon-like configuration of the entire circumpolar zone, the number of populations distributed over it longitudinally from the Kola Peninsula to Chukotka is 70, and latitudinally, from the Arctic Circle to the coast, is about 5. Thus a linear type of gene exchange between populations is being established, under the influence of which the genetic structure takes shape. This will be examined further, but first let us discuss the reproductive structure and the effective size of a population as an important parameter determining the level of random phenomena in the genetic composition of the circumpolar population.

Effective size of population

A possible approach to the assessment of the effective size of a population as Wright (1931) saw it, is the effective reproductive size, later designated as *Ne*. Its distinction from the reproductive size *Nr* depends on the scattering V_g of

Table 3.3. *Reproductive age limits in some circumpolar populations*

Ethnic group		Menarche (yr)	Age at birth of first Child (yr)		climacterium (yr)	Age at birth of last child (yr)		Authors
			Women	Men		Women	Men	
Aleuts of the	Mean		20		48			Rychkov &
Commander	Range		17–25		45–51			Sheremet'eva,
Island								1972a, 1973
Asiatic	Mean	14	20.8				42	Rychkov &
Eskimos	Range	13–16	18–24				31–54	Sheremet'eva,
								1972b
Chukchi	Mean	14	20.9		48		40	Rychkov &
	Range	12–17	16–28		45–50		30–52	Sheremet'eva,
								1972b
Yakuts of the	Mean	15.8	23.5	28.6	47	42	46	Gubenko, un-
Yana basin	Range	12–22	15–37	17–48	38–59	31–59	31–73	published data
Lapps of the	Mean		25	25	39			Khazanova, 1973
Kola	Range		17–33	17–33	32–50			

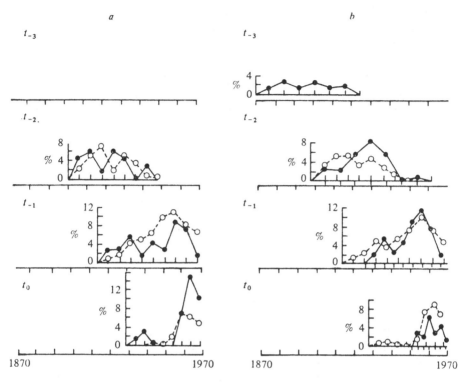

Fig. 3.3. Population structure according to generations, extention and overlapping of generations in time (in Rychkov & Sheremet'yeva, 1972b). *a*, —●—, Asian Eskimos; —○—, coastal Chukchi. *b*, —●—, taiga hunter and reindeer breeder Tungus–Evenks of Central Siberia; —○—, Arctic sea hunters: Asian Eskimos and coastal Chukchis.

Ordinate, percentage of total population and generation number; Abscissa, calendar year of birth for each member of population. Average year of birth of a generation is indicated as a point within 99% confidence limits.

the number of gametes g_i of the *i*th parent in relation to the average number of gametes $\bar{g}$ transmitted by the parent to the next generation. Reproductive size, as part of the total *Nt*, depends, in its turn, on the biological and social factors regulating the population's reproductive activity. The direct determinations of effective size have been carried out in several circumpolar populations (Rychkov & Sheremet'eva, 1972*a*, *b*). The biological and social data pertaining to the determination of reproductive age limits are given in Table 3.3. For assessing *Nr* we used the mixed biosocial limits of the beginning and the end of the reproductive period. When putting individuals in reproductive age categories we were not concerned with which generation in the population they belonged to. As genealogical data show, the overlapping of differently aged generations is very great indeed (*see* Fig. 3.3), whereas the

47

Table 3.4. *Effective population size and its equivalent reproduction time unit with account of generations overlap. From data on the Evenk population of mid-Siberia* (Rychkov *et al.*, 1974)

Generation and its size		Effective part of generation as percentage of total	Distribution of the Ne/Nt ratio of a population among generations	Overlapping of generations in effective part of population (%)
$t_n + 6$	89	2.25	0.003	1.15
$t_n + 5$	286	22.38	0.110	42.31
$t_n + 4$	165	51.51	0.140	53.85
$t_n + 3$	58	6.90	0.007	2.69
Totals	598		0.260	100.00

Unit of observation	Effective size	Unit of time (yr)
Population	27	$t = 25$
Generation	9	$3t = 75$

average intervals between generations is 25–8 yr. The estimates of Nr and Ne, derived without taking into account the overlapping of generations should refer to one generation. With the overlapping of generations taken into account, the effective population size turns out to be three times lower and should correspond to a unit of time equal to three generations, as had been determined for the Tungus populations of hunters and reindeer breeders in the taiga zone of Siberia (Table 3.4.) (Rychkov *et al.*, 1974).

Given the differences in the absolute values of Ne in the populations studied, attention is drawn to the extremely stable size of the Ne/Nt ratio of the total population size, fluctuating in the very narrow range of 0.20–0.36, among the Lapp populations of the Kola peninsula, the Yakuts of the Yana river basin, the Chukchis, Eskimos, Aleuts of the Kommander Islands and the Koryaks of Kamchatka (Table 3.5). These fluctuations are found to coincide with the values 0.21–0.34 established for the Indians of South America (Salzano, Neel & Maybury-Lewis, 1967).

Such stability of the Ne/Nt ratio in the populations studied can be used for the indirect determination of effective size Ne from the data on the distribution of the total size Nt of the total circumpolar population, as characterized in the preceding section. As shown in Table 3.2, the distribution of Nt values in a representative sample from the totality of circumpolar populations is characterized by the average value of $Nt = 253 \pm 20$ and the variance $V_{Nt} = 25\,098$. Taking into account fluctuations of total population size let us use for assessing the effective size a harmonic mean $H_{Nt} = 159$ /Table 3.2) and, by applying the established ratio $Ne/Nt = 0.28 \pm 0.01$ (Table 3.5) to it, we shall determine $Ne = 0.28 \times 159 \simeq 45$. This absolutely minute value of

Table 3.5. *Total* Nt, *reproductive* Nr, *reproductive-effective* Ne *size of population and the* Ne/Nt *ratio according to estimates in some circumpolar populations of Eurasia*

Ethnic group (no. of populations)	*Nt*	*Nr*	*Ne*	*Ne/Nt*
Aleuts, the Commander Islands (2)	223–72	75–86	70–81	0.27–0.31
Eskimos, the Chukotka peninsula (2)	249–99	78–144	54–98	0.22–0.34
Chukchi, the Chukotka peninsula (2)	180–209	70–92	50–76	0.28–0.36
Korjaks, the Kamchatka peninsula (2)	117–268	34–100	28–83	0.25–0.31
Yakuts, the Yana river basin (8)	258–843	108–344	68–173	0.20–0.34
Lapps, the Kola peninsula (2)	89–846	41–367	24–237	0.27–0.28

Average $Ne/Nt = 0.281 \pm 0.01$

an effective circumpolar population size points to the great role of random gene drift as the dominating background against which the process of genetic adaptation of populations takes place in the sub-Arctic zone. The measure in which the results of genetic adaptation may become the possession of all populations in the circumpolar zone depends on the intensity of gene exchange between the populations that counteracts random genetic drift.

The structure and intensity of gene migrations in circumpolar populations

To provide grounds for regarding a population as an integral whole, one needs first of all to determine whether there is any interaction between its constituent parts. On the plane of biological adaptation of human populations with which we are concerned, such interaction may rest on a genetic foundation and be implemented through gene exchange between populations. The directions and results of social adaptation are different for the different peoples and, due to gene migration, become the general possession of the population of a region provided that they have biological consequences for the population.

The existing methods of studying gene migration are applicable, and have in fact been applied, to small population groups only, in individual, rather limited, areas of Europe, America and Oceania (Cavalli-Sforza, 1962; Cavalli-Sforza & Bodmer, 1971; Morton, 1973).

Contrary to this, in our case it is necessary to trace gene migrations from the Kola peninsula of Fennoscandia to the Chukotka peninsula on the Bering

Table 3.6. Sampling data on the ethnic origin of 2939 persons belonging to the indigenous population of the Eurasian circumpolar zone

Ethnic group	1	2	3	4	5	6	7	8	9	10	11	12	13	Totals
1 Lapps	159	14	4											177
2 Komi		176	26	1										203
3 Nenets		21	387	17	3	11	1	2						442
4 Khants		2	16	200	2									220
5 Selkups			11	20	147									178
6 Enets								2						2
7 Nganasans			2			5	219	1						227
8 Dolgans									360	25				385
9 Yakuts									5	377				382
10 Evens									9	15	30	24		78
11 Yukaghirs									4	21	12	365	42	444
12 Chukchi												25		201
13 Eskimos													176	201
Totals	159	213	446	238	152	16	220	5	378	438	42	414	218	2939

Strait all along a circumpolar belt stretching over 160°. For this we shall make use of the ethnic differentiation of the population, due to which the genes exchanged by populations are marked by the ethnic origin of their carriers. This permits us to present a population as a totality of 'homozygotes' and 'heterozygotes' in regard to the ethnic origin of the genes, and take a direct count of the genes.

The initial sampling data on 2939 persons of established ethnic origin are given in Table 3.6. Diagonally indicated is the number of people originating from endogamous marriages within the given population; horizontally, the number of persons from exogamous marriages with one of the partners descending from another ethnic group; and vertically, the number of people in another ethnos, one of whose parents originates from the given ethnic group. In Table 3.7 the initial data are converted by the direct gene counting method into gene frequencies according to origin and indicate the frequency at which a given population gives off its genes (upper figure) and receives genes from outside (lower figure). The order of the ethnic groups given in Table 3.7 corresponds to their geographical situation.

In order to obtain a generalized gene migration index some additional information about the structure of the population is required. Thus, in respect to the entire North Asian subcontinent we used an assessment of gene migration intensity, directly proportional to the share of actually realized migration directions in regard to all possible directions linking a given population with other ones (Rychkov, 1973b; Rychkov $et\ al.$, 1973; Rychkov & Sheremet'eva, 1974).

$$m_i^{(u)} = \frac{n_i^{(u)}}{n_i^{(u)} + N_i} \cdot \frac{k_i^{(u)}}{K-1} \tag{1}$$

$$\bar{m} = \frac{1}{2K} \sum_{i(u,\,v)=1}^{K} m_i$$

where n_i is the number of migrations into the ith population, or from it; k_i, the number of realized migration directions; K, total number of possible migration directions; u, v, directions (to and fro) relative to the ith population.

This method, in which all the gene exchange directions between the populations are assumed to be equally probable, therefore proceeds from the concept of a population structure as a system of 'island' isolates. It has been established that the assessment of the migration coefficient $m = 0.0107 \pm 0.0018$, obtained by means of equation (1), agrees well with the assessment $m = 0.0106 \pm 0.0020$, derived through the gene frequency variance of the MN blood groups in northern Asia.

The application of equation (1) in the circumpolar zone gives an assessment of the migration coefficient $m = 0.0308 \pm 0.0068$, which turns out to

51

Table 3.7. Sample estimates of gene migration frequencies in circumpolar populations (top values refer to direction from given population to another along column: bottom values refer to direction into given population from others on line: diagonally, absence of migration frequency)

Group	1	2	3	4	5	6	7	8	9	10	11	12	13
1 Lapps	1.0000	0.0329	0.0045										
	0.9492	0.0395	0.0113										
2 Komi		0.9131	0.0291	0.0021									
		0.9335	0.0640	0.0025									
3 Nenets		0.0493	0.9340	0.0357	0.0099								
		0.0238	0.9378	0.0192	0.0034								
4 Khants		0.0047	0.0179	0.9202	0.0066								
		0.0045	0.0364	0.9546	0.0045								
5 Selkups			0.0123	0.0420	0.9835								
			0.0309	0.0562	0.9129								
6 Enets			0.0022				0.0023						
			0.0044				0.0011						
7 Nganasans						0.0124	0.9977	0.0023					
						0.0110	0.9824	0.0022					
8 Dolgans													
9 Yakuts									0.9762	0.0285			
									0.9675	0.0325			
10 Evens									0.0066	0.9304			
									0.0065	0.9935			
11 Yukaghirs									0.0119	0.0171	0.8572	0.0290	
									0.0577	0.0961	0.6924	0.1538	
12 Chukchi									0.0053	0.0240	0.1428	0.9408	0.0963
									0.0045	0.0236	0.0135	0.9111	0.0473
13 Eskimos												0.0302	0.9037
												0.0622	0.9378

be three times higher than for Siberia as a whole, its circumpolar zone included. Such a difference in the gene exchange intensity of the sub-Arctic belt and the continental regions of northern Asia would have been difficult to predict. This result probably points to a different structure of gene migrations and should be checked with the account of the close to linear situation of populations in the circumpolar zone, i.e. with the account of the unequal probability of gene exchange between different parts of the circumpolar population.

Let us analyze the data presented in Tables 3.6 and 3.7 proceeding from the linear stepping-stone model of population structure (Kimura & Weiss, 1964). The maximum number of steps through which direct gene exchange passes, as Table 3.7 shows, is five, i.e. it differs from the ideal gene exchange pattern only between adjacent populations, and for this reason the variance V_m is used in the calculation (migration in both directions relative to each population was taken into account). According to the authors of the model, the gene migration index is measured by the value

$$m(1 - r(1)) = \sigma_m{}^2 (1 - r(1)) \qquad (2)$$

where m is the migration rate between adjacent populations,

$r(1)$ – the genetic correlation between them,

$\sigma_m{}^2$ – variance of migration over distances larger than one step,

Table 3.8. *Assessment of gene migration intensity in a circumpolar population, based on the linear stepping-stone model of population structure*

Number of steps between the exchanging populations	Number of observations in the forward and reverse directions	Mean coefficient of gene migration
k		m_k
1	36	0.0342
2	32	0.0083
3	28	0.0008
4	24	0.00045
5	20	0.0001

$V_m = 0.0835$, $m_\infty = 0.0107$

$a = 1 - m - V_m = 0.9058$

$2\beta = V_m = 0.0835$

$R_1 = \sqrt{(1 + a)^2 - (2\beta)^2} = 1.9039$

$R_2 = \sqrt{(1 - a)^2 - (2\beta)^2} = 0.0436$

$1 - r(1) = \dfrac{1}{4d\beta} \left\{ R_1 R_2 - 1 + (1 - m_\infty{}^2) \right\} = 0.4079$

$m_1(1 - r(1)) = V_m(1 - r(1)) = 0.0341$

in which the multiplier $1 - r$ (1) depends on outside gene migration pressure m_∞ on the entire system of populations. We used for this figure the gene migration coefficient earlier derived for the sub-continent of northern Asia as a whole (Table 3.8). The chief results obtained from the sampling data and those concerning gene migration patterns in the circumpolar zone and in northern Asia are shown in the diagram (Fig. 3.4).

A comparison of the assessments of gene migration intensity, according to equations (1) and (2) indicates that they are sufficiently close (0.0308 and 0.0341 respectively). This confirms the very high level of gene exchanges between sub-Arctic zone populations that could hardly be predicted.

According to population genetics theory, a migration coefficient of even $m = 0.01$ is considered sufficiently high for regarding the corresponding totality of populations as an integral whole (Moran, 1962). There is, therefore, another reason to regard as a whole all the circumpolar populations relative to their genetic peculiarities.

Genetic markers in the circumpolar population

Summary sampling data on the distribution of blood groups, serum proteins and sensitivity to PTC in the circumpolar population of Eurasia are presented in Table 3.9.

The population is characterized within three borderlines from the Bering Strait: (1) to Scandinavia inclusive – total population; (2) to the Kola peninsula inclusive – population within the USSR; and (3) to the Arctic Urals – population within northern Asia. The sampling for the two latter regions sums up the data obtained from different points in the area throughout different years of research and by means of several differing methods of investigation. These differences concern first of all the number of sera used for determining MNSs and Rhesus blood groups, from which the sampling in the given review is described assuming a minimum set of phenotypes. The differences are also due to the methods of determining sensitivity to PTC used by different authors and by the same authors in different years. In recent years, a modified methodology with the use of a series of 30 (instead of 14) dilutions of a saturated solution and drop-by-drop administration (instead of large volumes), has been used for characterizing the peoples of the Bering Sea area and the Lapps of the Kola peninsula (Khazanova, Sheremet'eva & Spitsin, 1972; Rychkov & Sheremet'eva, 1972a, b). This technique allows one to distinguish four phenotypes in a sample (Rychkov & Borodina, 1969, 1973). However, in this case only two phenotypes are taken into account. The experience in studying Rhesus blood groups with the use of 6 or 5 sera showed that one should not ignore the presence of the Rh_z (CDE) haplotype in northern Asia, which rises to considerable frequencies by way of random gene drift among some Siberian populations. This experience was

Table 3.9. *Some genetic markers in the circumpolar population of Eurasia*

ABC Groups

From the Bering Strait to	Number tested		Phenotypes (%)				χ^2	Genes (%)		
			O	A	B	AB		O	A	B
The Urals	3595	OBS	37.86	32.57	23.45	6.12	6.52	62.09	21.77	16.14
		EXP	38.55	31.77	22.65	7.03				
The Kola peninsula	4295	OBS	34.85	34.83	23.80	6.52	14.48	59.84	23.54	16.61
		EXP	35.81	33.71	22.64	7.82				
Scandinavia	6413	OBS	32.87	42.13	18.73	6.27	25.18	58.20	28.32	13.48
		EXP	33.87	40.98	17.51	7.64				

MN Groups

			MM	MN	NN	χ^2	M	N
The Urals	2691	OBS	26.61	43.29	30.10	47.72	48.26	51.74
		EXP	23.29	49.94	26.77			
The Kola peninsula	2874	OBS	27.46	43.42	29.12	49.21	49.17	50.83
		EXP	24.18	49.98	25.84			
Scandinavia	3768	OBS	25.85	45.54	28.61	29.29	48.62	51.38
		EXP	23.64	49.96	26.40			

MNSs groups (samples tested with anti-M, anti-N and S)[a]

			MMS	MsMs	MNS	MsNs	NNS	NsNs	χ^2	MS	Ms	NS	Ns
The Urals	251	OBS	17.93	13.54	25.09	23.11	7.57	12.75	0.56	19.80	35.77	9.90	34.52
		EXP	18.08	12.79	24.67	24.70	7.82	11.92					
The Kola peninsula	271	OBS	18.45	13.65	25.46	22.14	8.49	11.81	0.56	20.02	35.88	10.99	33.11
		EXP	18.38	12.87	25.55	23.76	8.49	10.96					
Scandinavia	647	OBS	15.30	8.96	31.07	19.47	13.29	11.90	0.79	20.03	29.50	16.46	34.00
		EXP	15.83	8.70	29.92	20.06	13.90	11.56					

Table 3.9—contd.

P groups

		P+	P−	P$_1$	P$_2$
The Urals	258	41.09	58.91	23.25	76.75
The Kola peninsula	439	56.49	43.51	34.04	65.96
Scandinavia	1577	59.67	40.33	36.50	63.50

Rh groups (samples tested with anti-D only)

		Rh+	Rh−	D	d
The Urals	2285	98.99	1.01	89.95	10.05
The Kola peninsula	2468	98.86	1.14	89.32	10.68
Scandinavia	3187	98.62	1.38	88.25	11.75

Rh groups (samples tested with anti-C, anti-D and anti-E)[a]

			CDE	CDee	CddE	Cddee	ccDE	ccDee	ccddEe	ccddee	χ²	CDE	CDe	Cde	cDE	cdE	cDe	cde
The Urals	408	OBS	34.80	41.91	0.00	0.24	17.15	4.66	0.24	0.98	9.41	2.56	46.66	1.09	28.71	1.15	9.93	9.9
		EXP	32.92	41.51	0.66	0.23	20.51	2.95	0.24	0.98								
The Kola peninsula	428	OBS	35.28	41.59	0.00	0.23	16.82	4.91	0.23	0.93	10.78	2.79	46.44	1.07	28.73	1.13	10.21	9.64
		EXP	33.24	41.22	0.48	0.22	20.67	3.01	0.23	0.93								
Scandinavia	1181	OBS	20.58	59.10	0.00	0.17	12.45	3.89	0.09	3.72	2.74	0.35	54.07	0.43	17.94	0.23	7.69	19.2
		EXP	20.35	58.94	0.15	0.17	13.02	3.56	0.09									

Lewis Groups (samples tested with anti-Le^a and anti-Le^b)

		Le(a+b−)	Le(a−b+)	Le(a−b−)	Le(a+b+)	Le^a
The Urals	323	1.86	80.80	17.34	0	13.64
The Kola peninsula	502	3.59	78.69	17.72	0	18.95
Scandinavia	685	4.96	79.42	15.62	0	22.27

Table 3.9—contd.

Taste test PTC[a]

	tested	sensitive	non-sensitive	T	t
The Urals	1269	81.32	18.69	56.78	43.22
The Kola peninsula	1439	78.80	21.20	53.96	46.04
Scandinavia	1975	79.19	20.81	54.38	45.62

Haptoglobin

			Hp 1 – 1	Hp 2 – 1	Hp 2 – 2	χ^2	Hp^1	Hp^2
The Urals	842	OBS	11.16	45.60	43.23		33.96	66.03
		EXP	11.53	44.85	43.60	0.21		
The Kola peninsula	968	OBS	11.16	46.59	42.25		34.46	65.54
		EXP	11.87	45.17	42.95	0.85		
Scandinavia	1633	OBS	12.80	46.72	40.48		36.16	63.84
		EXP	13.08	46.17	40.76	0.25		

Transferrin

			C	CD_{chl}	$Bo – {}^1C$	χ^2	Tf^c	Tf^{Dchl}	$Tf^{Bo - 1}$
The Urals	850	OBS	96.82	0.94	2.24		98.41	0.47	1.12
		EXP	96.84	0.92	2.20	0.90			
The Kola peninsula	956	OBS	96.76	1.25	1.99		98.38	0.62	1.00
		EXP	96.79	1.22	1.97	1.16			

Gc Groups

			Gc (1 – 1)	Gc (2 – 1)	Gc (2 – 2)	χ^2	Gc^1	Gc^2
The Urals	529	OBS	37.24	47.64	15.12		61.01	38.99
		EXP	37.28	47.55	15.16	0.16		

[a] See p. 54 and Table 3.1.

57

taken into account when reducing the initial materials to their present shape, in which the method of assessing Rhesus gene frequencies differs somewhat from that suggested by Mourant (1954). Similarly, we recalculated the gene frequencies of Rhesus blood groups in the data for the Lapps of Scandinavia (Allison *et al.*, 1952, 1956; Beckman, 1959; Broman *et al.*, 1959). Apart from data by these authors, data by Eriksson *et al.*, 1967 and Eriksson, 1968 were used for characterizing the Lapp population outside the USSR.

The gene frequencies of different loci, calculated from the phenotype data of Table 3.9, represent the weighted mean assessments of gene frequencies in the circumpolar population. These were weighted by the different sample sizes from various parts of the historically and geographically differentiated total populations. No wonder, therefore, that for loci where an assessment of the Hardy–Weinberg equilibrium is possible, this equilibrium was found to be disrupted by a deficit of heterozygotic and an excess of homozygotic genotypes (ABO, MN blood groups). It is more startling that this equilibrium holds for the genotypes of a number of loci, despite the fact that the relevant samples were taken from a huge territory (MNSs, Hp, Tf, Gc, as well as Rh for the total population). Thus, proceeding from the distribution of a number of independent systems of genetic markers, the circumpolar population appears on some criteria homogeneous and panmictic and on others heterogeneous and inbred. The cause of this contradiction is hardly due to the different degree of sampling data representative of the different loci, since all the samples comprise a considerable proportion (0.5–8%) of the total population in the circumpolar zone. More probably, the reason lies in the different properties of the genotypes of different loci, whereas the conditions and rates of isolation and gene exchange between the parts of the circumpolar population are the same. Even in regard to the most representative data for ABO and MN blood groups one observes a different degree of deviation from the Hardy–Weinberg law, and, significantly, it appears to be in reverse to the representativeness of the samples for these two loci.

The same picture was observed in still greater contrast with regard to the total sampling from the indigenous population of Siberia within the boundaries from the Urals to the Pacific coast (Table 3.10). For 8633 persons examined in 70 populations the Hardy–Weinberg equilibrium for the ABO blood group locus is assessed by the value $\chi^2 = 0.009$. Of the 8633 persons blood typed for ABO, 6889 were simultaneously typed for MN blood groups. In the MN blood groups locus the Hardy–Weinberg equilibrium is characterized by the value $\chi^2 = 290.6$ with a high deficit of heterozygotes. This served as a preliminary indication of the different degree of selective neutrality of the two blood-group systems under Siberian conditions, which was confirmed by a measurement of the adaptive value of the genotypes of these loci (Rychkov, 1969, 1973*b*; Rychkov *et al.*, 1973; Rychkov & Sheremet'eva,

Table 3.10. *Distribution of ABO and MN blood groups in the total population of Siberia as an example of extreme states of Hardy–Weinberg equilibrium in independent loci*

ABO Groups

Number tested		Phenotypes (%)				χ^2 d.f.	Genes (%)		
		O	A	B	AB		O	A	B
8633	OBS	38.97	27.34	26.61	7.09	0.0090 1	62.41	19.02	18.57
	EXP	38.96	27.36	26.62	7.06				

MN Groups

Number tested		Phenotypes (%)			χ^2 d.f.	Genes (%)	
		MM	MN	NN		M	N
6889	OBS	37.96	38.86	23.18	290.6 1	57.39	42.61
	EXP	32.94	48.90	18.16			

The human biology of circumpolar populations

1974). A similar argument could be suggested in regard to an analogous picture revealed in the circumpolar population.

The data on gene frequencies in a circumpolar population represented in Table 3.9, though satisfactory in form, fall short of the essential purpose of the investigation. The formally correct assessments of gene frequencies as weighted mean frequencies in fact introduce a lot of chance into the genetic characteristic of the population. The statistical procedure of weighting gene frequencies according to the sizes of samples obviously contradicts the purpose of such characteristics, since the differences in the sizes of samples drawn from these or other populations in the circumpolar zone are by no means related either to the biological or the social characteristics of these populations, and fully depend on chance factors in the organization of expeditionary work over so vast a territory. Let us move, therefore, from characterizing the circumpolar population according to summary sampling data, to a description based on non-weighted mean gene frequencies $\bar{q}$ in the populations concerned (Table 3.11).

However small the number of populations studied in the Soviet part of the circumpolar zone (10–45, according to different loci), it nevertheless comprises 3–14% of the total number of populations existing within these boundaries. Because of this, highly significant non-weighted mean gene frequencies can be established, which may provide a reliable characteristic of the circumpolar populations along the belt between 30° E and 170° W of Greenwich. 90% of the length of the belt to which the distributions of the studied populations refer lies within northern Asia, and about 90% of the population originates from the Asian part of the circumpolar area. In this connection special attention should be given to a comparison of the mean gene frequencies of the Asian part of the circumpolar population with similarly calculated non-weighted mean gene frequencies in the Siberian populations south of the Arctic Circle (Table 3.11). A characteristic of the total Siberian population, excepting its circumpolar part, is cited from other works (Rychkov & Sheremet'eva, 1972a, b; Sheremet'eva, 1973).

Significant differences between the bulk of the indigenous population of Siberia and the population of its circumpolar zone are established according to a number of genes of different loci. Taking into account only the significant differences in gene frequencies (for example, A, B, N, P_2), the circumpolar populations of Siberia differ from other Siberian populations in the same measure in which they approximate genetically to the Lapps and Eskimos outside Asia. In any case, the significant differences observed point to the reality of the genetic specificity of the circumpolar population not only within the boundaries of Europe and America, but also within Asia, thus indicating a special direction in the evolution of human populations in the circumpolar zone.

However, neither the gene frequencies *per se*, nor the established fact of

Table 3.11. *Mean gene frequencies in the populations of the circumpolar zone within three border lines from the Bering Strait to the Urals, the Kola peninsula and Scandinavia, and comparative data on the populations of Siberia south of the Arctic Circle*

Locus	Gene	No. of populations	$\bar{q}$	$s_{\bar{q}}$	No. of populations	$\bar{q}$	$s_{\bar{q}}$	t	d.f.	P
		Circumpolar populations			The populations of Siberia south of the Arctic Circle (Rychkov & Sheremet'eva, 1972a, b; 1973; Sheremet'eva, 1973)			Comparison between Siberian populations north and south of the Arctic Circle		
ABO	A	39	0.2164	0.0117	147	0.1880	0.0074	1.8164	184	<0.05
		45	0.2298	0.0127						
		58	0.2676	0.0144						
	B	39	0.1449	0.0117	147	0.1907	0.0083	2.6466	184	<0.01
		45	0.1529	0.0110						
		58	0.1324	0.0104						
	O	39	0.6387	0.0153	147	0.6292	0.0094	0.4396	184	
		45	0.6172	0.0160						
		58	0.5999	0.0142						
MNSs	N	30	0.4729	0.0261	128	0.3560	0.0185	2.8948	156	<0.0025
		34	0.4652	0.0256						
		39	0.4686	0.0229						
	S	9	0.3416	0.0468	66	0.4186	0.0240	1.1414	73	
		10	0.3699	0.0506						
		12	0.3762	0.0423						
	Ms	9	0.3132	0.0425	66	0.3308	0.0164	0.3740	73	
		10	0.3130	0.0380						
		12	0.2995	0.0348						
	MS	9	0.2262	0.0428	66	0.2576	0.0194	0.5696	73	
		10	0.2324	0.0388						
		12	0.2302	0.0323						
	Ns	9	0.3452	0.0349	66	0.2505	0.0156	2.1324	73	<0.025
		10	0.3170	0.0421						
		12	0.3242	0.0358						
	NS	9	0.1154	0.0205	66	0.1610	0.0159	1.0378	73	
		10	0.1375	0.0287						
		12	0.1460	0.0244						
Rhesus	d	26	0.0527	0.0191	122	0.0350	0.0083	0.9436	146	
		30	0.0713	0.0193						
		33	0.0788	0.0181						
	c	10	0.4734	0.0408	62	0.5711	0.0150	2.3856	70	<0.01
		11	0.4625	0.0386						
		18	0.4439	0.0260						
	e	10	0.6674	0.0544	62	0.5967	0.0101	2.1524	70	<0.025
		11	0.6841	0.0504						
		18	0.7517	0.0376						

Table 3.11—contd.

Locus	Gene	Circumpolar populations			The populations of Siberia south of the Arctic Circle (Rychkov, & Sheremet'eva, 1972a, b; 1973; Sheremet'eva, 1973)			Comparison between Siberian populations north and south of the Arctic Circle		
		No. of populations	$\bar{q}$	$s_{\bar{q}}$	No. of populations	$\bar{q}$	$s_{\bar{q}}$	t	d.f.	P
Rhesus	R_z	9	0.1002	0.0201	62	0.1165	0.0118	0.5074	69	
		10	0.1097	0.0203						
		17	0.0744	0.0164						
	R_0	9	0.1694	0.0212	62	0.2792	0.0184	2.2260	69	<0.025
		10	0.1645	0.0261						
		17	0.1202	0.0181						
	R_2	9	0.2324	0.0407	62	0.2865	0.0146	1.3060	69	
		10	0.2321	0.0365						
		17	0.1951	0.0254						
	R_1	9	0.4260	0.0463	62	0.3121	0.0180	2.2523	69	<0.025
		10	0.4289	0.0415						
		17	0.4834	0.0319						
P	P_2	7	0.7618	0.0214	89	0.6098	0.0243	1.7530	94	<0.05
		11	0.6364	0.0555						
		15	0.6278	0.0470						
Lewis	Le^a	7	0.0822	0.0529	59	0.2019	0.0228	1.7361	64	<0.05
		11	0.1135	0.0494						
		12	0.1289	0.0472						
PTC	t	13	0.4610	0.0371	36	0.4258	0.0311	0.6230	47	
		18	0.5032	0.0319						
		20	0.4921	0.0311						
Hp	Hp^1	17	0.3579	0.0243	31	0.3389	0.0240	0.5125	46	
		19	0.3630	0.0221						
		21	0.3647	0.0206						
Tf	Tf^c	17	0.9840	0.0046	21	0.9860	0.0048	0.2873	36	
		18	0.9839	0.0034						
Gc	Gc^1	15	0.6292	0.0237	8	0.6348	0.0563	0.1078	21	

their difference from the gene frequencies in the more southerly areas of Eurasia, contain any indications as to whether these gene frequencies and gene differences are the result of neutral or adaptive evolution on the part of the circumpolar population. In the light of this question let us turn to the genogeography of the circumpolar territory.

Genogeography of the circumpolar population of Eurasia

The entire circumpolar zone of Eurasia is permeated by migration of genes which in about 70 generations (expected number of populations according to longitude) may penetrate from the extreme western point to the extreme eastern, or vice versa. In this connection it would be useful to take a look, if only as a preliminary step, at the actual picture of gene frequency distribution in the geographical space of the circumpolar zone. We shall examine the appropriate data at the level of the ethnic organization of the circumpolar population. The mean frequencies $\bar{q}$ and, for some inadequately studied ethnic groups and genes, the single populational frequencies q_i have been related to the centers of the territories occupied by the ethnic groups. Each such center has one coordinate in longitudinal degrees east of Greenwich (Table 3.12). Data from Table 3.12 are presented in graphic form in Fig. 3.4, which reveals that in the majority of cases gene frequencies of different loci represent most diverse functions of geographic distance. This visual impression of gene frequency dependence on distance is confirmed by a correlative assessment (Table 3.13). According to the genes of different loci, the average absolute value of the correlation coefficient is 0.433 ± 0.060, while the absolute value of the gene frequency regression coefficient per degree of geographic distance is $b_{q/d} = 0.00091 \pm 0.00027$. However, the gene frequency distribution curves given in Fig. 3.5 indicate that the revealed dependencies are by no means of a linear order and, consequently, their assessment through correlation and rectilinear regression is rather approximate. Nevertheless, given such a preliminary examination, the existence of gene frequency gradients related to geographic separation in the circumpolar region comes to light.

The existence of such gradients in biogeography usually indicates the dependence of a character on natural factors, which is true in the case of a character's clinal variability oriented to geographic latitude, when the gradients of changing environmental factors are apparent. In this case, however, we were dealing with the sufficiently homogeneous natural conditions of the circumpolar belt, which hardly allows the assumption of environmental longitudinal heterogeneity as the cause of the correlations detected. Longitudinally oriented gene migrations constitute a necessary, but obviously insufficient, condition for the emergence of such correlations and such gradients, for this would also necessitate the assumption of a certain 'difference of potentials' at the end points of distribution, without an explanation of its causes, i.e. in Fennoscandia and in Chukotka. Moreover, if such correlations appeared only because of gene migration in the neutral sub-Arctic natural environment, then all the empirical geographic distributions of the gene frequencies at different loci would have to reflect one and the same type of function. In reality, the picture is quite different and the gene frequencies at

63

Table 3.12. *Gene frequencies in some geographical points of the circumpolar belt of Eurasia. (The number of populations studied is given in parentheses)*

Locus	Gene	25° E	35° E	55° E	65° E	73° E	83° E	90° E	95° E	110° E	126° E	144° E	155° E	172° E	175° W
ABO	A	0.3985	0.3459	0.1666	0.1845	0.2200	0.1112	0.1534	0.2575	0.1720	0.2408	0.3160	0.3494	0.1990	0.2348
	B	0.0615	0.1909	0.2116	0.2506	0.1880	0.0752	0.0822	0.1214	0.2310	0.1498	0.0780	0.0624	0.1003	0.1654
	O	0.5400	0.4632	0.6218	0.5649	0.5920	0.8136	0.7644	0.6211	0.5970	0.6094	0.6060	0.5881	0.7007	0.5998
		(13)	(3)	(2)	(6)	(1)	(3)	(1)	(2)	(1)	(10)	(1)	(1)	(7)	(4)
MNSs	N	0.4921	0.3056	0.6085	0.5979	0.5400	0.4254	0.6170	0.6860	0.3535	0.2200	0.3085	0.4075	0.4271	0.4382
		(5)	(3)	(2)	(5)	(1)	(3)	(1)	(2)	(1)	(1)	(1)	(2)	(8)	(3)
	Ms	0.2318	0.3116	0.3614	0.2756	0.2241	0.0931							0.3126	0.4131
	MS	0.2194	0.2884	0.2496	0.2244	0.3315	0.4838							0.1858	0.1249
	Ns	0.3604	0.0635	0.2158	0.4772	0.3202	0.2992							0.3414	0.3706
	NS	0.1884	0.3365	0.1732	0.0228	0.1242	0.1239							0.1602	0.0914
		(2)	(1)	(1)	(1)	(1)	(1)							(2)	(3)
Rhesus	d	0.1533	0.1999	0.2892	0.1256	0.1291	0.0000	0.0000	0.0470	0.0000	0.0262			0.0000	0.0000
		(3)	(3)	(2)	(4)	(1)	(3)	(1)	(2)	(1)	(9)			(2)	(3)
	R_z	0.0215	0.1951		0.0516	0.0359	0.0349							0.1238	0.1602
	R_o	0.0568	0.1201		0.1896	0.2313	0.2047							0.1586	0.1307
	R_2	0.1422	0.2299		0.1848	0.2158	0.0599							0.2287	0.3297
	R_1	0.5458	0.4549		0.3839	0.2501	0.7005							0.4889	0.3794
		(7)	(1)		(2)	(1)	(1)							(2)	(3)

Longitude from Greenwich

		25°E	35°E	55°E	65°E	95°E	126°E	172°E	175°W
P	P_2	0.6043 (4)	0.3910 (3)	0.2462	0.6708	0.7428 (1)		0.7683 (2)	0.7941 (3)
Lewis	Le^a	0.2982 (1)	0.1616 (3)	0.4324	0.2886	0.2865 (1)		0.0000 (2)	0.0000 (3)
PTC	t	0.3925 (2)	0.5633 (3)	0.7385	0.5252 (2)	0.3996 (1)	0.5340 (7)	0.2179 (2)	0.5024 (2)
Hp	Hp^1	0.3800 (2)	0.3630 (3)	0.4260 (2)	0.3985 (2)	0.2470 (1)	0.4680 (4)	0.2688 (5)	0.3558 (4)
Tf	Tf^c		0.9810 (3)	0.9840	0.9800 (2)	0.9647 (1)	0.9881 (4)	0.9941 (5)	0.9743 (4)
Gc	Gc^1						0.6305 (4)	0.6628 (5)	0.6132 (5)

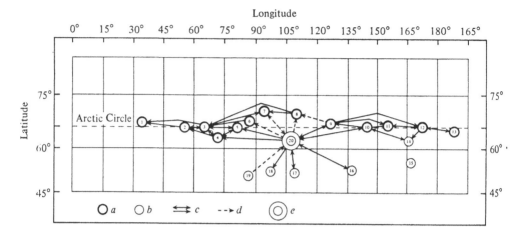

Fig. 3.4. Structure of gene migration in indigenous population of the circumpolar zone of Eurasia and Siberia, established from sampling data (modified according to Rychkov & Sheremet'eva, 1974). *a*, circumpolar populations examined in this work; *b*, other populations of northern Asia; *c*, established directions of migration; *d*, directions not reflected in sampling data; *e*, nucleus of gene migration structure. Scale 1:100 000 000. 1, Kola Lapps; 2, Komi; 3, Nenets; 4, Khants; 5, Selkups; 6, Enets; 7, Nganasans; 8, Dolgans; 9, Yakuts; 10, Evens; 11, Yukaghirs; 12, Chukchi; 13, Eskimos; 14, Koryaks; 15, Aleuts; 16, Amur and Sakhalin peoples; 17, Buryat; 18, Peoples of the Sayan plateau; 19, Peoples of the Altai highlands; 20, Evenks.

different loci are related in different ways to the geographic longitude of the locality. By way of example the approximations of empirical distributions are cited by mathematical functions of different types.

The shaping out dependencies of the periodic type (genes Tf^c, Hp^1, t, B, A), pointing to the cyclic nature of gene frequency in spatial dynamics, are of special interest as they may reflect the cyclic character of the interaction of a population with the environment, usually taking place in well balanced ecological systems. Such a dependence may be interpreted as the result of a stationary process of genetical reorganization of the generations of a circumpolar population.

No less indicative are the shaping out dependencies of the parabolic type (genes O, d, R), which show that, apart from the geographical longitude of a population's habitat, gene frequency may be influenced also by the square and the cube of the location's longitude, corresponding to which may be the area and volume of the circumpolar environment, which thus becomes significant in the process of a population's genetic adaptation.

A deeper analysis is not possible in this review and with the materials available today, but we shall put forward one more argument in favour of gene frequency dependence on the natural conditions of the sub-Arctic area.

As was noted above, there is no ground for assuming environmental

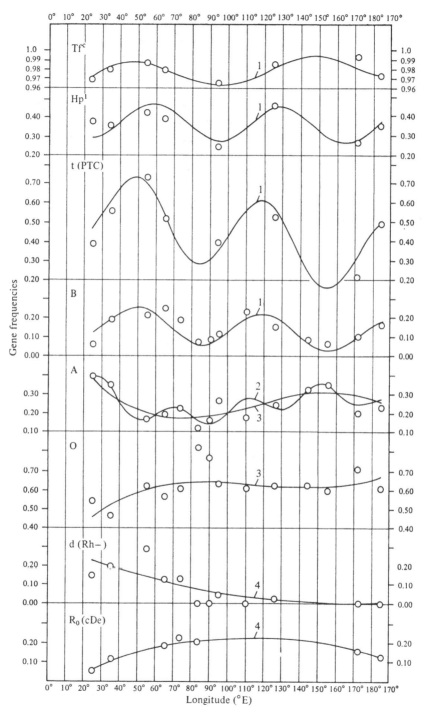

Fig. 3.5. Geographical longitude gene frequency distribution in some loci of the circumpolar belt of Eurasia.

Approximating functions: 1, $y = a + bx + c \sin k\,x + d \cos k\,x$; 2, $y = a + bx + cx^2 + dx^3 + e \sin kx + f \cos kx$; 3, $y = a + bx + cx^2 + dx^3$; 4, $y = a + bx + cx^2$.

heterogeneity in the circumpolar belt. Yet there is another characteristic of this belt, which, just as gene frequencies, turns out to be related to geographic longitude. This is the area of the sub-Arctic climato-geographic belt, at its minimum in the European part and reaching its maximum in the Chukotka–Kamchatka region in Asia. Such unequal development of areas with a tundra landscape and sub-Arctic climate is the consequence of different degrees of extremity of the same natural conditions and it is, therefore, quite probable that the law-bound alteration of longitudinal environmental extremity along the Arctic Circle is precisely what underlies the connections of gene frequencies with longitude. The different type of dependence of gene frequencies in different loci on the geographic coordinate, revealed in this case, points to the different degree of neutrality and adaptive value of the studied loci in the sub-Arctic natural environment. In this connection let us turn to an analysis of gene frequency variances, from which it is possible to derive new and more direct indications about the different degree of the susceptibility of the gene loci under review to pressure on the part of the sub-Arctic natural environment.

An analysis of gene frequency variances in the circumpolar zone of Eurasia

Since an assessment of the role of genetic factors in the process of human population adaptation in the circumpolar zone is an end purpose of the work undertaken in the IBP–HA section, we shall evaluate the degree of neutrality of the loci studied in the sub-Arctic natural environment.

Let us assume a neutral evolution, as a process of genetic differentiation of the initial population, proceeding on the basis of random gene drift counteracted by gene migration. In the conditions described above of small circumpolar populations ($Ne = 45$) and the considerable level of gene migration ($m = 0.448$) characteristic of the circumpolar zone, one may neglect the recurrent mutation pressure and possible differences in the mutation rates in the loci concerned. Under these conditions the gene frequency variances V_{qi} of the different loci, being standardized according to obviously different average gene frequencies $V_q/\bar{q}_i (1 - \bar{q}_i) = f_i$, must be similar within limits allowed by the accidents of sampling.

Conversely, in the case of the process of adaptive evolution, the variances will differ because not all the polymorphic gene loci concerned are equally involved in this process.

However, a direct comparison of gene variances in different loci, similar to what has been done for the Indian populations of the Yanomama, Makiritare and Xavante (Neel & Ward, 1972) is inadvisable in view of the fact that the pressure of natural selection on the different loci may vary not only in magnitude, but also in regard to type, from stabilizing to disruptive.

Proceeding from the hypothesis of neutral evolution, comparison of variances according to genes can be adequately replaced by comparison with some standard theoretical variance value f_e. For this it must, in turn, be assumed that the number of generations through which the process developed was sufficient for the establishment of a stable equilibrium between random gene drift and gene migrations, as opposing forces affecting variability of gene frequencies. In this case, according to Wright (1951)

$$f_e = \frac{1}{4Nm - 1} \qquad (3)$$

The data already given on the age of the successive development of the circumpolar population, going back to pre-Neolithic times, and in the long run to the final stage of the Upper Paleolithic period, can now be adopted as a foundation for assuming a stable equilibrium between random and systematic factors in the neutral evolution of the population.

The gene migration coefficient in equation (3) has the following equivalent in the stepping-stone model of populational structure (Kimura & Weiss, 1964)

$$m = m \infty - m_1 (1 - r (1)) \qquad (4)$$

where $m_1 = \sigma_m^2$ and $r(1)$ is the genetic correlation between adjacent gene-exchanging groups.

Thus, the theoretical value of the variance in the conditions of a step gene migration structure equals:

$$f_e^1 = \frac{1}{4N [m_\infty - \sigma_m^2 (1 - r (1))] + 1} \qquad (5)$$

In the concrete case of the circumpolar zone of Eurasia, which comprises a limited number of populations, the theoretical value of the variance undergoes the following correction (Kimura & Weiss, 1964)

$$f_e = f_e^1 \left\{ 1 - \frac{1}{n} - \frac{2}{n^2} \sum_{i=1}^{n} (n - i) r (i) \right\}, \qquad (6)$$

where $r(i)$ is the genetic correlation between populations separated by i steps ($i = 1, 2, \ldots n$). By using demographic data on the effective population size and on the step structure of gene migrations described above, let us calculate the theoretical value of the variance (Table 3.13).

The found value f_e can now be used for comparison with the empirical values of gene frequency variances of different loci. To be more precise, taking part in the comparison is not the empirically observed gene frequency variance V_{q_o} as such, but the part of it which remains after deducting the

Table 3.13. *Coefficients of correlation and of regression of gene frequency* q *on geographic distance* d, *expressed in* °E

Gene	Number of geographical points	r_{qd}	$b_{q/d}$
A	13	0.735†	0.00120
B	13	−0.665*	−0.00065
O	13	0.168	0.00033
N	13	−0.468	−0.00113
Ms	7	0.354	0.00062
MS	7	−0.548	−0.00108
Ns	7	0.431	0.00103
NS	7	−0.347	−0.00057
d	11	−0.674*	−0.00150
R_z	6	0.207	0.00024
R_0	6	−0.290	0.00021
R_2	6	0.506	0.00072
R_1	6	−0.0001	−0.0000003
P_2	6	0.697	0.00276
Le^a	6	−0.714	0.00215
t	7	−0.570	0.00168
HP^1	7	−0.267	−0.00040
Tf^c	7	0.152	0.00003
Mean and SD		0.433 ± 0.060	0.00091 ± 0.00027

* 95% level of significance
† 99% level of significance

sampling variance V_{qs}, which represents the genetic variance $V_{qg} = V_{qo} - V_{qs}$. The value V_{qg}, being standardized to gene frequency, now becomes comparable for the genes of different loci and the theoretically expected value. The observed assessments of genetic variances are given in Table 3.14. The mean magnitude of the observed genetic variances, according to the average for different loci, taking the number of degrees of freedom equal to the sum of the genes minus the sum of loci with multiple alleles, is $f_o = 0.0761 \pm 0.0131$, $t = 5.80$, d.f. $= 16$, $P < 0.001$. A comparison of this magnitude with the theoretically expected one gives the value of Student's criterion $t = 0.364$, $P = 0.721$ and the value of Fischer's criterion $F = 1.06$. $P \gg 0.25$.

Thus, the degree of genetic differentiation of the circumpolar population, assessed on the average per locus, corresponds with a high degree of significance to the value expected according to the hypothesis of neutral evolution of the population over a sufficiently long stretch of time.

In reality, however, we checked by this procedure not the hypothesis of neutral evolution, but merely the applicability of the theoretical assessment

Table 3.14. *Assessment of expected gene frequencies variance according to data on the structure and intensity of gene migration and the effective size of the circumpolar population, proceeding from the linear stepping-stone model of population structure*

Number of steps between the populations (k)	Genetic correlation ($r(k)$)
1	0.5919
2	0.3590
3	0.2175
4	0.1319
5	0.0799
6	0.0484
7	0.0293
8	0.0178
9	0.0108
10	0.0065
11	0.0039
12	0.0024

$N = 45$, $m_\infty = 0.0107$, $m_1 (1 - r(1)) = 0.0341$

$$f_e^1 = \frac{1}{1 + 4N[m_\infty + m_1(1 - r(1))]} = 0.1104$$

$$k = 13, \quad \sum_{i=1}^{k} (k - i) \, r(i) = 15.7294$$

$$f = f_e^1 \left\{ 1 - \frac{1}{k} - \frac{2}{k^2} \quad \sum_{i=1}^{k} (k - i) \, r(i) \right\} = 0.08135$$

of the gene frequency variance by comparing against it the variances observed in the sample of genes studied. This does not rule out the possibility that, although they correspond on an average to a neutral state of the environment, the loci when examined separately will reveal an essential deviation from this state. It follows from Table 3.15, which presents the results of such a comparison that indeed behind the mean gene neutrality are significant deviations from this state in some of the loci involved.

The most essential deviations from the expected level toward a drastic inhibition of genetic differentiation in the circumpolar zone were established for the loci of ABO blood groups, P (within northern Asia), transferrins Tf and the group-specific component Gc (within northern Asia).

The most essential deviations from the expected level toward a steep intensification of genetic differentiation of the circumpolar population were established for the Dd locus of Rhesus (Rhesus-negative) blood groups and Lewis.

The other of the studied loci (Hp, MN, sensitivity to PTC) prove really

Table 3.15. *Empirical gene frequencies variances in the circumpolar population, classified within three boundaries from the Bering Strait: to the Urals, the Kola peninsula and Scandinavia*

Locus	Gene	Number of populations K	Average number of genes studied in the population $\bar{n}$	Observed variance V_q	χ^2	Genetic variance V_{qg}	Standardized genetic variance f_0
ABO	A	39	184	0.0054	224.03	0.0045	0.0264
		45	190	0.0073	344.16	0.0064	0.0359
		58	213	0.0121	748.86	0.0112	0.0569
	B	39	184	0.0054	303.40	0.0047	0.0377
		45	190	0.0055	352.50	0.0048	0.0369
		58	213	0.0063	667.72	0.0058	0.0503
	O	39	184	0.0091	276.96	0.0078	0.0340
		45	190	0.0115	405.71	0.0102	0.0433
		58	213	0.0117	592.18	0.0106	0.0441
MNSs	N	30	179	0.0205	427.13	0.0191	0.0767
		34	168	0.0222	494.84	0.0207	0.0833
		39	193	0.0204	600.44	0.0191	0.0767
	S	9	62	0.0197	43.53	0.0161	0.0716
		10	60	0.0256	59.41	0.0218	0.0934
		12	113	0.0215	113.87	0.0194	0.0828
	Ms	9	60	0.0162	36.26	0.0127	0.0590
		10	60	0.0144	36.28	0.0110	0.0506
		12	113	0.0145	85.05	0.0124	0.0594
	MS	9	60	0.0165	47.31	0.0144	0.0820
		10	60	0.0151	51.24	0.0140	0.0783
		12	113	0.0125	87.94	0.0110	0.0618
	Ns	9	60	0.0109	23.23	0.0072	0.0318
		10	60	0.0177	44.15	0.0141	0.0652
		12	113	0.0154	87.37	0.0134	0.0611
	NS	9	60	0.0038	17.81	0.0021	0.0205
		10	60	0.0082	37.54	0.0063	0.0530
		12	113	0.0072	71.41	0.0060	0.0483
Rhesus	d	26	174	0.0095	830.20	0.0092	0.1851
		30	163	0.0112	800.52	0.0108	0.1632
		33	192	0.0109	918.86	0.0105	0.1443
	c	10	94	0.0167	56.63	0.0140	0.0561
		11	89	0.0164	58.64	0.0136	0.0545
		18	99	0.0122	83.20	0.0097	0.0393
	e	10	94	0.0296	118.16	0.0273	0.1286
		11	89	0.0279	114.99	0.0254	0.1176
		18	99	0.0254	229.30	0.0235	0.1261
	R_z	9	91	0.0036	29.36	0.0026	0.0291
		10	86	0.0041	32.57	0.0030	0.0304
		17	95	0.0046	101.07	0.0038	0.0559
	R_0	9	91	0.0040	20.95	0.0025	0.0178
		10	86	0.0068	38.29	0.0052	0.0378

Table 3.15—contd.

Locus	Gene	Number of populations K	Average number of genes studied in the population $\bar{n}$	Observed variance V_q	χ^2	Genetic variance V_{qg}	Standardized genetic variance f_0
	R_0	17	95	0.0056	80.72	0.0045	0.0425
	R_Z	9	91	0.0149	61.01	0.0129	0.0722
		10	86	0.0133	57.67	0.0112	0.0627
		17	95	0.0110	106.44	0.0093	0.0593
	R_1	9	91	0.0193	57.38	0.0166	0.0678
		10	86	0.0172	54.32	0.0143	0.0585
		17	95	0.0172	104.40	0.0145	0.0580
P	P_2	7	74	0.0032	7.87	0.0008	0.0042
		11	80	0.0339	117.24	0.0310	0.1340
		15	210	0.0265	333.87	0.0254	0.1088
Lewis	Le^a	7	63	0.0196	98.37	0.0184	0.2444
		11	72	0.0269	192.28	0.0255	0.2532
		12	97	0.0268	254.84	0.0257	0.2285
PTC	t	13	118	0.0179	102.01	0.0158	0.0638
		18	142	0.0183	187.38	0.0175	0.0700
		20	182	0.0193	280.54	0.0190	0.0759
Hp	Hp^1	17	122	0.0100	174.42	0.0091	0.0812
		19	122	0.0093	178.48	0.0084	0.0731
		21	175	0.0089	268.92	0.0082	0.0711
Tf	Tf^c	17	100	0.0004	36.20	0.0002	0.0126
		18	106	0.0002	22.75	0.00005	0.0032
Gc	Gc^1	15	70	0.0084	35.61	0.0051	0.0219

	f_0	t	d.f.	P
Average for gene sample:	0.0648 ± 0.0141	4.59	16	>99.9%
	0.0761 ± 0.0134	5.68	15	>99.9%
	0.0776 ± 0.0116	6.69	14	>99.9%

close to the state of neutrality, since gene frequency variances do not significantly differ from values corresponding to the neutral status in the environment.

We find sufficiently contradictory results in the cases of linked loci MNSs and CcDEe, whose individual haplotypes have significantly low gene frequency variances (NS and R^0 or cDe), while others do not differ from the neutral value, and still others, for example d, regarded as the sum of r, r' and r'' haplotypes, differ by their significantly higher variability of frequencies.

The human biology of circumpolar populations

It is interesting to note that among the genes whose frequency variability is significantly restricted, we found first of all those that are known as characteristic typological markers of the Lapps as classical representatives of the circumpolar population. In other words, according to these genes (A, P_2, NS, Rs) the circumpolar zone of Eurasia manifests great unity, despite the obvious differences existing in regard to these genes between Europe and Asia south of the Arctic Circle. It follows, therefore, that the source of stabilizing pressure on these loci should be placed in the natural environment of the circumpolar zone as that which alone unites the populations being compared.

On the contrary, in the case of genes subjected to excessive differentiating pressure (d and Le[a]) the source of such pressure cannot be equally associated with the natural environment, since the external gene migrations m_∞ taken into account by this analysis, are directed to the western and eastern parts of the circumpolar area from sources in Europe and in Siberia, clearly differing sharply as to the frequency of these genes. Since the analytical method used did not take into account the possibility of genetic differences in an 'infinitely remote' external source of migrations, the significantly higher heterogeneity of the population according to genes d and Le[a] can be attributed to external pressure, and the loci concerned can be regarded in this case as neutral.

Similar considerations may be put forward also in regard to those loci that were earlier classified as neutral on the basis of a comparison of variances. If the differences between continental Europe and continental Asia were accurately assessed (which goes far beyond the present review), the neutral loci should be placed among those subjected to the stabilizing pressure and united with loci of doubtless adaptive significance.

Thus, the main result of this part of the analysis of population genetics data is that it becomes possible to discover in the circumpolar zone of Eurasia an additional systematic pressure, differing from the pressure of gene migrations and pointing to the adaptive significance of the majority of the investigated genetic properties of the circumpolar population. We therefore cite the assessed values of such stabilizing pressure (Table 3.16), irrespective of the degree of certainty with which they had been earlier assumed on the basis of an analysis of variances in Table 3.15. This additional systematic factor was singled out on the assumption that the empirical gene frequency variance f_o, in distinction from the theoretical variance f_e, may include this additional factor in the form shown in equation (7) (Kimura & Weiss, 1964; Cavalli-Sforza & Bodmer, 1971).

$$f_o = \frac{1}{1 + 4\,Nb} \tag{7}$$

where $b = m_\infty - m_1\,(1 - r\,(1)) - S - u - v$, where u, v are the direct and reverse mutation frequencies, which may be ignored.

Table 3.16. *A comparison of observed standardized genetic variances* f_o *with the value* f_e *expected under the hypothesis of neutral evolution of the circumpolar population and an estimate of the stabilizing pressure* S *of the natural environment on the gene loci involved*

Locus	Gene	Variances ratio f_e/f_o (expected value of $f_e = 0.03135$)	D.F.	P	Observed value of 4 Nb (expected value 11.29)	Stabilizing pressure (S)
		F				
ABO	A	3.08	38	<0.001	36.8788	0.1421
		2.27	44	<0.001	26.8550	0.0865
		1.43	57	<0.05	16.5739	0.0293
	B	2.16	38	<0.005	25.5252	0.0791
		2.20	44	<0.001	26.1003	0.0823
		1.62	57	<0.01	18.8807	0.0422
	O	2.39	38	<0.001	28.4118	0.0951
		1.88	44	<0.025	22.0954	0.0600
		1.84	57	<0.005	21.6762	0.0577
MNSs	N	1.06	29		12.038	0.0041
		0.98	33		11.005	−0.0016
		1.06	38		12.038	0.0041
	S	1.14	8		12.9684	0.0093
		0.87	9		9.7101	−0.0088
		0.98	11		11.0831	−0.0012
	Ms	1.38	8		15.9492	0.0259
		1.61	9		18.6078	0.0415
		1.37	11		15.8805	0.0252
	MS	0.99	8		11.1951	−0.0005
		1.04	9		11.7681	0.0027
		1.32	11		15.1734	0.0216
	Ns	2.56	8		30.4494	0.1064
		1.25	9		14.3374	0.0169
		1.33	11		15.3693	0.0226
	NS	3.97	8	<0.025	44.7805	0.2027
		1.53	9		17.8679	0.0365
		1.68	11		19.7039	0.0467
Rhesus	d	0.44	25	<0.01	4.4025	−0.0383
		0.50	29	<0.025	5.1274	−0.0342
		0.56	32	<0.025	5.9300	−0.0298
	c	1.45	9		16.8253	0.0307
		1.49	10		17.3486	0.0336
		2.07	17		24.4452	0.0730
	e	0.63	9		6.7779	−0.0251
		0.69	10		7.5041	−0.0211
		0.64	17		6.9302	−0.0243
	R_z	2.80	8		33.3643	0.1226
		2.68	9		31.8947	0.1145
		1.46	16		16.8891	0.0310
	R_o	4.57	8	<0.025	55.1801	0.2438
		2.15	9		25.4537	0.0787
		1.91	16		22.5849	0.0624

75

Table 3.16—contd.

Locus	Gene	Variances ratio f_e/f_o (expected value of $f_e = 0.03135$	D.F.	P	Observed value of $4\,Nb$ (expected value 11.29)	Stabilizing pressure (S)
	R_2	1.13	8		12.8504	0.0087
		1.30	9		14.9489	0.0203
		1.37	16		15.8634	0.0254
	R_1	1.20	8		13.7487	0.0136
		1.39	9		16.0932	0.0267
		1.40	16		16.2414	0.0275
P	P_2	19.37	6	<0.001	237.100	1.2545
		0.61	10		6.463	−0.0268
		0.75	14		8.191	−0.0172
Lewis	Le^a	0.33	6		3.092	−0.0457
		0.32	10	<0.025	2.9494	−0.0464
		0.36	11	<0.05	3.3764	−0.0440
PTC	t	1.28	12		14.6986	0.0188
		1.16	17		13.2857	0.0111
		1.07	19		12.1752	0.0049
Hp	Hp^1	1.00	16		11.3153	0.0001
		1.11	18		12.6799	0.0077
		1.14	20		13.0645	0.0098
Tf	Tf^c	6.46	16	<0.001	78.3629	0.3726
		25.42	17	<0.001	311.5000	1.6678
Gc	Gc^1	3.71	14	<0.005	44.6621	0.1854

	$\bar{S}$	t	d.f.	P
Average for gene sample:	0.1275 ± 0.0780	1.63	16	<90%
	0.1087 ± 0.1071	1.01	15	<80%
	0.0183 ± 0.0097	1.89	14	<95%

S is the index of the stabilizing pressure on a polymorphic locus at gene frequencies in the locus close to the points of equilibrium. If gene frequencies are near the equilibrium points, the S coefficient approximately corresponds to the standard deviation σ_{wt} of genotype adaptive values W_t. Proceeding from a comparison of equation (7) with the theoretical value of gene frequency variance equation (6),

$$S = \frac{f_e - f_o}{4\,N f_e f_o} \tag{8}$$

As Table 3.15 shows, the assessments of value S are, indeed, much higher than the expected order of mutation rate and are frequently higher than the established gene migration order. Only in cases when f_o is negligibly small compared to f_e, do S values turn out to be more than unity.

As could be expected, the mean value S, according to the totality of genes, does not significantly differ from zero, which agrees with the result of a similar comparison of variances. Within the limits of this overall result the value S varies not only according to the genes of different loci, but also depending on whether we are examining the circumpolar population within the boundaries from the Bering Strait to Scandinavia, or to the Kola peninsula, or only to the Urals. The S values increase in this order.

It has already been pointed out when reviewing genogeography, that it is not so much the heterogeneity of natural conditions throughout the length of the circumpolar zone as the different degree of extremity of those same natural conditions, less in the west and greater in the east, that is the dominant factor in the formation of the genetic structure of the population and the geographical distribution of genes. Directly corresponding to this is the value S, which, within the boundaries of the northern Asian part, proved nearly seven times higher than throughout the entire circumpolar area of Eurasia.

Let us recall that in the northern Asian part of the area the heterogeneity of the social component of the environment is much higher than in the European part: there is greater sparsity of the population, more ethnic groups with deeper linguistic differentiation and greater diversity of economic–cultural structures. The existence of man in the sub-Arctic zone apparently goes farther back into antiquity in the Siberian part (particularly in the Far East and the Bering Sea area), than in the European part, which remained longer under the glacier. Nevertheless, the genetic differentiation of the circumpolar population in Siberia proves much lower, from which it follows that the source of such a high stabilizing pressure lies precisely in the natural component of the environment.

As was noted during the description of genogeography, the chief characteristic of the sub-Arctic natural environment is the degree of its extremity, which increases from west to east, from Scandinavia to the Bering Sea area. It is from west to east that population density in the sub-Arctic zone drops.

Gene concentrations change also in the same direction, though with different intensity and according to different types of functions from geographical coordinates. Lastly, increasing from west to east is the degree of the selective stabilizing pressure on polymorphic gene loci.

Thus, we may conclude that the intensity of the process of the genetic adaptation of human populations correlates with the extremity of natural conditions in the circumpolar zone of Eurasia.

References

Allison, A. C., Broman, B., Mourant, A. E. & Ryttinger, L. (1956). The blood group of the Swedish Lapps. *J. Roy. Anthrop. Inst.*, **86**, 87–94.
Allison, A. C., Hartmann, O., Brendemoen, O. & Mourant, A. E. (1952). The blood groups of the Norwegian Lapps. *Acta path. microbiol. Scand.*, **31**, 334–8.
Beckman, L. (1959). *A contribution to the physical anthropology and population genetics of Sweden.* Lund.
Broman, B., Jonsson, B., Beckman, L. & Mellbin, T. (1959). In *A contribution to the physical anthropology and population genetics of Sweden*, ed. L. Beckman. Lund.
Cavalli-Sforza, L. L. (1962). The distribution of migration distances: models and applications to genetics. In *Human Displacements*, ed. L. L. Cavalli-Sforza & W. F. Bodmer (1971).
Cavalli-Sforza, L. L. & Bodmer, W. F. (1971). *The genetics of human populations.* San Francisco: W. F. Freeman and Company.
Chernetsov, V. N. (1973). Ethno-cultural areas in forest and subarctic zones of Eurasia in the Neolithic epoch. In *Problems of archaeology of the Ural and of Siberia.* (In Russian.) 10–17. Moscow: Nauka Publishing House.
Chersky, N. V., Kondakov, K. G. & Morov, A. P. (ed.). (1968). *Economy and culture of the peoples of Yakutia.* (In Russian.) Moscow: Nauka Publishing House.
Dikov, N. N. (1960). *After the traces left by ancient camp fires.* (In Russian.) Magadan: Magadan Book House.
Eriksson, A. W. (1968). Serologisk populationsgenetik. *Nordisk Medicin*, **79**, 419.
Eriksson, A. W., Fellman, J., Forsius, H., Harris, H., Hopkinson, D., Lehman, W., Mäkeläo, O., Nevanlinna, H., Robson, E & Tiilikainen, A. (1967). Population genetic studies on the skolt Lapps. *Scand. J. clin. lab. Invest.*, **19**, suppl. 95, 73.
Khazanova, A. B. (1973). Anthropogenetic surveys of Saamy (Lapps) of the Kola peninsula related to the problem of origin of lapanoid anthropological type. (In Russian.) Dissertation, Moscow University, Moscow.
Khazanova, A. B. & Shamlyan, N. P. (1970). The anthropology and the population genetics of the Kola peninsula Lapps. (In Russian.) *Vop. Anthrop.*, **34**, 71–8.
Kimura, M. & Weiss, G. H. (1964). The stepping stone model of population structure and the decrease of genetic correlation with distance. *Genetics*, **49**, 561–76.
Lashuk, L. P. (1968). 'Sirtja' – ancient inhabitants in Subarctic. In *Problems of anthropology and of historic ethnography of Asia.* (In Russian.) 178–93. Moscow: Nauka Publishing House.
Levin, M. G. (1958). Blood groups among the Chukchi and Eskimos. (In Russian.) *Sovetskaia Etnografiia*, **5**, 113–16.
Levin, M. G. (1959). New materials on the blood groups among Eskimos and Lamuts. (In Russian.) *Sovetskaia Etnografiia*, **3**, 98–9.
Mochanov, Yu. A. (1973). North-eastern Asia in IX–X millenniums B.C. (Sumnaghin culture). In *Problems of archaeology of the Ural and of Siberia.* (In Russian.) 29–43. Moscow: Nauka Publishing House.
Mochanov, Yu. A & Fedoseeva, S. A. (1973). Archaeology of Arctic and Bering Sea ethnocultural bonds between the Old and New Worlds in the Holocene. (In Russian.) Conference: *The Bering Land Bridge and its role for the history of Holarctic Floras and Faunas in the late Cenozoic*, Khabarovsk, 197–201.
Moran, P. A. P. (1962). *The statistical processes of evolutionary theory.* Oxford: Clarendon Press.
Morton, N. E. (1973). Population Structure of Micronesia. In *Methods and Theories*

of Anthropological Genetics (eds. M. H. Crawford & P. L. Workman), pp. 333–66. Albuquerque: University of New Mexico Press.

Mourant, A. E. (1954). *The distribution of the human blood groups*. Oxford: Blackwell.

Movsesyan, A. A. (1973). Some aspects of genetics of contemporary and of ancient Siberian populations. (In Russian.) *Vop. Anthrop.*, **45**, 77–84.

Neel, J. V. & Ward, R. H. (1972). The genetic structure of a tribal population, the Yanomama Indians. VI. Analysis by F-statistics (including a comparison with the Makiritare and Xavante). *Genetics*, **72**, 639–66.

Okladnikov, A. P. (1968). Siberia in the paleolithic epoch. In *The history of Siberia*, ed. A. P. Okladnikov. (In Russian) vol. I, Ancient Siberia, pp. 37–93. Leningrad: Nauka Publishing House.

Okladnikov, A. P. & Beregovaya, N. A. (1971). Ancient settlements of cape Baranov. (In Russian.) Leningrad: Nauka Publishing House.

Okladnikov, A. P., Dikov, N. N., Derevyanko, A. P., Kozyreva, R. V., Larichev, V. E., Maximenkov, G. A. & Fedoseeva, S. A. (1968). Siberia in the Neolithic epoch. In *The history of Siberia*, ed. A. P. Okladnikov. (In Russian) vol. I, Ancient Siberia, pp. 94–150. Leningrad: Nauka Publishing House.

Rychkov, Yu. G. (1965). Peculiarities of serological differentiation in Siberian peoples. (In Russian.) *Vop. Anthrop.*, **21**, 18–34.

Rychkov, Yu. G. (1969). Some approaches to Siberia anthropology based on population genetics. (In Russian.) *Vop. Anthrop.*, **33**, 16–33.

Rychkov, Yu. G. (1973*a*). Genetic and anthropologic aspects of the problem 'Man and Environment' in the region of Beringia. (In Russian.) Conference: 'The Bering Land Bridge and its role for the history of Holarctic Floras and Faunas in the late Cenozoic', Khabarovsk, 186–9.

Rychkov, Yy. G. (1973*b*). The System of ancient human isolates in Northern Asia in genetic aspects of evolution and stability of populations. (In Russian.) *Vop. Anthrop.*, **44**, 3–22.

Rychkov, Yu. G. & Borodina, S. R. (1969). Hypersensitivity to phenylthioures (PTC) in one of isolated populations of Eastern Siberia. A possible hypothesis of inheritance. (In Russian.) *Genetics*, **5**, 116–23.

Rychkov, Yu. G. & Borodina, S. R. (1973). Further research in genetics of man hypersensitivity to phenylthiocarbamide (experimental, population, family evidence). (In Russian.) *Genetics*, **9**, 141–52.

Rychkov, Yu. G. & Movsesyan, A. A. (1972). A genetical–anthropological analysis on the distribution of cranial anomalies in the Mongoloids of Siberia in connection with the problem of their origin. (In Russian.) *Transactions of the Moscow Society of Naturalists*, **58**, 114–32.

Rychkov, Yu. G., Rusakova, O. L., Rappoport, M. P. & Sheremet'eva, V. A. (1973). Factors of genetic differentiation of population system of the Northern Asia aboriginal stock. I. Estimation of the integrative adaptive values of ABO blood groups genotypes in the Siberian native population. (In Russian.) *Genetics*, **9**, 136–45.

Rychkov, Yu. G. & Sheremet'eva, V. A. (1972*a*). Population genetics of the Commander Islands Aleuts related to the problems of history and of adaptation of the ancient Beringian peoples. I. (In Russian.) *Vop. Anthrop.*, **40**, 45–70.

Rychkov, Yu. G. & Sheremet'eva, V. A. (1972*b*). Genetics of the North Pacific populations related to problems of history and adaptation of peoples. III. The Eskimos and Chukchi populations of Bering sea coast. (In Russian.) *Vop. Anthrop.*, **42**, 3–30.

The human biology of circumpolar populations

Rychkov, Yu. G. & Sheremet'eva, V. A. (1973). Population genetics of the Commander Islands Aleuts and its relevance to problems of adaptation of the ancient Beringias aborigines. *IX Intern. Congress of Anthropological and Ethnological Sciences.* August–September, 1973. USA: Chicago.

Rychkov, Yu. G. & Sheremet'eva, V. A. (1974). Factors of genetic differentiation of population system of the Northern Asia aboriginal stock. II. Structure of gene migration in Siberia and adaptive landscape of the Northern Asia in space of ABO blood groups gene frequencies. (In Russian.) *Genetics*, **10**, 147–59.

Rychkov, Yu. G., Tauzik, T., Tauzik, N. E., Zhukova, O. V., Borodina, S. R. & Sheremet'eva, V. A. (1974). Genetics and Anthropology of Siberian Taiga hunters and reindeer-breeders population. (Events of the Middle Siberia.) II. Effective size, time and spatial structure of population and intensity of gene migration. (In Russian.) *Yop. Anthrop.*, **48**, 3–18.

Salzano, F. M., Neel, J. & Maybury-Lewis, D. (1967). Further Studies on the Xavante Indians. I. Demographic Data on two Additional Villages: Genetic Structure of the Tribe. *Amer. J. hum. Genet.*, **19**, 463.

Sheremet'eva, V. A. (1971). The population genetics of indigenous peoples of Siberia (Western Siberia). (In Russian.) *Vestnik Moscow University*, **2**, 97–8.

Sheremet'eva, V. A. & Spitsin, V. A. (1972). Anthropogenetical surveys of Lapps in the Kola peninsula. In *Human Adaptability.* (In Russian.) 42–5. Leningrad: Nauka Publishing House.

Sheremet'eva, V. A. (1973). Population genetics of the North-East Asiatic peoples related to the problems of ethnic anthropology. (In Russian.) Dissertation, Moscow University, Moscow.

Shluger, S. A. (1940). Anthropological type of the Nenets in relevance to ethnogenesis of the Northern peoples. (In Russian.) Dissertation, Moscow University, Moscow.

Simchenko, Yu. B. (1968). Some data on the ancient ethnic substrate in the North Eurasian peoples. In *Problems of anthropology and of historic ethnography of Asia.* (In Russian.) 194–213. Moscow: Nauka Publishing House.

Spitsin, V. A. (1967). Distribution of haptoglobin and of some other inheritable factors in the North-Eastern Siberia. (In Russian.) *Vop. Anthrop.*, **25**, 62–9.

Spitsin, V. A. (1973). Polymorphism of blood serum proteins and some blood enzymes in the aboriginal population of Siberia and some other adjacent territories. *IX Intern. Congress of Anthropological and Ethnological Sciences.* August–September 1973. USA: Chicago.

Vasiliev, V. I., Gurvich, I. S. & Simchenko, Yy. B. (ed.) (1967). *Modern Life of Northern Peoples.* (In Russian.) Moscow: Nauka Publishing House.

Wright, S. (1931). Evolution in Mendelian populations. *Genetics*, **18**, 97–159.

Wright, S. (1951). The genetical structure of populations. *Ann. Eugenics*, **15**, 322–54.

Zolotareva, I. M. (1968). Blood group distribution of the peoples of Northern Siberia, pp. 502–8. *VII Intern. Congress of Anthropological and Ethnological Sciences*, August, 1964. Moscow: Nauka Publishing House.

4. Genetic studies on circumpolar populations

A. W. ERIKSSON, W. LEHMANN & N. E. SIMPSON

This chapter deals with comparisons of the frequencies of genetic markers in circumpolar populations. The data are derived from studies which were undertaken as part of the IBP and from more recently published studies. The populations consist of the Ainu living on the island of Hokkaido in Japan; Eskimos living in Alaska, Canada and Greenland; the Lapps from Norway, Sweden and Finland and some of the peoples living in Siberia. These populations were chosen because they have lived as nomads in relative isolation from other populations for several thousands of years until recently; their origins were somewhat ambiguous and they lived for the most part north of the Arctic Circle. The Ainu, of course, live considerably south of the Arctic Circle (just south of the 45th parallel) but are included because of their isolation and their ambiguous origin. Most of the sites of the specific IBP studies are shown in Figs. 4.1, 4.2 and 4.3.

Comparisons of genetic markers and traits in the circumpolar peoples were made for several reasons. Historical and archaeological evidence is ambiguous about the origins of these peoples. By determining genetic markers, so-called genetic affinities may be measured and expressed in genetic distances and it was hoped that, at least, more conclusive evidence than previously could be obtained for the origins of these peoples. The studies in the past have depended on polymorphisms of the red-cell antigens and anthropometric measurements. For the purpose of this chapter the following definition of polymorphisms was arbitrarily chosen: if there were two or more alleles at a gene locus, each with a frequency ≥ 0.01 in at least one of the populations that gene locus was considered to be polymorphic. Most of the anthropometric data did not have a simple genetic determination, which limited their use. Although the erythrocyte antigens were useful more data were needed. Additional studies seemed very worthwhile after the discovery of polymorphic loci for many serum proteins and enzymes in blood, particularly in red blood cells, together with the advent of simple methods of screening for them using electrophoresis. Like those of the red-cell antigens, the enzyme and other protein polymorphisms are the result of mutations at the loci which code for them. Harris & Hopkinson (1972) have reviewed 71 enzymes which have been surveyed by electrophoresis in at least one population and although only 20 of these were polymorphic in Europeans, several more were polymorphic in other populations. There are at least ten proteins in blood other than the red-cell enzymes, which are polymorphic in some populations and which may be detected either by electrophoretic or by immunochemical methods. In addition to the above systems, those which are most useful for determining origins and genetic affinities are the Gm and HL–A systems which have many polymorphic haplotypes and are also coded for by specific loci. Consequently, populations often have unique haplotypes or haplotype frequencies more variable than other systems.

Other reasons for comparing the gene frequencies were either to study the possibilities of natural selection which might have acted in the past or because they may have adaptive functions now. Questions about natural selection and adaptation are difficult to answer, however, from the study of isolates. The variation in gene frequencies can be due to 'founder effect' or 'genetic drift' that is completely determined by chance. Additional variation is known to have arisen from some mixing of the populations with their neighbors. Although there are examples of selective advantage for the heterozygote against malaria from some of the alleles in man such as G–6–PD deficiency and some forms of haemoglobin (Harris, 1970), and in the past, against the heterozygote in cases of Rh incompatibility (Stern, 1973), the selective advantage, if any, is not known for other systems. In fact, there is much controversy as to whether loci are polymorphic because of natural selection or selective neutrality (Kimura, 1968; Harris, 1970; Yamazaki & Maruyama, 1974). An adaptive gene giving protection against the harshness of the Arctic climate or lowered amount of an essential nutrient, might be recognized if the frequencies of some of the alleles were increased consistently around the pole.

GENETIC MARKERS IN BLOOD
N. E. SIMPSON, W. LEHMANN & A. W. ERIKSSON

The Populations

The Ainu

The Ainu are thought to be a unique population and both a Caucasoid origin has been suggested and, more recently, an affinity to Australian aborigines (reviewed by Omoto & Misawa, 1974). Recent studies on monogenic markers and dental morphology indicate that the Ainu have affinities to the Mongoloid race (Omoto, 1972, 1973). It is known, however, that in recent times and particularly in the last 100 years there has been considerable gene flow from the Japanese to the Ainu (Omoto & Harada, 1972). The oldest written record of the Ainu dates from 1323 (Omoto, 1972). It is thought, however, that there was contact with the Japanese in earlier times and there is archaeological evidence for this at least as far back as the eighth to ninth centuries A.D. (*see* Omoto, 1972).

The 120 Ainu whose blood groups are included in the present review lived

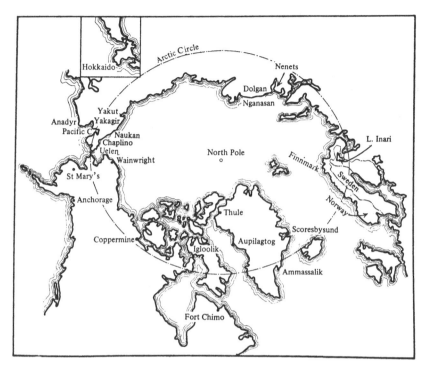

Fig. 4.1. Polar map showing the principal areas of the study.

in Niikappu, in the district of Hidaka on the Hokkaido Island, Japan, and the data were published by Misawa & Hayashida (1968, 1970). They were considered to have a mean degree of admixture with the Japanese of about one half. The sample consisted of 69 Ainu students in a junior high school and 51 adults presumably selected at random and unrelated.

The red-cell enzymes and serum-protein groups were determined for the Ainu living in Piratori, Niikappu, Shizunai and Urakawa in the district of Hidaka on the island of Hokkaido, Japan, by Omoto & Harada (1972). The mean admixture with Japanese was estimated to be about 40% in these populations. The estimate was made from official family records; limited information came from reliable local residents in the older age group and finally some information originated from pedigrees made in earlier anthropological studies.

The Eskimos

The data summarized in this chapter on the markers for the Eskimos living in Alaska are not all taken from the same area of Alaska which, of course, is a much larger geographical area than Hokkaido. The data for blood-group antigens come from the Eskimos living in Wainwright which is located on the north shore of Alaska (*see* Fig. 4.1) and the beginnings of the settlement occurred with the founding of a school in 1904 (Milan, 1970). Blood was analyzed from about 200 of the 333 inhabitants of the village. The time of sampling was August 1968. Unfortunately, other marker data for this village were not available and, therefore, gene frequencies have been included for the enzymes and serum proteins which were determined for Eskimos from various areas of Alaska by Scott *et al.* (1966), Scott, Weaver & Wright (1970), and Scott & Powers (1974). The frequencies of the alleles at the first locus (E_1) for the enzyme cholinesterase were determined by surveying Eskimos from both northern and southern Alaska when they were hospital patients in Anchorage. The finding of an extraordinary number of Eskimos who were homozygous for the $E_1{}^s$ allele (silent) (Gutsche, Scott & Wright, 1967) stimulated a subsequent survey of 322 unrelated individuals living in the three villages of St Mary's (Fig. 4.1), Mountain and Pilot Station which were in the area from which the homozygotes came.

Subsequently, additional villages were surveyed for both the first and second locus for cholinesterase (Scott *et al.*, 1970). The remainder of the enzyme and serum proteins were surveyed in hospital patients who were classified as northern or southern depending on the district from which they came (Scott *et al.*, 1966; Duncan, Scott & Wright, 1974).

The sample from the village of Igloolik (Fig. 4.1) in Canada consisted of about 400 of the 560 Eskimos living there. The Canadian Government initiated a program to encourage the Eskimos to move into this community in the 1940s and they have gradually been settling there for about the last 30 years. Although the population consisted of family groups, the relationships between the individuals were known from careful interviewing independently by a number of investigators and subsequent pooling of the information. In addition, data were used from the vital statistics on births, marriages and deaths which go back to 1946. The gene frequencies were calculated from the complete sample of nuclear or extended families. The number of unrelated individuals was small, and in fact, for those frequencies calculated both for the total population and the unrelated individuals there was virtually no difference (McAlpine *et al.*, 1974). By using the gene frequencies for Gm, acid phosphatase and 6-phosphogluconate dehydrogenase the Caucasian admixture was estimated to be about 7% for the Igloolik population (McAlpine *et al.*, 1974). Unfortunately, there were no data available for the blood group antigens in this population.

The sample of about 300 Eskimos in Fort Chimo, Canada (Fig. 4.1), represented about 50% of the population in the village. Auger & Guevin (personal communication) assumed that there was considerable Caucasian–Indian admixture. The population again was made up of complete family groups excluding the children under five years of age, and the gene frequencies were calculated from the total population.

A great many populations of Eskimos in Greenland have been studied as part of the IBP and when interpreting their gene frequencies some background information needs to be considered. The studies are reviewed by Persson (1970) and Gürtler, Gilberg & Tingsgård (1974). Although it is assumed that the first people on Greenland came there about 4000 years ago, probably none of the original population has descendants. The ancestors of the present Eskimo population probably arrived on Greenland around the time of the birth of Christ. These ancestors founded the Dorset culture which was followed by the Thule Eskimo culture. The latter are thought to have come from northeast Asia via the Bering Strait and North America. Probably few Dorset people were left by A.D. 1000. It is a controversial matter whether the remaining Eskimo population and new waves of Eskimo immigrants from North America did mix with the remnants of the mediaeval Scandinavian (Icelandic) colonists. In the last 300–400 years the Eskimos have increasingly mixed with Nordic (Danish) people.

It is now assumed that the Polar Eskimos in Thule (Fig. 4.1) are relatively pure and that they came to Greenland after the Thule culture period. The Eskimos on the west and southwest coasts are assumed to be mixed in varying degrees with mainly Norwegians, Danes and Netherlanders and especially in the last century with Danes.

Other pure Eskimo tribes that have remained in isolation are thought to be those living in Ammassalik and Scoresbysund on the east coast of Greenland (*see* Figs. 4.1, 4.2); the latter was founded by 70 immigrants from Ammassalik in 1925 with subsequent integration of eight Eskimos from West Greenland (Fernet *et al.*, 1971*b*).

Almost all of the Eskimos living on the island of Aupilagtoq (Fig. 4.2) were studied. The majority of these Eskimos came from the northwest of Greenland and the population studied is thought to be representative of the neighbouring coastal area and not an isolate of pure Eskimos. The population in Aupilagtoq consisted of complete family groups. The samples were collected during an expedition led by Dr Balslev J. Jørgensen, Copenhagen, and the results were published in a preliminary report by Eriksson, Kirjarinta & Gürtler (1970).

The Danish studies of the settlements along the west coast of Greenland consist of unrelated adults, representing about 10% of the total population. They are classified according to their birth place; the individual areas are shown in Fig. 4.2. The material from Thule was divided into mixed and pure Polar Eskimos by the use of the pedigrees of Gilberg and Holm (Gürtler *et al.*, 1974).

The blood-group markers and most of the serum markers were determined for about 70% of the Ammassalik population in East Greenland in the

Fig. 4.2. Detailed map of the areas studied in Greenland.

Danish studies (Persson & Tingsgård, 1966; Gürtler *et al.*, 1974). The population was considered to be almost pure Eskimo as there has been little mixing of this population until 1945 and all of the individuals known from pedigree studies to have admixture with Caucasians were excluded from the survey (Persson & Tingsgård, 1966). A smaller sample was tested for Gm types (Nielsen *et al.*, 1971) and red-cell enzymes (Dissing *et al.*, 1974). Blood groups, Gm, Inv and haptoglobin types were determined in an additional sample of 130 individuals over 12 years of age not previously tested in Ammassalik (Fernet *et al.*, 1971*a*). Fernet *et al.* (1971*b*) also determined blood groups, Gm, Inv and haptoglobins in individuals over 12 years of age for nearly 60% of the population in Scoresbysund.

The Lapps

On the basis of evidence from place names and archaeological findings it is assumed that until about 2000 years ago the Lapps were spread over the northern parts of Fennoscandia including almost the whole of present Finland. Fig. 4.3 shows how the Lapps were forced to retreat to the northernmost areas in Finland by the Finns (whose ancestors arrived in Finland about A.D. 300) and to the less fertile northern areas in the Scandinavian peninsula by the Swedes and Norwegians. Because the Lappish areas have always been sparsely populated (< 1 person/km²) and because the travel communications have always been difficult, the Lapps have until recently lived in relative isolation. Their origin is unknown, one hypothesis being that they are of Mongolian–Asiatic descent and another that they are an ancient Nordic population which survived the glacial period on the coastal area of northern Scandinavia (Beckman, 1959). It has also been claimed that the Lappish ancestry consisted of different marginal populations, some of whom, at least, had rather extreme gene frequencies which are still reflected among the majority of the present Lapp populations (Eriksson, 1973; Eriksson *et al.*, 1976).

Some Norwegian scientists have studied the genetic markers in populations of Lapps living in the county of Finnmark in northern Norway (Fig. 4.3). The populations consisted of school children and adults from institutions for the sick and aged. An attempt was made to exclude known non-Lappish admixture by using the criteria that parents were Lappish and that the Lappish language was chiefly spoken at home. The individuals were also questioned about their relatives in other institutions in an attempt to sample a population which was unrelated. The enzymes and serum proteins were determined on the same population or parts thereof (Monn, 1969) and the blood groups on a separate sample taken in the same manner from institutions in the same area by Kornstad (1972).

The studies of Swedish Lapps were not part of the IBP but are included for

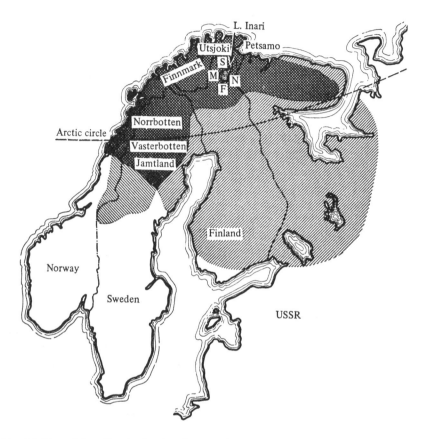

Fig. 4.3. Map of Scandinavia showing the areas in detail in the study. Around Lake Inari are M (Inari Mountain Lapps), F (Inari Fisher Lapps), S (Svettijärvi Skolt Lapps) and N (Nellim Skolt Lapps). The Svettijärvi and Nellim Skolt Lapps around Lake Inari came from Petsamo after the Second World War. The shaded areas show how the Lapps have retreated north in about the last 2000 years. The approximate borderlines and years 1200, 1500 and 1800 show the extent of Lappish (settlement) territory in Finland at the indicated time. —▨, Lapp territory at about 200 BC. ▧—, Lapp territory at A.D. 1970.

comparison. The blood groups, haptoglobin and transferrin types were determined in children attending nomad schools in the county of Norrbotten (Beckman & Holmgren, 1961). More recently, the red-cell enzymes were determined by Beckman *et al.* (1971) in school children. About 65% of the children in the latter population were born in the county of Norrbotten, and the remainder in the counties of Västerbotten and Jämtland (*see* Fig. 4.3). No more than two siblings from one sibship were included and 172 of the 210 individuals were unrelated.

The Lapps who were studied in Finland live around Lake Inari (Fig. 4.3). Those living on the eastern shore were Skolt Lapps who had lived in isolation in the Petsamo region (Fig. 4.3) until the end of the second World War when this territory became part of the USSR. The Skolt Lapps from the village of Suenjel were evacuated to Svettijärvi to the northeast of Lake Inari, and the Skolt Lapps from Pasvik to Nellim (which is southeast of Lake Inari). Over 95% of the Skolt Lapps living in the parish of Inari are included in the populations which were studied. There is evidence for a considerable flow of genes to the Lapps, and particularly to the Skolt Lapps, from the neighboring populations. The Fisher and Mountain Lapps who live mainly on the west side of Lake Inari had not mixed with the Finns until 1800 when only 4% of the population in the parish of Inari was Finnish and the percentage of Finns living in the area gradually increased with considerable immigration after the second World War. Particularly in the present century the Lapps and the Finns have intermarried, resulting in considerable Finnish gene flow to the Fisher and Mountain Lapps. The populations of Fisher and Mountain Lapps were, however, selected by pedigree and linguistic criteria for lack of Finnish and other non-Lappish admixture. Over 50% of the Fisher and Mountain Lapps living in the parish of Inari were included in the studies of genetic markers. The four populations of the Inari region consisted of family groups and the relationships were determined by interview and checked in parish registers. Further details of these populations are described in Eriksson *et al.* (1971*a*, *b*, *c*) and Steinberg *et al.* (1976).

The Utsjoki Lapps live in the most northerly part of Finland and the population which was studied consisted of a random sample between the ages of 20 and 60 years (Eskola *et al.*, 1976).

The Siberians

Data are published for the ABO and MN blood groups for a considerable number of populations living in Siberia. Three groups of Chukchi living in the northeast near the Pacific were studied by Levin (1958): those living in the Anadyr river basin, the village of Uelen and a number of villages on the Pacific coast (Fig. 4.1). In addition, there were two groups of Siberian Eskimos living in the northeast near the Arctic ocean, those in Naukan (Levin, 1958) and in Chaplino (Levin, 1959). The data for the two blood groups are given for seven other northeastern Siberian populations (Zolotareva, 1965) and four northwestern populations. For all of these populations, whose number was at least 65, Szathmáry (1974) has recalculated the frequencies using the computer program MAXLIK (Reed & Schull, 1968) and the frequencies are listed in Table 4.6. For further studies on Siberians and on the Aleut islanders see Rychkov & Sheremet'eva (chapter 3 in this volume).

Methods

The phenotyping for all of the markers was done using established methods which are given in the original papers and it is beyond the scope of this chapter to reiterate them. For the unpublished data, the methods were also similar to those in the published works.

The genetic distances among circumpolar populations and the construction of the genetic networks uniting these populations were calculated by Dr Emöke Szathmáry, Trent University, Peterborough, Ontario, Canada, using a computer program based on the method of Cavalli-Sforza & Edwards (1967). The program uses gene frequency data as input. All frequencies in Tables 4.1 and 4.2 used in the calculation of distances, unless obtained by counting, were calculated by Dr Szathmáry using the computer program MAXLIK, which gives the maximum likelihood estimate of Reed & Schull (1968). The gene frequencies for a few systems would not converge using MAXLIK and these frequencies were calculated using computer program ALLTYPE and are labelled in the tables. Only those data for populations which had been typed for the same array of systems or genes could be used in this method of calculating distances. One additional Eskimo population which was otherwise not included in this review, the Copper Eskimos described by Chown & Lewis (1959), was included in the distance calculations.

The Data

The gene frequencies for the genetic markers in blood are summarized in Tables 4.1, 4.2, 4.3, and 4.4. An attempt has been made to include all of the studies that were carried out under the auspices of the IBP and most of the other relevant studies in the last decade or so. There were data from a few IBP studies that were not available. Some published studies may have been inadvertently omitted but an attempt was made to make the review as complete as possible.

Table 4.1 summarizes the data by the individual population studies for all of the markers in blood, except Gm and HL–A types from Ainu, Eskimos and Lapps. The data for the mixed West Greenland populations are pooled under the heading of West Greenland. Table 4.2 gives the blood-group data for the individual regions of West Greenland which were studied. Data for the same areas of West Greenland for Gc, Hp and Gm may be found in Persson (1970).

In Table 4.3 those populations are mentioned for which the gamma-globulin haplotype frequencies are based on tests for Gm (1, 2, 3, 5, 6, 13, 14 and 21). The Gm haplotype frequencies in Table 4.3 were estimated by a maximum likelihood method using the computer program of MAXIM (Kurczynski & Steinberg, 1967). The frequencies were taken from the individual publications. Additional Gm data when other or fewer reagents were used are given in Table 4.4.

The phenotypic frequencies for the HL–A histocompatibility antigens are tabulated in Table 4.5 for one Ainu, three Eskimo and three Lapp populations. Maximum likelihood estimates for the gene frequencies were not available for all of the populations and therefore it seemed more reliable to compare the phenotype frequencies.

Table 4.6 summarizes the gene frequencies for the ABO and the MN blood group antigens which were calculated using the MAXLIK program (Szathmáry, 1974) from the data of Levin (1958) and Zolotareva (1965) for some of the Siberian populations.

The Frequencies

Genes with frequencies similar in the Ainu and Eskimos but different in Lapps

There were eight alleles that probably were not Ainu or Eskimo genes in the original populations although they occur at present in the Eskimo populations known to have considerable admixture with Caucasians. They were the alleles A_2, K, R^w (C^wDe), and R^o (cDe) and the P^c allele for red-cell acid phosphatase, AK^2 for adenylate kinase, E_1^a for cholinesterase and SOD^2 for cytoplasmic superoxide dismutase. In addition, there were two haplotypes, $Gm^{3, 5, 13, 14}$ and $Gm^{1, 2, 21}$ which were probably neither Ainu nor Eskimo haplotypes but were polymorphic in the Lapps. The HL–A antigens A3 and A7 had relatively low frequencies in the Ainu and Eskimos and exceptionally high ones in the Lapps while A5 had a higher frequency in the Ainu and Eskimos than in the Lapps.

The gene for the A_2 red-cell antigen was non-existent in the Japanese and Ainu and probably was absent in the Eskimos. As can be seen from Fig. 4.4 the A_2 gene was not present in the pure Polar Eskimos of Thule and had a very low frequency in the Ammassalik and Fort Chimo Eskimos. It had a frequency of less than 0.02 in the Aupilagtoq and other West Greenland Eskimos who were stated to have considerable Caucasian admixture particularly from the Danes. Also it is clear from Fig. 4.4 that there was a north–south cline for the frequencies of this gene among the Europeans. Finland had the highest A_2 frequency of the Europeans (0.10) (Nevanlinna, 1972). It is striking that all Lapp populations so far tested have the highest frequencies for the A_2 gene of any population in the world; A_2 frequencies for the Skolt Lapps are 0.18–0.21 and for other Lapps 0.22–0.37, i.e. two to five times higher than in the neighboring populations (Eskola *et al.*, 1976).

The allele for the K antigen is not likely an Ainu or Eskimo gene (Table 4.1). The allele was not found in the Ainu or in the Japanese population (Omoto & Misawa, 1974). It only occurred in the Fort Chimo and West

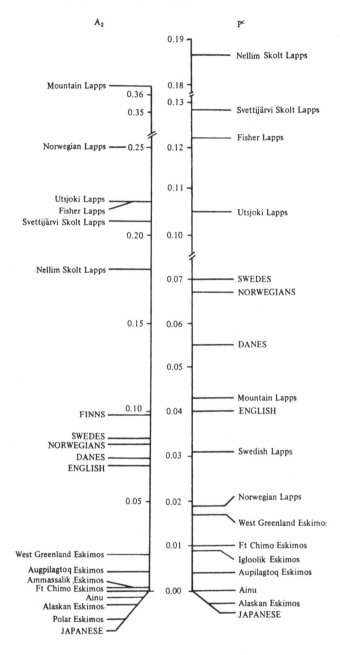

Fig. 4.4. Frequencies of the A_2 blood group allele and the P^c allele for red-cell acid phosphatase in circumpolar peoples and their neighbors.

Greenland Eskimos; both populations are known to have considerable admixture and the *K* allele is polymorphic in Scandinavians with a frequency of about 0.04. The allele was polymorphic in the Mountain Lapps of Inari and Utsjoki and the Sevettijärvi Skolts and occurred in low frequencies in the Fisher Lapps and the Nellim Skolts (Table 4.1). Also the Norwegian and Swedish Lapps have low frequencies of *K* (Beckman *et al.*, 1959; Kornstad, 1972). An exception are the Norwegian Lapps investigated by Allison *et al.* (1952) with a *K* of 0.062 which is the highest noted *K* gene frequency in Scandinavia.

The alleles R^w (C^wDe) and R^o (cDe) were also probably neither Ainu nor Eskimo genes (Table 4.1). Antiserum for detecting R^w was not used in testing the Ainu or the Japanese. The R^o gene was not found in the Ainu although it occurs in the Japanese with a frequency of 0.01 (Omoto & Misawa, 1974). The R^w gene appeared to be present in Eskimos only in Fort Chimo. However, few Eskimo populations were tested for this antigen. The R^o gene was found in the West Greenland Eskimos, who have considerable Caucasian admixture, but also in Eskimos in Ammassalik and Scoresbysund. Both the R^w and R^o genes are polymorphic in the British although with low frequencies (R^w = 0.013 and R^o = 0.026) (Race & Sanger, 1968), R^w has a somewhat higher frequency in Danes (0.016), Swedes (0.020) and Finns (0.020). R^o has a rather high frequency in Finns (0.037) (Nevanlinna, 1972). In Swedes it is 0.019 (Heiken & Rasmuson, 1966) and in Norwegians 0.021 (Kornstad, 1959). The two genes R^w and R^o are polymorphic in most of the Lapp populations (*see* Table 4.1) with some of the Finnish Lapps having higher frequencies than the Finns. Some Lapp populations have the highest known R^w frequencies (Eriksson, 1973). It should be noted that MAXLIK would not converge for the Inari Mountain Lapps for the Rh system. Therefore, the gene frequencies which were calculated by hand assuming that R^z is absent have been included in Table 4.1.

The P^c gene for the enzyme red-cell acid phosphatase (Fig. 4.4) was absent in the Ainu and Japanese and its frequencies in the few Eskimo populations for which there are data available were < 0.02. Among the Norwegian, Swedish and Inari Mountain Lapps the frequencies of P^c were between 0.019 and 0.043 with the frequency for English falling in this range (Harris, 1970). In the Scandinavians the frequencies of P^c were slightly higher than in the above populations, and in the Finnish Lapps (Kirjarinta, 1976) other than the Inari Mountain Lapps the frequencies were the highest of those populations compared in Fig. 4.4. The P^c gene may not be an Ainu and Eskimo gene and have entered the Eskimo populations by gene flow from Caucasian admixture.

Recently, Kirjarinta (1976) has noted a negative correlation between red-cell acid-phosphatase activity and glutathione-reductase (GSSG–R) activity. The P^c allele produces an acid phosphatase type with an activity twice as high

Table 4.1. *Gene frequencies for genetic markers in blood of circumpolar peoples*

		Ainu							Green
		Japan	USA	Canada					
		Hok-		Igloo-	Fort			Aupi-	
System	Gene	kaido	Alaska	lik	Chimo	Thule[M]	Thule[P]	lagtoq	West
Red cell antigens									
ABO[a]	N	523 [1]	191 [2]		278 [3]	119 [4]	152 [4]	152 [5]*	1995 [4]
	O	0.5871	0.5616		0.7763	0.8010	0.9105	0.6699	0.6583
	A_1	$\{$ 0.2511	$\{$ 0.3728		0.2087	0.1650	0.0895	0.2151	0.2579
	A_2				0.0023	0.0256	0.0000	0.0123	0.0189
	B	0.1618	0.0656		0.0127	0.0084	0.0000	0.1027	0.0649
Diego[a]	N	77 [1]	191 [2]						
	Di^a	0.0398	0.0026						
	Di	0.9602	0.9974						
Duffy[a]	N	523 [1]	189 [2]		229 [3]			152 [5]	
	Fy^a	0.9020	0.7818		0.7808			0.7435	
	Fy^b	0.0980	0.2182		0.2192				
Kell[a]	N	523 [1]	191 [2]		273 [3]	118 [4]	151 [4]	152 [5]	1428 [4]
	K	0.0000	0.0000		0.0166	0.0000	0.0000	0.0000	0.0088
	k	1.0000	1.0000		0.9834	1.0000	1.0000	1.0000	0.9912
Kidd[a]	N	313 [1]	191 [2]		273 [3]				
	Jk^a	0.3312	0.4442		0.7505			0.1991	
	Jk^b	0.6688	0.5558		0.2495			0.8009	
Lutheran	N		191 [2]		273 [3]			152 [5]	
	Lu^a		0.0000		0.0010			0.0033	
	Lu		1.0000		0.9990			0.9967	
MNSs[a]	N	280 [1]	191 [2]		278 [3]†	119 [4]	152 [4]	152 [5]	1994 [4]
	MS	0.0628	0.1549		0.0980	$\{$ 0.6597	$\{$ 0.6118	0.2134	$\{$ 0.7410
	Ms	0.3980	0.6907		0.6070			0.5399	
	NS	0.1032	0.0074		0.0243	$\{$ 0.3403	$\{$ 0.3882	0.1334	$\{$ 0.2590
	Ns	0.4360	0.1470		0.2707			0.1133	
P[a]	N	522 [1]	182 [2]		273 [3]	119 [4]	152 [4]	152 [5]	1980 [4]
	P_1	0.1468	0.1647		0.2417	0.0879	0.0932	0.2348	0.1946
	P_2	0.8532	0.8353		0.7583	0.9121	0.9068	0.7652	0.8054
Rh[a]	N	414 [1]	191 [2]		278 [3]	118 [4]	151 [4]	152 [5]*	1987 [4]
CDe	R^1	0.5311	0.4553		0.5378	0.6737	0.7252	0.5428	0.5618
C^{w}De	R^{1w}	0.0000	0.0000		0.0018	0.0000	0.0000	0.0000	0.0000
cDE	R^2	0.2213	0.5182		0.3885	0.2548	0.2748	0.4210	0.3059
cDe	R^o	0.0000	0.0000		0.0000	0.0000	0.0000	0.0000	0.0051
CDE	R^z	0.0032	0.0080		0.0000	0.0000	0.0000	0.0000	0.0013
Cde	r'	0.0462	0.0000		0.0000	0.0000	0.0000	0.0000	0.0186
cde	r	0.0600	0.0185		0.0719	0.0715	0.0000	0.0362	0.1072
cdE	r''	0.1382	0.0000		0.0000	0.0000	0.0000	0.0000	0.0000
CdE	r^y	0.0000	0.0000		0.0000	0.0000	0.0000	0.0000	0.0000

Population									
Eskimos			Lapps						
land			Norway	Sweden	Finland				
					Utsjoki		Inari	Skolt	
Ammassalik	Ammassalik	Scoresbysund	Finnmark	North	Mountain	Fisher	Mountain	Svettijärvi	Nellim
739 [4]	130 [6]	246 [7]	423 [8]		170 [9]	259 [10]	133 [10]	348 [10]	150 [10]
0.4580	0.4373	0.6299	0.4623		0.5342	0.6933	0.4559	0.3807	0.4058
0.4366	{0.4058	0.3558	0.1738		0.1423	0.0826	0.1035	0.2710	0.0912
0.0024		0.0000	0.2563		0.2246	0.2202	0.3660	0.2085	0.1830
0.1030	0.1569	0.0143	0.1076		0.0989	0.0039	0.0746	0.1398	0.3200
			423 [8]	220 [11]		159 [10]	48 [10]	227 [10]	134 [10]
			0.0000	0.0000		0.0000	0.0000	0.0000	0.0000
			1.0000	1.0000		1.0000	1.0000	1.0000	1.0000
		228 [7]	423 [8]		170 [9]	247 [10]	130 [10]	348 [10]	143 [10]
		0.8853	0.5413		0.4971	0.5368	0.5794	0.7663	0.6985
		0.1147							0.3015
739 [4]	130 [6]	246 [7]	423 [8]	419 [11]	170 [9]	256 [10]	131 [10]	348 [10]	147 [10]
0.0000	0.0000	0.0000	0.0060	0.0012	0.0238	0.0020	0.0590	0.0174	0.0068
1.0000	1.0000	1.0000	0.9940	0.9988	0.9762	0.9980	0.9410	0.9826	0.9932
		246 [7]	423 [8]			58 [10]	45 [10]	113 [10]	112 [10]
		0.7221	0.5311			0.1906	0.3333	0.2129	0.3121
		0.2779	0.4689			0.8094	0.6667	0.7871	0.6879
739 [4]	130 [6]	246 [7]*	359 [8]		170 [9]	246 [10]	130 [10]†	345 [10]*	140 [10]‡
{0.9350	{0.9077	0.1182	0.2643		0.1448	0.3283	0.2451	0.3147	0.2880
		0.6643	0.2594		0.3670	0.1839	0.2010	0.3766	0.3156
{0.0650	{0.0923	0.0160	0.1578		0.2135	0.1067	0.1011	0.0476	0.0513
		0.2015	0.3185		0.2747	0.3811	0.4528	0.2611	0.3451
738 [4]	130 [6]	246 [7]	336 [8]		170 [9]	259 [10]	133 [10]	348 [10]	150 [10]
0.0923	0.1962	0.3721	0.3056		0.5089	0.4339	0.3869	0.2592	0.3367
0.9077	0.8038	0.6279	0.6944		0.4911	0.5661	0.6131	0.7408	0.6633
738 [4]	130 [6]	243 [7]‡	423 [8]		170 [9]	259 [10]§	133 [10]b	348 [10]‡	150 [10]‡
0.6474	0.5960	0.6509	0.5447		0.5676	0.5830	0.6135	0.6552	0.7933
0.0000	0.0000	0.0000	0.0390		0.0177	0.0521	0.0150	0.0115	0.0000
0.3483	0.3575	0.2250	0.2541		0.2206	0.2490	0.1917	0.2212	0.0700
0.0000	0.0354	0.0227	0.0000		0.0000	0.0685	0.0193	0.0412	0.0400
0.0010	0.0040	0.0219	0.0000		0.0000	0.0000	0.0000	0.0000	0.0000
0.0000	0.0000	0.0000	0.0073		0.0000	0.0000	0.0745	0.0000	0.0000
0.0033	0.0071	0.0795	0.1549		0.1941	0.0473	0.0859	0.0709	0.0966
0.0000	0.0000	0.0000	0.0000		0.0000	0.0000	0.0000	0.0000	0.0000
0.0000	0.0000	0.0000	0.0000		0.0000	0.0000	0.0000	0.0000	0.0000

95

Table 4.1—contd.

Enzymes									
Red cell	N	489 [12]	264 [13]	383 [14]	152 [3]			126 [10]	144 [15]
acid phos-	P^a	0.2249	0.5644	0.4530	0.3553			0.2817	0.3472
phatase	P^b	0.7751	0.4356	0.5379	0.6349			0.7143	0.6354
	P^c	0.0000	0.0000	0.0091	0.0099			0.0040	0.0174
Adenylate	N	200 [12]	66 [18]	362 [14]		122 [15]	147 [15]	153 [19]	144 [15]
kinase	AK^1	1.0000	1.0000	1.0000		0.9800	1.0000	0.9837	0.9751
	AK^2	0.0000	0.0000	0.0000		0.0000	0.0000	0.0163	0.0069
	AK^3	0.0000	0.0000	0.0000		0.0000	0.0000	0.0000	0.0174
	AK^5	0.0000	0.0000	0.0000		0.0200	0.0000	0.0000	0.0000
Adenosine	N	125 [12]	170 [18]	362 [14]		122 [15]	147 [15]	152 [21]	144 [15]
deaminase	ADA^1	0.9720	1.0000	1.0000		0.9835	1.0000	0.9737	0.9896
	ADA^2	0.0280	0.0000	0.0000		0.0165	0.0000	0.0263	0.0104
Cholinester-	N	125 [12]	322 [23]	362 [14]				146 [24]	
ase (1st	$E_1{}^u$	1.0000	0.8851	1.0000				1.0000	
locus)	$E_1{}^a$	0.0000	0.0000	0.0000				0.0000	
	$E_1{}^s$	0.0000	0.1149	0.0000				0.0000	
Cholinester-	N	195 [12]	1603 [55]	298 [14]					
ase^a (2nd	$E_2{}^-$	0.9714	0.9763	0.9323					
locus)	$E_2{}^+$	0.0286	0.0237	0.0677					
Esterase D	N	86 [26]		145 [27]					
	EsD^1	0.6649		0.7083					
	EsD^2	0.3351		0.2917					
Glutamic-	N		124 [18]	145 [28]					
pyruvate	Gpt^1		0.5766	0.6207					
transamin-	Gpt^2		0.4234	0.3690					
ase	Gpt^5		0.0000	0.0103					
6-Phospho-	N	200 [12]	100 [18]	136 [14]		122 [15]	147 [15]		144 [15]
gluconate	PGD^A	0.9575	1.0000	0.9926		0.9877	1.0000		0.9861
dehydro-	PGD^C	0.0425	0.0000	0.0074		0.0123	0.0000		0.0139
genase									
Phospho-	N	191 [12]	299 [13]	397 [14]	152 [3]	119 [15]	143 [15]	152 [32]	144 [15]
glucomutase	$PGM_1{}^1$	0.8220	0.8244	0.7582	0.9704	0.8950	0.9266	0.6546	0.7500
	$PGM_1{}^2$	0.1780	0.1756	0.2418	0.0296	0.1050	0.0734	0.3454	0.2500
Superoxide	N			359 [14]				152 [10]	
dismutase	SOD^1			1.0000				1.0000	
	SOD^2			0.0000				0.0000	
Serum types									
Ag serum	N					111 [35]		153 [36]	101 [35]
lipoprotein^a	Ag^x					0.3563		0.5958	0.4965
	Ag					0.6437		0.4042	0.5035
Lp serum^a	N							153 [36]	
lipoprotein	Lp^a							0.1637	
	Lp							0.8363	
C'3 com-	N							63 [39]	
ponent^a	$C3^1$							0.0635	
	$C3^2$							0.9365	
Gc com-	N	467 [12]	214 [13]	145 [27]		147 [42]	150 [42]	153 [5]	1180 [43]
ponent^c	Gc^1	0.7495	0.6963	0.6655		0.7585	0.6704	0.6438	0.6941
	Gc^2	0.2505	0.3037	0.3276		0.2415	0.3296	0.3562	0.3059
	Gc^{Esk}	0.0000	0.0000	0.0069		0.0000	0.0000	0.0000	0.0000

	105 [16]	210 [17]	171 [9]	218 [10]	129 [10]	330 [10]	142 [10]
	0.4381	0.5071	0.3480	0.1583	0.2907	0.2985	0.2324
	0.5429	0.4619	0.5468	0.7179	0.6667	0.5727	0.5810
	0.0190	0.0310	0.1053	0.1239	0.0426	0.1288	0.1866
159 [15]	273 [20]	210 [17]	171 [9]	231 [19]	133 [19]	343 [19]	147 [19]
0.9906	0.9872	0.9976	0.9912	0.9762	0.9774	0.9985	1.0000
0.0031	0.0128	0.0024	0.0088	0.0238	0.0226	0.0015	0.0000
0.0063	0.0000	0.0000	0.0000	0.0000	0.0000	0.0000	0.0000
0.0000	0.0000	0.0000	0.0000	0.0000	0.0000	0.0000	0.0000
159 [15]	300 [22]	187 [17]	171 [9]	217 [21]	129 [21]	291 [21]	133 [21]
1.0000	0.8483	0.8930	0.8246	0.8433	0.9031	0.8660	0.8835
0.0000	0.1517	0.1070	0.1754	0.1567	0.0969	0.1340	0.1165
				143 [25]	124 [25]	349 [25]	189 [25]
				0.9930	0.9798	0.9871	0.9762
				0.0070	0.0202	0.0129	0.0238
				0.0000	0.0000	0.0000	0.0000
				143 [25]	124 [25]	349 [25]	189 [25]
				0.9752	0.9672	0.9410	0.9201
				0.0248	0.0328	0.0590	0.0799

	198 [29]		171 [30]				
	0.6111		0.5029				
	0.3889		0.4971				
	0.0000		0.0000				
159 [15]		210 [17]	171 [31]	153 [31]	116 [31]	250 [31]	91 [31]
1.0000		0.8690	0.8918	0.8726	0.8836	0.9680	0.9725
0.0000		0.1310	0.1082	0.1275	0.1164	0.0320	0.0275

159 [15]	303 [33]	210 [17]	171 [9]	224 [32]	132 [32]	339 [32]	138 [32]
0.6879	0.5132	0.5857	0.5175	0.4509	0.4848	0.6608	0.7935
0.3121	0.4868	0.4143	0.4825	0.5491	0.5152	0.3392	0.2065
			171 [9]	224 [10]	217 [10]	339 [34]	272 [34]
			1.0000	0.9978	1.0000	1.0000	0.9761
			0.0000	0.0022	0.0000	0.0000	0.0239

54 [35]	294 [37]			214 [38]	121 [38]	191 [38]	113 [38]
0.2546	0.4168			0.6073	0.4039	0.4303	0.4201
0.7454	0.5832			0.3927	0.5961	0.5697	0.5799
	252 [37]						
	0.1933						
	0.8067						
	198 [40]	148 [41]		165 [39]	116 [39]	208 [39]	100 [39]
	0.0631	0.0270		0.0424	0.0215	0.0889	0.0750
	0.9369	0.9730		0.9576	0.9784	0.9111	0.9250
583 [44]	713 [37]	190 [45]	170 [9]	225 [46]	130 [46]	333 [46]	139 [46]
0.6102	0.7861	0.8447	0.8735	0.9756	0.8885	0.7793	0.6655
0.3898	0.2139	0.1553	0.1265	0.0244	0.1115	0.2207	0.3345
0.0000	0.0000	0.0000	0.0000	0.0000	0.0000	0.0000	0.0000

Table 4.1—contd.

Haptoglobin	N	467 [12]	220 [13]	356 [14]	246 [3]	143 [42]	150 [42]	152 [5]	1241 [43]
	Hp^1	0.1623	0.3250	0.3441	0.3496	0.3042	0.3723	0.4211	0.3360
	Hp^2	0.8377	0.6750	0.6559	0.6504	0.6958	0.6277	0.5789	0.6640
Inv[a]	N	346 [49]		365 [14]			29 [50]	144 [51]	101 [50]
	Inv 1	0.1689		0.2691			0.2122	0.1380	0.1621
	Inv	0.8311		0.7309			0.7878	0.8620	0.8379
Alpha$_1$ anti-	N	238 [12]		145 [27]				151 [52]	
trypsin	Pi^M	0.9790		1.0000				0.9801	
	Pi^S	0.0021		0.0000				0.0000	
	Pi^Z	0.0000		0.0000				0.0199	
	Pi^F	0.0189		0.0000				0.0000	
Transferrin	N	466 [12]		356 [14]				153 [5]	1277 [43]
	Tf^C	0.9829		1.0000				1.0000	0.9984
	Tf^B	0.0021							
	Tf^B0–1			0.0000				0.0000	0.0000
	Tf^B2			0.0000				0.0000	0.0004
	Tf^D1	0.0000		0.0000				0.0000	0.0012
	Tf^D	0.0150		0.0000				0.0000	0.0000

[a] The gene frequencies were calculated from the raw data given in the references after the number in the population by Dr E. Szathmáry using the program MAXLIK (Reed & Schull, 1968) for most of the systems in which at least one of the alleles were dominant or when there were multiple alleles and are marked [a]. For the codominant allele systems the gene frequencies were estimated by counting.
[b] MAXLIK would not converge, given frequencies calculated by hand, assuming that $R^Z = 0.0000$.
[c] Gc^{Esk} is a tentative designation: see text.
* Not in Hardy–Weinberg equilibrium: $P < 0.05$.
† Not in Hardy–Weinberg equilibrium: $P < 0.01$.
‡ Not in Hardy–Weinberg equilibrium: $P < 0.001$.
§ Not in Hardy–Weinberg equilibrium: $P < 0.001$ and not likely to be reliable as program took 31 iterations to find P.
[1], Omoto & Misawa (1974); [2], F. Pauls & F. H. Allen (personal communication); [3], F. Auger & P. M. Guevin (personal communication); [4], H. Gürtler, A. Gilberg & P. Tingsgård (personal communication); [5], Eriksson, Kirjarinta & Gürtler (1970); [6], Fernet, Langaney & Robbe (1971); [7], Fernet, Mortensen, Langaney & Robert (1971); [8], Kornstad (1972); [9], Eskola *et al.* (1976); [10], Eriksson (1973); [11], Beckman (1959); [12], Omoto & Harada (1972); [13], Scott,

737 [47]	127 [6]	245 [7]		329 [48]	168 [9]	231 [46]	129 [46]	346 [46]	148 [46]
0.4817	0.4488	0.4612		0.3100	0.2861	0.3420	0.2442	0.4697	0.3514
0.5183	0.5512	0.5388		0.6900	0.7139	0.6580	0.7558	0.5303	0.6486
438 [50]	130 [6]	243 [7]			172 [9]	254 [51]	132 [51]	345 [51]	153 [51]
0.2058	0.2715	0.1988			0.0882	0.1535	0.1835	0.2540	0.1482
0.7942	0.7285	0.8012			0.9118	0.8465	0.8165	0.7460	0.8518
			302 [53]			98 [52]		154 [52]	126 [52]
			0.9917			1.0000		0.9742	1.0000
			0.0000			0.0000		0.0228	0.0000
			0.0083			0.0000		0.0033	0.0000
			0.0000			0.0000		0.0000	0.0000
			301 [37]	329 [54]	168 [9]	234 [5]	127 [5]	337 [5]	141 [5]
			0.9967	0.9818	0.9911	1.0000	0.9921	1.0000	0.9965
			0.0017	0.0000	0.0000	0.0000	0.0000	0.0000	0.0035
			0.0000	0.0000	0.0000	0.0000	0.0000	0.0000	0.0000
			0.0017	0.0000	0.0000	0.0000	0.0000	0.0000	0.0000
			0.0000	0.0182	0.0089	0.0000	0.0079	0.0000	0.0000

Duncan, Ekstrand & Wright (1966); [14], McAlpine *et al.* (1974); [15], J. Dissing, A. Gilberg, H. Gürtler, J. B. Knudson & P. Tingsgård (1974); [16], Camoens, Monn & Berg (1973); [17], Beckman, Beckman & Cedergren (1971); [18], Duncan, Scott & Wright (1974); [19], Eriksson *et al.* (1971*a*); [20], Berg (1969); [21], Eriksson *et al.* (1971*b*); [22], Camoens, Monn & Berg (1972); [23], Gutsche, Scott & Wright (1967); [24], Singh *et al.* (1974); [25], Singh *et al.* (1971); [26], Omoto, Aoki & Harada (1974); [27], D. W. Cox & N. E. Simpson (personal communication); [28], Chen *et al.* (1972); [29], Olaisen & Teisberg (1972); [30], Virtaranta *et al.* (1976); [31], R. Nyholm (personal communication); [32], Eriksson *et al.* (1971*c*); [33], Monn (1969); [34], Kirjarinta *et al.* (1969); [35], Persson & Swan (1971); [36], Berg & Eriksson (1971); [37], Monn *et al.* (1971); [38], Berg & Eriksson (1973*a*); [39], Arvilommi, Berg & Eriksson (1973); [40], Teisberg (1971); [41], Brönnestam, Beckman & Cedergren (1971); [42], Gilberg & Persson (1967); [43], Persson (1968); [44], Persson & Tingsgård (1966); [45], Hirschfeld & Beckman (1961); [46], Eskola *et al.* (1971); [47], Persson & Tingsgård (1968); [48], Beckman & Mellbin (1959); [49], Steinberg & Kageyama (1970); [50], Persson *et al.* (1972); [51], Steinberg *et al.* (1974); [52], Fagerhol (personal communication); [53], Fagerhol, Eriksson & Monn (1969); [54], Beckman & Holmgren (1963); [55], Scott, Weaver & Wright (1970).

as that of the P^a and one and a half times higher than that of the P^b allele (Spencer, Hopkinson & Harris, 1964). Since it is known that riboflavin-5-phosphate (FMN) is the best known natural substrate for acid phosphatase, and FMN is necessary for the production of flavin adenine dinucleotide (FAD) (the prosthetic group for glutathione reductase) it could be expected that high acid-phosphatase activity (subjects with the P^c allele) would reduce glutathione-reductase activity. It is known that dietary deficiency of riboflavin (vitamin B_2) also reduces glutathione-reductase activity (Gustafsson *et al.*, 1971). This was suggested as a cause for the Skolt Lapps having a considerably lower glutathione-reductase activity than the other Inari Lapps (Eriksson *et al.*, 1967; Waller *et al.*, 1972). This may indeed be true because quantitative and qualitative nutritional studies have shown that the riboflavin intake, at least, in some Skolt Lapp families is rather low (Hasunen, unpublished observations). From the point of view of selection it seems paradoxical that the Skolt Lapps have one of the highest known frequencies of the P^c allele and at the same time the lowest glutathione-reductase activity. However, it may be that the riboflavin intake was nearer an optimal one in the ancient Skolt Lapp society.

It had been proposed that the red-cell acid-phosphatase allele P^b is in some way advantageous for survival under tropical conditions. As a matter of fact, the frequency of the P^b allele is among the lowest on record in some Arctic populations, e.g., in the Swedish Lapps 0.46, and in Alaskan Eskimos 0.44. However, some of the circumpolar populations appear to have rather high P^b frequencies: i.e. Greenland Eskimos, 0.71; Icelanders, 0.61; Inari Fisher Lapps, 0.72 and Inari Mountain Lapps, 0.67. Thus a selective disadvantage for the P^b allele in an Arctic environment is only one possible alternative hypothesis to genetic drift (for references see Eriksson, 1973).

The AK^2 gene for adenylate kinase is also not likely to be an Ainu or Eskimo gene. It was not detected at all in 223 Japanese (Omoto & Harada, 1972) nor in the Ainu, North American or Polar Eskimos (see Table 4.1). It occurs in low frequencies among the Greenland Eskimos, probably reflecting their relative admixture with Caucasians. It has low but polymorphic frequencies in most of the Lapp populations. The AK^2 allele, however, was almost absent in the Skolt and Swedish Lapps. Although the AK^2 gene is polymorphic in Caucasians, its frequency is low; 0.05 in the English and 0.04 in the Finns (Giblett, 1969; Eriksson *et al.*, 1971a). The AK^3 gene occurred in West Greenland and Ammassalik Eskimos and since this allele is nonpolymorphic in Caucasians it is not clear whether the gene originated from an Eskimo or Caucasian mutation.

The E_1^a gene for serum cholinesterase (Table 4.1) is another gene which does not appear to be present in the Ainu or Eskimos. It is polymorphic in Caucasians, again with a low frequency of about 0.02 to 0.06 (Giblett, 1969), and the frequency in the Lapps is similar to that of Caucasians (Singh *et al.*,

1971, 1974). An extraordinarily high frequency of the E_1^8 (silent) allele for cholinesterase (0.12) was found in the area around St Mary's in Alaska (Fig. 4.1) (Gutsche *et al.*, 1967). This gene is non-polymorphic in other populations and indeed is quite rare. Probably it has gained its high frequency due to a mutation or gene flow from a Caucasian who happened to have the allele with subsequent inbreeding in this isolated area.

The SOD^2 allele for cytoplasmic superoxide dismutase (Table 4.1) was not found in the two Eskimo populations which were tested, i.e. in Igloolik and Aupilagtoq. There are no comparable data for the Ainu or Japanese. One heterozygote (SOD 2-1) was found in 224 Fisher Lapps giving a SOD^2 frequency of only 0.002. In 272 Nellim Skolt Lapps there were 13 heterozygotes in two, not closely related, families. The SOD^2 allele does not seem to exist in a polymorphic frequency in any Lapp population other than the Nellim Skolt Lapps, and this may be a consequence of Finnish admixture. The SOD^2 allele has a relatively high frequency among all Finns so far studied both in Finland and in Sweden (Kirjarinta *et al.*, 1969; Beckman & Pakarinen, 1973; Eriksson, 1973). The SOD^2 allele has not been found in thousands of other Caucasians tested (Brewer, 1967) and appears to be rather exclusive to Fennoscandian peoples, and particularly the Finns, although there are not many reports on this enzyme. On the other hand, the isozymes may be seen as achromatic bands in a number of electrophoretic systems that are commonly used for screening other enzymes such as that for phosphoglucomutase. Consequently, variants are likely to have been absent in the populations tested for phosphoglucomutase in Table 4.1, i.e. all of the circumpolar peoples not mentioned above in this study even though the negative results for SOD have not always been reported.

The $Gm^{3, 5, 13, 14}$ haplotype is generally considered to be a Caucasian haplotype and Table 4.3 shows that it did not occur in the Ainu. The haplotype was detected neither in a population of 87 Japanese (Steinberg & Kageyama, 1970) nor in an earlier investigation of 748 Ainu (Steinberg & Matsumoto, 1964). The $Gm^{3, 5, 13, 14}$ haplotype has probably come into the Eskimo populations from admixture and occurred in the Igloolik Eskimos with a frequency of 0.04 (McAlpine *et al.*, 1974) and in the Ammassalik Eskimos with a frequency of 0.005 (Nielsen *et al.*, 1971), although this population was not tested for Gm (14) (Table 4.4). The Aupilagtoq population is known to have substantial admixture, and because the $Gm^{3, 5, 13, 14}$ haplotype has a frequency of 0.65 in Caucasians (Steinberg, 1969) admixture probably accounts for the polymorphic frequency of 0.258 in the Aupilagtoq Eskimos. The $Gm^{3, 5, 13, 14}$ haplotype has high frequencies among the Lapps (see Table 4.3) with the Inari Mountain Lapps and the Svettijärvi Skolt Lapps having frequencies as high as Caucasians and the Nellim Skolts with a frequency of 0.755 (Table 4.3). The $Gm^{1, 2, 21}$ haplotype is also not likely to be an Ainu or an Eskimo one. This haplotype had a frequency of 0.077

101

Table 4.2. *Gene frequencies for genetic markers in Eskimos[a] by the community*

System	Gene	Uper-navik	Umanak	K'utd-ligssat	Godhavn	Vajgat	Jakobs-havn	Christi-anshåb
Red-cell antigens								
ABO	N	134	101	97	56	42	98	57
	O	0.6443	0.6909	0.6071	0.6675	0.4635	0.6438	0.6174
	A_1	0.2554	0.2172	0.2825	0.2338	0.4739	0.2862	0.2171
	A_2	0.0150	0.0460	0.0295	0.0240	0.0000	0.0442	0.0230
	B	0.0853	0.0458	0.0809	0.0747	0.0625	0.0258	0.1425
Kell	N	93	84	88	39	22	77	44
	K	0.0108	0.0000	0.0230	0.0260	0.0000	0.0065	0.0465
	k	0.9892	1.0000	0.9770	0.9740	1.0000	0.9935	0.9535
MN	N	134	101	97	56	42	98	57
	M	0.7873	0.7277	0.7629	0.7411	0.7262	0.6990	0.6930
	N	0.2127	0.2723	0.2371	0.2589	0.2738	0.3010	0.3070
P	N	133	98	96	55	42	95	57
	P_1	0.1439	0.1566	0.1665	0.2065	0.1736	0.1857	0.2164
	P_2	0.8561	0.8434	0.8335	0.7935	0.8264	0.8143	0.7836
Rh	N	133	101*	97	55*	42	97[b]	57
CDe	R^1	0.6165	0.5693	0.6340	0.5363	0.5714	–	0.3957
cDE	R^2	0.3009	0.3490	0.2715	0.3622	0.2775	–	0.3747
CDE	R^z	0.0000	0.0000	0.0000	0.0000	0.0000	–	0.0000
Cde	r'	0.0000	0.0000	0.0000	0.0000	0.0000	–	0.0516
cde	r	0.0826	0.0817	0.0945	0.1014	0.1511	–	0.1308
cDe	$R°$	0.0000	0.0000	0.0000	0.0000	0.0000	–	0.0471

[a] The gene frequencies in this table were calculated by Dr E. Szathmáry using the program MAX-LIK (Reed & Schull, 1968) from the raw data supplied by Gürtler, Gilberg & Tingsgård (1974).

of birth in West Greenland

Egedes-minde	Kangat-siak	Holsteins-borg	Sukker-toppen	Godthåb	Frederiks-håb	Narssak	Juliane-håb	Nanor-talik
128	96	204	184	147	135	77	165	274
0.7355	0.5961	0.6652	0.6742	0.6141	0.6697	0.7281	0.6533	0.6620
0.1165	0.2177	0.2759	0.2843	0.3240	0.2658	0.2114	0.2621	0.2719
0.0045	0.0205	0.0034	0.0000	0.0203	0.0419	0.0000	0.0252	0.0176
0.1436	0.1657	0.0555	0.0415	0.0416	0.0226	0.0605	0.0594	0.0485
81	79	133	117	99	88	62	117	205
0.0062	0.0127	0.0075	0.0043	0.0051	0.0057	0.0000	0.0086	0.0049
0.9938	0.9873	0.9925	0.9957	0.9949	0.9943	1.0000	0.9914	0.9951
127	96	204	184	147	135	77	165	274
0.7244	0.7500	0.6250	0.7391	0.7789	0.7185	0.7987	0.7909	0.7792
0.2756	0.2500	0.3750	0.2609	0.2211	0.2815	0.2013	0.2091	0.2208
127	95	204	183	146	133	77	165	274
0.2285	0.3041	0.2197	0.2530	0.2381	0.1822	0.1626	0.1847	0.1345
0.7715	0.6959	0.7803	0.7470	0.7619	0.8118	0.8374	0.8153	0.8655
128	96	204	183	146	133	77	164	274
0.5664	0.5433	0.5361	0.5367	0.5274	0.5902	0.5909	0.5407	0.6189
0.3565	0.3181	0.2786	0.2801	0.3760	0.2747	0.3129	0.2965	0.2861
0.0000	0.0000	0.0000	0.0000	0.0000	0.0000	0.0000	0.0107	0.0029
0.0000	0.0453	0.0301	0.0179	0.0000	0.0000	0.0000	0.0401	0.0406
0.0771	0.0933	0.1555	0.1653	0.0966	0.1351	0.0961	0.1120	0.0515
0.0000	0.0000	0.0000	0.0000	0.0000	0.0000	0.0000	0.0000	0.0000

[b] MAXLIK would not converge.
* Not in Hardy–Weinberg equilibrium: $P < 0.01$.

Table 4.3. *Gm haplotype frequencies in circumpolar peoples based on tests for Gm (1, 2, 3, 5, 6, 13, 14, 21)*

		Ainu	Eskimos		Lapps		Skolt Lapps	
		Japan	Canada	Greenland	Inari, Finland			
Allotypes tested	Haplotype	Hokkaido	Igloolik	Aupilagtoq	Fisher	Mountain	Svettijärvi	Nellim
1, 2, 3, 5, 6, 13, 14, 21	N	159 [1]	365 [2]	144 [3]	254 [3]	132 [3]	345 [3]	153 [3]
	3, 5, 13, 14	0.000	0.041	0.258	0.453	0.625	0.671	0.755
	1, 21	0.522	0.782	0.656	0.497	0.267	0.269	0.179
	1, 2, 21	0.000	0.005	0.018	0.044	0.105	0.016	0.036
	1, 13	0.273	0.171	0.060	0.006	0.004	0.043	0.029
	1, 3, 5, 13, 14	0.026	?	0.008	0.000	0.000	0.000	0.000
	2, 21	0.179	0.000	0.000	0.000	0.000	0.000	0.000

Haplotype frequencies were calculated by a maximum likelihood method and were taken from the individual publications. [1], Steinberg & Kageyama (1970); [2], McAlpine *et al.* (1974); [3], Steinberg *et al.* (1974).

Table 4.4. *Gm haplotype frequencies in circumpolar peoples based on a variety of tests for Gm*

Population	Number tested	Allotypes tested	Haplotype frequencies									
			1, 21	1, 17, 21	1, 13	1, 13, 15, 16	1, 4, 5, 13	1, 2, 21	1, 2, 17, 21	3, 5, 11, 13, b⁵	3, 5, 11, 13, 23, b⁵	2, 21
Ainu												
Hokkaido	407[1]ᵃ	1, 2, 4, 5, 13, 15, 16, 21	–	ᵇ 0.563	0.000	0.252	0.043	–	0.093	–	–	0.050
	407[1]ᶜ	1, 2, 4, 5, 13, 15, 16, 21	–	0.571	0.000	0.252	0.043	–	0.000	–	–	0.134
Eskimos												
Thuleᵈ	150[2]	1, 2, 3, 5, 11, 13, 15	–	0.879	0.121	–	–	–	–	–	–	–
Ammas-salikᵈ	283[2]	1, 2, 3, 5, 11, 13, 15	–	0.804	0.191	–	–	–	–	–	–	–
*Lapps*ᵉ												
Utsjoki Mountain	176[3]	1, 2, 3, 5, 11, 13, 15, 16, 17, 21, 23, b⁵	–	0.261	0.000	0.006	–	–	0.108	0.280	0.345	0.000

ᵃ Frequencies if 1, 2, 21 present.
ᵇ Would not be detected. Gm 1, 11, 13, 15, 16, 17, b⁵.
ᶜ Frequencies if 1, 2, 21 not present.
ᵈ Thought to be pure.
ᵉ Haplotype frequencies calculated by A. G. Steinberg using MAXIM.
[1], Matsumoto & Miyazaki (1972); [2], Nielsen *et al.* (1971); [3], Eskola *et al.* (1976).

among the Ainu which, as Steinberg & Kageyama (1970) have shown, is not significantly different from zero and, based on this assumption, the authors estimated the haplotype frequencies for the Ainu which are given in Table 4.3. $Gm^{1, 2, 21}$ occurs in the Japanese, however, with a frequency of about 0.22 (Steinberg & Matsumoto, 1964; Steinberg & Kageyama, 1970). The $Gm^{1, 2, 21}$ haplotype has a low frequency in the Igloolik Eskimos (0.005) and a somewhat higher one in the Aupilagtoq Eskimos (0.018). Again, these frequencies may be explained by the admixture with Caucasians. The $Gm^{1, 2, 21}$ haplotype has a frequency of 0.11 in Caucasians (Steinberg, 1969) which is considerably lower than that for $Gm^{3, 5, 13, 14}$ in Caucasians (0.65) and also lower than $Gm^{3, 5, 13, 14}$ in the Aupilagtog Eskimos (0.26).

In contrast to the genes and haplotypes which are unlikely to be Ainu or Eskimo genes but which are polymorphic in Lapps, the $Gm^{3, 5, 13, 14}$ haplotype (Table 4.3) appeared to be polymorphic in the Ainu and possibly in the Eskimos although there are few conclusive data for this haplotype in Eskimos. The haplotype was non-existent in all of the Lapp populations studied. This haplotype is considered to be a Mongolian one (Steinberg, 1969) and its polymorphic frequency of 0.026 in the Ainu is probably due to the Japanese admixture.

Three alleles, P_1, PGM_1^2 and Gc^2 and two haplotypes, $Gm^{1, 21}$ and $Gm^{1, 13}$ were polymorphic in the three types of populations, i.e. the Ainu, Eskimos and Lapps. The alleles and haplotypes had frequencies which were on the average alike in the Ainu and Eskimos and different in the Lapps.

The blood-group allele P_1 (Fig. 4.5) had lower frequencies in the Ainu and Eskimos than in the Lapps with the exception of the Scoresbysund Eskimos. The Utsjoki Lapps had the highest frequencies among the Lapps, which may be due to their admixture with Norwegians and Finns.

The polymorphism of the phosphoglucomutase PGM_1 locus seems to be fairly stable; the PGM_1^2 gene frequency is between 0.20 and 0.30 in Caucasians (Giblett, 1969; Eriksson et al., 1971c). The data available so far indicate that the PGM_1^2 allele is in general less frequent in Mongolians than in Caucasians. With the exception of the Skolt Lapps all of the Lapp populations had PGM_1^2 frequencies ranging from 0.41 to 0.55, i.e. about twice as high as in the neighboring Scandinavian populations and considerably higher than in the Ainu and Eskimo populations (Fig. 4.5). The frequency of the PGM_1^2 gene seems to increase from south to north both in Norway and in Finland. The increase may be interpreted as the result of increasing Lappish influence (Monn, 1969; Eriksson et al., 1971c).

The Gc^2 allele (Table 4.1 and Fig. 4.5) has a high frequency in the Ainu and Eskimos and a low, although more variable one, among the Lapp populations. The frequencies for the Skolt Lapps were in the range of those for the Ainu and Eskimos with that for the Nellim Skolts being the highest among the Lapp populations. The Inari Fisher Lapps had the lowest frequency of the

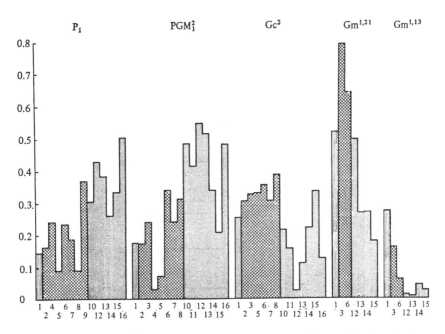

Fig. 4.5. Frequencies of the P_1 blood group allele; the $PGM_1{}^2$ allele for phosphogluco-mutase, the Gc^2 allele for Gc component and the $Gm^{1, 21}$ and the $Gm^{1, 13}$ haplotypes in Ainu from 1. Hokkaido: in Eskimos from 2. Alaska, 3. Igloolik, 4. Ft Chimo, 5. Thule, 6. Aupilagtoq, 7. West Greenland, 8. Ammassalik, 9. Scoresbysund: and Lapps from 10. Norway, 11. Sweden, 12. Inari Fisher Lapps, 13. Inari Mountain Lapps, 14. Svettijärvi Skolt Lapps, 15. Nellim Skolt Lapps, 16. Utsjoki Mountain Lapps.

populations reviewed (Fig. 4.5). This Gc^2 frequency (0.024) is the lowest noted in Europeans. The Gc^2 frequencies were next lowest in the Inari and Utsjoki Mountain Lapps and the Swedish Lapps; in all three the frequencies were lower than those in the Norwegian and Skolt Lapps (for references, *see also* Table 4.1; Reinskou & Kornstad, 1965; Cleve, 1973). The rather low frequency of Gc^2 in Finns (0.20) may be indicative of Lappish or eastern immigrants' influence (Eriksson, 1973).

The $Gm^{1, 21}$ haplotype has a high frequency in the Ainu and Eskimos (Table 4.3, Fig. 4.5). This haplotype is both a Caucasoid and Mongoloid characteristic and if tests for the antigen Gm (17) had been performed the haplotypes would probably have been $Gm^{1, 17, 21}$ in all cases (Steinberg *et al.*, 1961; Steinberg, 1969). On the other hand, the $Gm^{1, 13}$ haplotype is a Mongo-loid but not a Caucasoid one. The few available data for this haplotype show a cline (i.e. Fig. 4.5) with the frequency decreasing from east to west (from Mongolians to Caucasians) with very low frequencies among the Lapps. Again if the antigen Gm (17) had been used this haplotype would probably have been $Gm^{1, 13, 17}$ (Steinberg, 1969).

107

r (cde)   ADA²

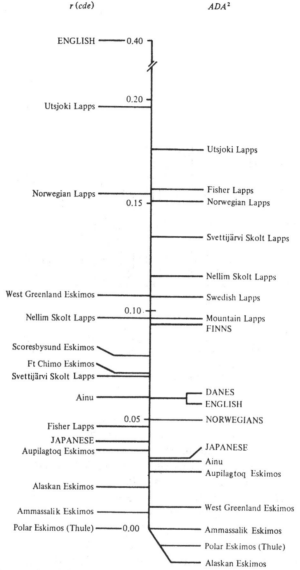

Fig. 4.6. The frequencies of the r(cde) allele for the Rh blood group and the ADA² allele for adenosine deaminase in circumpolar peoples and their neighbors. All of the r frequencies in the Caucasian neighbors are similar to those of the English.

There were two alleles which are probably not originally Eskimo ones and which have low but polymorphic frequencies in the Ainu and even higher frequencies in the Lapps than in either the Ainu or Eskimos. The alleles are *ADA*² for the red-cell enzyme adenosine deaminase and r(cde) for the Rh antigen system. Fig. 4.6 demonstrates that the frequencies for the *ADA*² allele were low but polymorphic (0.03), in the Ainu and that there was a similar frequency in the Japanese (Omoto & Misawa, 1974). The populations of Eskimos that were assumed to have relatively little admixture with Caucasians did not have ADA^2 i.e. those in Thule, Ammassalik, Alaska and Igloolik. The Aupilagtoq and West Greenland Eskimos who were assumed to have considerable admixture had low frequencies of ADA^2. The frequencies of ADA^2 for the English and Scandinavians are between 0.05 and 0.06 (Ageheim & Bergström, 1972). On the other hand, the Finns have a higher frequency of ADA^2 (0.10) than other Scandinavians, and the frequencies among the Lapps are the highest (0.10–0.18) of all the world populations hitherto studied (Eriksson *et al.*, 1971*b*).

The Rh-negative gene *r* (*cde*) occurred in low frequencies (Table 4.1) in the Ainu and Eskimos. Since *r* is not generally regarded as a Mongolian gene the polymorphism may have originated in the Ainu rather than in the Japanese whose frequency (0.04) is less than that in the Ainu (0.06); see Omoto & Misawa (1974). The gene does not seem to be present in the so-called pure Eskimos in Thule and was rare (0.002–0.007) in Ammassalik (Table 4.1). The Lapps have higher but more variable frequencies of the *r* gene than have the Ainu and Eskimos but still not as high as the usually quoted frequency of nearly 0.40 for Caucasians. There is a known selective disadvantage for the heterozygotes for the Rh-negative and Rh-positive genes (i.e. those complexes with *D*). When the frequency of the less frequent gene is low the selection will be faster in reducing the frequency of the less frequent gene (Stern, 1973). Therefore, it is not surprising that the frequencies are so much lower in all of the circumpolar peoples than in Caucasians assuming that they had fewer Rh-negative alleles originally. The Rh-negative gene was polymorphic in the Eskimos known to have considerable admixture (i.e. in Alaska, Fort Chimo, Aupilagtoq, and West Greenland), and these frequencies as well as those of the Lapps, may have reached these levels due to gene flow from Caucasians. However, the frequency of *r* is relatively low (about 0.35) among Finns (Nevanlinna, 1947, 1972).

The HL-A antigen A3 which is common among Caucasians with a phenotypic frequency of about 30% (Int. Workshop, 1973) and uncommon in Mongolians (e.g. 3% in Japanese) had a low frequency in the Ainu and Eskimos but a surprisingly high frequency in the Lapps (47.2–57.9%, Table 4.5). The Lapps (around 50%) and the Finns (42%) have the highest of European frequencies of HL-A3 (Tiilikainen *et al.*, 1973).

The A7 antigen which has a low phenotype frequency of 10% or less in Mongolians was comparatively rare in the Ainu and Eskimos (Table 4.5) with the exception of the Eskimos in Igloolik (19%) but had a frequency in Lapps (about 40%) which was higher than those in their Caucasian neighbors whose frequencies were about 30% (Int. Workshop, 1973; Tiilikainen *et al.*. 1973). The antigen A5, on the other hand, had a frequency > 20% in

Table 4.5. *Phenotypic frequencies for the histocompatibility antigens (HL–A) of circumpolar peoples*

		Population							
		Ainu	Canada	Eskimos			Lapps		Skolt Lapps
		Japan		Canada	Greenland		Norway	Finland	Finland
HL–A Sublocus	Gene	Hokkaido	MacKenzie	Igloolik	West and South	East	Finnmark	Utsjoki	Inari
First	N	125[1]	236[2]	320[2]	120[3]	62[3]	176[6]	140[5]	178[4]
	A1	0.008	0.055	0.000	0.050	0.000	0.130	0.114	0.045
	A2	0.548	0.356	0.506	0.375	0.065	0.449	0.386	0.584
	A3	0.040	0.055	0.003	0.033	0.000	0.472	0.486	0.579
	A9	0.536	0.886	0.877	0.867	0.952	0.398	0.465	0.213
	A10	0.357	0.000	0.000	–	–	0.034	0.050	0.073
	A11	0.081	0.000	0.000	0.042	0.000	0.068	0.093	0.140
	A28	0.079	–	0.089	0.233	0.323	0.097	0.064	0.062
	W19	0.386	–	0.003	–	–	0.063	0.007	0.022
	W23	0.008	–	–	–	–	–	0.000	–
	W25	0.065	–	–	0.008	0.000	–	0.000	–
	W26	0.352	–	–	0.008	0.000	–	0.051	–
	W29	0.087	–	–	0.008	0.000	–	0.000	–
	W30	0.000	–	–	0.033	0.000	–	0.000	–
	W31	0.183	–	–	–	0.000	–	0.007	–
	W32	0.016	–	–	0.000	0.000	–	0.050	–

Second

B5	0.246	0.229	0.274	0.200	0.403	0.023	0.085	0.202
B7	0.024	0.110	0.193	0.042	0.000	0.409	0.393	0.489
B8	0.008	0.068	0.000	0.033	0.000	0.085	0.057	0.011
B12	0.081	0.000	0.000	0.050	0.000	0.119	0.078	0.006
B13	0.000	0.000	0.000	0.000	0.000	0.017	0.007	0.006
B14	0.024	0.000	0.006	0.008	0.000	0.000	0.000	0.000
B17	0.071	–	–	0.025	0.065[b]	0.085	0.071	0.000
B27	0.000	–	–	0.310[a]	0.054	0.244	0.236	0.045
W5	–	0.013	0.107	0.149	0.468	0.119	0.164	0.292
W10	0.228	0.758	0.801	0.358	0.500	0.171	0.150	0.073
W15	0.492	0.110	0.083	0.408	0.000	0.250	0.308	0.579
W16	0.276	–	–	0.017	0.000	–	0.121	0.017
W18	0.032	0.004	–	0.008	–	–	0.050	0.039
W20	–	–	0.003	–	–	–	0.206	0.034
W21	0.024	–	–	0.008	0.000	0.080	0.000	0.000
W22	0.039	0.229	–	0.008	0.000	0.017	0.100	0.006
407*	–	–	–	0.000	0.000	0.017	–	–
FJH*	–	–	–	–	–	–	–	–
JA	–	–	–	0.250	0.160	–	–	–

[a] FJH = 0.250; FJH – AJ = 0.008.
[b] FJH = 0.065; FJH – AJ = 0.

[1], Mittal et al. (1973); [2], Dossetor et al. (1973b); [3], Kissmeyer-Nielsen et al. (1973); [4], Tiilikainen et al. (1973); [5], E. van den Berg-Loonen & M. Kort-Bakker (personal communication); [6], Thorsby, Bratlie & Teisberg (1971).

the Ainu and Eskimos and was lower in the Lapps although the Skolts had a frequency of 20%. The rather high frequency of A5 in Finnish Lapps may be due to gene flow from the Finns whose frequency of 13% (Tiilikainen *et al.*, 1973) was almost twice that of Norwegians 8% (Thorsby, Bratlie & Teisberg, 1971).

Genes with frequencies similar in the Eskimo and Lapps but different in the Ainu

There were two alleles with frequencies which were similar in the Eskimos and Lapps and different from those in the Ainu; the Fy^a gene for the Duffy blood group and the Hp^1 gene for the serum protein haptoglobin. Three of the HL–A antigens (A10, W19, at the first and W5 at the second sublocus) were more common among the Ainu than in the Eskimos and Lapps. In addition, there were seven alleles which occurred in the Ainu and probably did not occur at all in the original Eskimo and Lapp populations or have been lost through drift or selection. These alleles in the Ainu were Di^a for the Diego system, r'' (*cdE*) and r' (*Cde*) for the Rh antigen system, Pi^S and Pi^F for alpha$_1$ antitrypsin, Tf^DChi for transferrin, and the $Gm^{2,\ 21}$ haplotype. There is genealogical evidence for the assumption that the Tf^DChi allele in the Mountain Lapps in Inari and Utsjoki may be a consequence of gene flow from Finns. The same is the case for the Tf^B0–1 allele found in the Nellim Skolt Lapps.

The frequency of the Fy^a gene was particularly high in the Scoresbysund Eskimos (0.89) and the Ainu (0.90) (Fig. 4.7) and the Japanese (Omoto & Misawa, 1974) compared to the other Eskimos, the Lapps and the Caucasians. Generally speaking it had a high frequency in both Eskimos and Lapps but not as high as in the Ainu. Fig. 4.7 shows that the Eskimos with the exception of those in Aupilagtoq, Greenland, had higher frequencies of Fy^a than those in the Lapp populations although not many Eskimo populations were tested for the Duffy system.

The frequencies of the Fy^a allele for all of the Eskimo and Lapp populations were higher than those (about 0.45) for the English (Race & Sanger, 1968) and Scandinavians (Heiken, 1962; Berg, 1973), and vary considerably. We do not yet know to what extent the phenotype Fy (a − b−) (or *Fy*) allele, that is common in Negroes, is present outside Africa. The Nellim Skolt Lapp population was tested also for the Fy^b antigen. The Fy^b allele appeared to be relatively rare in comparison with other European data.

The frequencies for the Hp^1 gene were similar in the Eskimos and Lapps although with considerable variability between the subpopulations (Fig. 4.7) and the frequency was substantially lower (0.16) in the Ainu than either in the Eskimo or Lapp populations. The frequency in the Ainu was also lower than in their Japanese neighbors whose frequency was 0.25 (Osmoto & Harada,

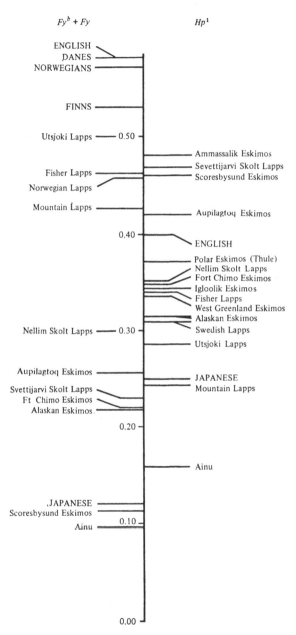

Fig. 4.7. The frequencies of the Fy^b + Fy for the Duffy blood group and the Hp^1 allele for haptoglobin in circumpolar peoples and their neighbors.

113

1972). The Hp^2 allele codes for a partial duplication of the Hp^1 molecule (Giblett, 1969) and therefore, codes for a larger molecule than Hp^1 does. It has been suggested that the Hp^1 allele confers a disadvantage in cases where there is red-cell destruction because the Hp^1 form of the protein passes through the kidney glomerular membrane more readily than that of the Hp^2 type of protein. Hp^1 is smaller than Hp^2 and thereby Hp^1 would make it more difficult to retain the smaller HpHb complex (Berggård & Bearn, 1962). Knowledge of the history of haemolytic diseases in the Ainu population in the past would be of interest as a possible selective factor having caused the unusually low frequency of the Hp^1 allele in the Ainu.

The Di^a allele for the Diego antigen was polymorphic in the Ainu (0.04). It did not occur in over 1000 investigated Lapps in Sweden, Norway and Finland (Beckman, 1959; Kornstad, 1972; Eriksson, 1973), and had a frequency of 0.003 in the Alaskan Eskimos (Table 4.1). Since this allele is considered to be a Mongolian one it is not surprising that it is polymorphic only in the Ainu. It had a frequency of 0.03 in their Japanese neighbors (Omoto & Misawa, 1974). The allele may not have been present in the original Eskimos or may have been lost by drift or selection. Of the Eskimo populations unfortunately only the population in Coppermine in the western Arctic of Canada was tested for the Di^a antigen (Chown & Lewis, 1959). The allele was absent in 320 Copper Eskimos.

The r'' (*cdE*) allele for the Rh antigen had an exceptionally high frequency (0.14) in the Ainu (Table 4.1). This is considerably higher than in their Japanese neighbors whose frequency was 0.05 (Omoto & Misawa, 1974). Most of the other populations were tested for this gene and it was completely absent. The r' (*Cde*) allele occurred in the Ainu with a frequency of 0.05. The West Greenland Eskimos and Norwegian Lapps probably got their few r' alleles from Caucasians, the frequency in Caucasians being about 0.01 (Race & Sanger, 1968).

The Pi^F allele for alpha$_1$-antitrypsin was polymorphic in the Ainu (0.02) and absent in Igloolik and Aupilagtoq Eskimos, Norwegian and Finnish Lapps (Table 4.1). The Pi^S allele occurred with a low frequency in the Ainu but was absent in Eskimos and in the majority of Lapp populations. Pi^S was polymorphic only in the Svettijärvi Skolts. Since there are few data for the Eskimos it is not clear whether the Pi^S allele was originally exclusive to the Ainu as far as circumpolar peoples are concerned. The exception is that the Skolt Lapps have the allele. However, the three Skolt subjects who have the Pi phenotype MS are closely related. It may be that the ancestors of these Skolt Lapps got the gene from the Norwegians who have a frequency of 0.023. The Finns, on the other hand, do not seem to have the gene in a high frequency (Fagerhol, Eriksson & Monn, 1969; unpublished observations).

The Tf^DChi allele for transferrin occurred only in the Ainu and in the Swedish and Finnish Mountain Lapps. Tf^DChi is generally polymorphic in

Mongolian populations and occurs with a frequency of 0.01 in the Japanese (Omoto & Harada, 1972). Surprisingly, the *Tf*Chi^D allele also exists in Finns with a frequency of 0.01 (Nevanlinna, 1972). Beckman & Holmgren (1961) reported a variant in Swedish Lapps which was later considered to be *Tf*^{D}Chi by W. C. Parker and A. G. Bearn (*see* Beckman & Holmgren, 1963).

The *Gm*$^{2,\,21}$ haplotype was polymorphic in the Ainu (Table 4.3) and was absent in the Eskimo and Lapp populations and in the Japanese (Steinberg & Matsumoto, 1964; Steinberg & Kageyama, 1970). It appears, therefore, to be exclusively an Ainu haplotype.

The HL–A antigen A10 had a very low frequency in all of the Eskimo populations which were studied (Table 4.5). The antigen had a rather low phenotypic frequency in Lapps like those in Norwegians (0.07) (Thorsby *et al.*, 1971) and Finns (0.10) (Tiilikainen *et al.*, 1973).

On the other hand the Ainu had a high frequency of the A10 antigen (0.36) which is considerably higher than the average in Japanese (0.26) (Int. Workshop, 1973). Similarly W19 antigen had a high frequency in the Ainu (0.39) as it did in the Japanese (0.42) (Int. Workshop, 1973). W19 was possibly absent in the original Eskimos in Igloolik (0.003) (Dossetor *et al.*, 1973*a*) and had low frequencies in the Lapps in whom it has a somewhat lower frequency than in the neighboring Norwegians (0.13) (Thorsby *et al.*, 1971) and Finns (0.16) (Tiilikainen *et al,*, 1973). The W5 antigen at the second sublocus had a higher frequency in the Ainu (0.32) than in the Eskimos and Lapps (Table 4.5) with the Lapps generally having a frequency slightly higher than in the Eskimos. The W5 antigen was also more frequent in Japanese (0.27) (Mittal *et al.*, 1973; Int. Workshop, 1973; Bodmer & Bodmer, 1973) than in Norwegians (0.13) (Thorsby *et al.*, 1971) and in Finns (0.19) (Tiilikainen *et al.*, 1973).

*Genes with frequencies similar in the Lapps and Ainu
and different in the Eskimos*

There were three alleles, *Jk*^{a} for the Kidd, R^2 (*cDE*) for the Rh antigens and *PGD*^{C} for 6-phosphogluconate dehydrogenase which had frequencies which were alike in the Lapps and Ainu but which were different in the Eskimo population and three alleles which occurred only in the Eskimo populations, R^Z (*CDE*) for the Rh antigen, *Gc*^{Esk}* for group specific component and *Gpt*^{1} for glutamic-pyruvate transaminase. The frequencies of the HL–A antigens A9 and W10 were higher in the Eskimo populations which were tested than in the Lapp and Ainu populations (Table 4.5)

The phenotypic frequency of W18 was lower in the Eskimos than in the Ainu and Lapps.

The allele *Jk*^{a} was on the average twice as frequent in the Eskimos (0.60)

* Arbitrarily designated as *Gc*^{Esk}; the identification of the variant is at present under study.

as in both the Ainu and Lapp populations (Fig. 4.8) whose frequency on the average was about 0.30. The frequencies in the Lapps were quite variable. The frequency of the allele Jk^a in the Finnish Lapps is about 0.25, in the Norwegian Lapps 0.53; the latter frequency is similar to those in other European populations.

The Lu^a allele for the Lutheran antigen is rare in Eskimos but antiserum for Lu^a was not used in testing the Ainu (Table 4.1). Unfortunately, the only data available for Lapps are probably not exact (Allison *et al.*, 1952; 1956).

The R^2 (cDE) allele frequency (Fig. 4.8) appears to be on the average higher in the Eskimos (0.36) than in the Ainu and Lapps both of which had a frequency around 0.20. Also North American Indians have a high frequency of R^2, which is in general higher than is found anywhere outside America (Mourant, Kopec & Domaniewska-Sobczak, 1975). The Nellim Skolt Lapps had an exceptionally low frequency of 0.07 for the R^2 allele which can probably be explained by genetic drift.

The PGD^C allele for phosphogluconate dehydrogenase (Table 4.1) was probably present in the original Ainu population (0.04) and may have come from the Japanese who have a frequency of 0.05 (Omoto & Harada, 1972). It was probably not present in the original Eskimo populations or it may have got lost. It could have come from Caucasian sources, in which it has a low but polymorphic frequency of about 0.02 (Giblett, 1969).

The PGD^C allele had the highest frequency of the circumpolar populations in the Swedish Lapps and Inari Fisher Lapps (0.13). Among Inari and Utsjoki Mountain Lapps this gene also occurred in high frequencies, 0.12 and 0.11 respectively, but had only the frequency of 0.03 in the Skolt Lapps (Beckman, G. *et al.*, 1971; Nyholm, R. *et al.*, in preparation).

The R^z (CDE) allele occurred in low frequencies in the Ainu and in some of the Eskimo populations (Table 4.1) and was absent in the Lapp populations. It is uncommon in Caucasians, and the Eskimo populations in which it was present were not particularly contiguous to each other. The gene occurred in both the East and some of the West Greenlanders but also occurred in Alaskan Eskimos in Wainwright.

Gc^{Esk} may be similar to a variant found in the Ammassalik Eskimos reported by Persson & Tingsgård (1965), a study which is not included in Table 4.1. It was only detected in Igloolik in the present survey with a frequency of 0.007. Although it was not possible to retest Gc in the Igloolik Eskimos, the plasma was drawn off the samples soon after taking the blood and frozen in liquid nitrogen until typing. The result, therefore, was not likely to be due to poor handling which is known to produce artifacts in this system (Persson & Tingsgård, 1965). The variant is described by Cox *et al.* (1978).

Another hitherto undescribed allele found in Igloolik was a new one, the glutamic-pyruvic transaminase allele Gpt^5 (Chen *et al.*, 1972). Only three

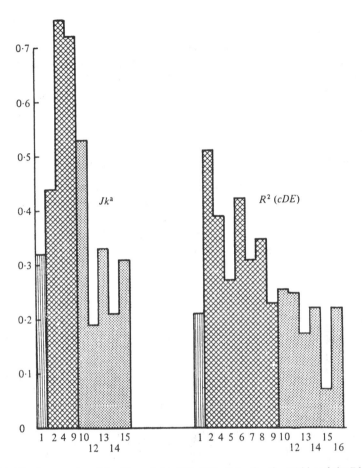

Fig. 4.8. The frequencies of the *Jk*ᵃ and the *R*² (*cDE*) alleles for the Kidd and the Rh blood groups, respectively, of the Ainu (——) in 1. Hokkaido: the Eskimos (——) in 2. Alaska, 4. Ft Chimo, 5. Thule, 6. Aupilagtoq, 7. West Greenland, 8. Ammassalik, 9. Scoresbysund: the Lapps (——) in 10. Norway, 12. Inari Fisher Lapps, 13. Inari Mountain Lapps, 14. Sevettijärvi Skolt Lapps, 15. Nellim Skolt Lapps, 16. Utsjoki Mountain Lapps.

populations were tested for this system and it is too soon to know if this is a typical Eskimo allele.

The Eskimos had exceptionally high phenotypic frequencies for the histocompatibility antigens A9 and W10 (Table 4.5). The frequencies were higher than those of the Japanese and Ainu. The Japanese frequencies of 60% for A9 and 36% for W10 (Int. Workshop, 1973) were, however, higher than the 18% for A9 in Norwegians and Finns, the 21% for W10 in Norwegians

The human biology of circumpolar populations

(Thorsby *et al.*, 1971) and 15% in Finns (Tiilikainen *et al.*, 1973). The Lapp frequencies for A9 were higher than those of their neighbors. On the other hand, the Ainu frequencies for both A9 and W10 were somewhat lower than the Japanese frequencies. The exceptional frequencies in the Eskimos suggest that these antigens may have some adaptive significance. The W22 antigen also had a high frequency in the MacKenzie Delta Eskimos. The Japanese frequency for this antigen is 18% (Int. Workshop, 1973) and is low in the Ainu. The W18 antigen occurred less frequently in the Eskimos than in either the Ainu or Lapps. The phenotypic frequency of W18 is 5% in Finns (Tiilikainen *et al.*, 1973).

Genes with frequencies which suggest several clines around the Pole

There were two blood-group systems for which there were data for populations around the North Pole, the ABO and MN systems. In both systems there was considerable variation in the frequencies of individual populations but in general the frequencies of the *A*, *B* and *O* alleles and the *M* and *N* alleles had four clines around the Pole (Fig. 4.9).

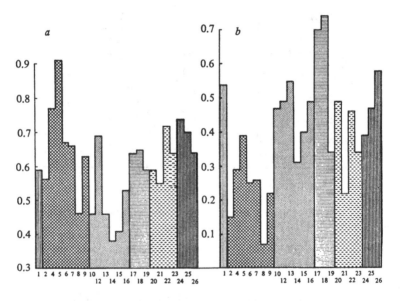

Fig. 4.9. The frequencies of alleles for the O antigen (*a*) and the N antigen (*b*) in circumpolar peoples; the Ainu (———) in 1. Hokkaido; the Eskimos (———) in 2. Alaska, 4. Ft Chimo, 5. Thule, 6. Aupilagtoq, 7. West Greenland, 8. Ammassalik, 9. Scoresbysund; the Lapps (———) in 10. Norway, 12. Inari Fisher Lapps, 13. Inari Mountain Lapps, 14. Svettijärvi Skolt Lapps, 15. Nellim Skolt Lapps, 16. Utsjoki Mountain Lapps; the northwest Siberians (———), 17. Nenets, 18. Nganasans, 19. Dolgans: northeast Siberians (———) 20. Tundra Yukagirs, 21. Tundra Yakuts; Eskimos in 22. Chaplino, 23. Naukan: the Chukchi (———) in 24. Uelen, 25. Pacific, 26. Anadyr.

The frequencies for the *O* allele increased from 0.56 in the Alaskan Eskimos to 0.91 in the pure Polar Eskimos in Thule, Greenland. With the exception of the Fisher Lapps (*O* = 0.69) and the Inari Mountain Lapps (*O* = 0.53) so far all the other Lapp populations studied have *O* gene frequencies below 0.50, being as low as 0.38 in the Svettijärvi Skolt Lapps. Similarly, the Finns have lower *O* gene frequencies than other Scandinavians i.e. 0.56 and 0.62, respectively. The frequency is higher in the Siberian populations with a drop in that for the eastern Tundra Yakuts. Another increase occurred in the most northerly Chukchi whose frequencies then declined in a southerly direction to 0.59 in the Ainu which is a similar frequency to that of the Alaskan Eskimos (Fig. 4.9). Generally speaking, the *B* allele had a relatively low frequency where the *O* was high (note that *B* is completely absent in the pure Polar Eskimos). Conversely, the *B* allele was high where the frequency of *O* was low (*see* Table 4.1). The Norwegian and Swedish frequencies of the *B* allele of about 0.07 and the Finnish of about 0.13 were higher than was generally found in the Eskimos. It was lower, however, than that in the Lapps, whereas the frequency of 0·16 in the Ainu was typical of Mongolian populations. The antisera for the subtypes A_1 and A_2 were not used in studies of all the populations.

The distribution of the frequencies for A_2 has already been discussed. Until now it has been supposed that the gene A_2 was missing or had a very low frequency among the Siberian populations. Recently, however, a relatively high frequency of the A_2 allele has been found in the North Altayan group, 0.012 in Kumandintsi and 0.010 in Chelnantsi (R. Sukernik, unpublished observations, 1974).

The distribution of frequencies of the alleles for the N and O antigens followed a similar pattern in the North American and Greenland Eskimos (Fig. 4.9). The highest frequency of N was found in the pure Polar Eskimos of Thule and the lowest in Ammassalik. The frequency of the *N* allele was on the average higher in the Lapps than in the Eskimos and increased to the highest frequency in the Nenets and Nganasans of western Siberia. The frequency distribution was lower in the eastern Siberian Tundra Yakuts and higher in the Chukchi increasing in a southerly direction along the Pacific coast. The frequency in the Chukchi living in the Anadyr valley was higher than that in the Ainu. Although there was considerable variation between the populations within the six groups (in Fig. 4.9), the distribution of frequencies appeared to be generally a bimodal curve with two peaks; one for the Lapps and western Siberians and another for the Chukchi and Ainu. The two nadirs were for the North American and Greenland Eskimos and for the eastern Siberian Eskimos and Yakuts.

The Siberian populations were not tested for the S antigens but the Ainu, some of the North American Eskimos and the Lapp populations were (Table 4.1). The *NS* complex had a low frequency in Alaskan and Fort Chimo

Table 4.6. *The gene frequencies for the ABO and MN blood groups in some Siberian populations*

Population	Number tested	Gene frequencies				
		O	A	B	M	N
Northwest						
Nenets	235	0.6373	0.2008	0.1619	0.2979	0.7021
Nganasan	230	0.6472	0.2200	0.1328	0.2609	0.7391
Dolgan	242	0.5880	0.1825	0.2295	0.6467	0.3533
Northeast						
Tundra Yukagir	75	0.5884	0.3492	0.0623	0.5133	0.4867
Tundra Yakut	137, 132	0.5486	0.2116	0.2398	0.7803	0.2197
Chaplino Eskimo	87	0.7191	0.0903	0.1906	0.5345	0.4655
Naukan Eskimo	65	0.6369	0.3080	0.0551	0.6615	0.3385
Chukchi						
Uelen	69	0.7276	0.1655	0.1069	0.6087	0.3913
Pacific	76	0.6979	0.2053	0.0968	0.5263	0.4737
Anadyr	71	0.6390	0.3101	0.0509	0.4225	0.5775

Gene frequencies calculated by MAXLIK from Szathmáry (1974).

Eskimos and in the Skolt Lapps but a frequency of ≥ 0.10 in the Ainu, and Aupilagtoq Eskimos. With the exception of the Skolt Lapps in the east, the Lapp populations have frequencies of *NS* which are the highest in Europe. The frequencies of *NS* in the Lappish core area are higher than those of any other populations, except the Ainu. For a sample of Ainu an *NS* frequency as high as 0.28 has been reported (Mourant, 1954). However, recent studies of the Ainu by Misawa & Hayashida (1968) show an *NS* frequency of only 0.10. Among the Lapps, the frequency of *NS* reaches 0.26 in Kautokeino, in Norway, close to the Finnish border (Kornstad, 1972) and 0.21 in Utsjoki (Eskola *et al.*, 1976). In other Scandinavians the frequency of *NS* is considerably lower than in the Lapps: in Finns 0.08 (Nevanlinna, 1972), in Swedes 0.08 (Heiken, 1965) and in Icelanders 0.06 (Bjarnason *et al.*, 1973). The English have a frequency of 0.08 (Race & Sanger, 1968) and the Japanese 0.10 (Omoto & Misawa, 1974).

In addition, the frequencies for two of the Gm haplotypes, *Gm*[1, 13] and *Gm*[1, 21] formed apparent clines around the North Pole (Tables 4.3 and 4.4) although there are still only a few data available. *Gm*[1, 13] had the highest frequency in the Ainu (0.27), generally decreased from west to east in the Eskimo populations and had low frequencies in the Lapps. The *Gm*[1, 13] haplotype is considered to be a Mongolian one and Steinberg *et al.* (1974) suggest that its presence in the Lapps may be because of common ancestry of Eskimos and Lapps. The *Gm*[1, 21] frequency was highest in Igloolik (0.79) of all the circumpolar populations and was lowest in the Lapps. The frequency in the Ainu (0.52) was intermediate between the Eskimos and the Lapps.

Genes which had similar frequencies in all circumpolar populations

Five alleles and nine HL–A antigens probably did not occur in any of the original circumpolar populations. They were $Tf^{B}0$-1, $Tf^{B}2$, $Tf^{D}1$, Pi^{Z}, r^{y} (*CdE*), HL–A antigens A1, A12, A8, A11, A28, A13, A14, A17 and W21. Two alleles, E_2 + for cholinesterase and Inv^1 and the HL–A antigen A2 probably had much the same frequency in all of the original circumpolar populations.

The $Tf^{B}0$-1 allele for transferrin was absent in the Ainu and Eskimos but was present in two Lapp populations, i.e. the Nellim Skolts and Norwegian Lapps in low frequencies (Table 4.1). The $Tf^{B}0$-1 allele is primarily an American Indian one (Giblett, 1969) but is polymorphic also in some Siberian and Finno-Ugrian peoples (Spitsin, 1972; for references *see* Eriksson, 1973). The frequency is 0.012 in Finns (Nevanlinna, 1972) although $Tf^{B}0$-1 has not been found in most other European populations and may have entered the Lapp populations by Finnish admixture. Since the Norwegian Lapps who were studied live just north of the Finnish border it is not unreasonable to suppose that there is Finnish admixture in them as well as in the Skolts. The $Tf^{B}2$ is a western European allele (Beckman & Holmgren, 1961; Braend *et al.*, 1965) but it is completely missing in the Lapp populations studied so far. It probably entered West Greenland (Table 4.1) from European sources. $Tf^{D}1$ is essentially a South Pacific (Omoto, 1973) and Negro allele but it has been reported in Swedes (Beckman & Holmgren, 1963) and in Norwegian Lapps (Monn *et al.*, 1971). It could have entered the Greenland and Lapp populations from Scandinavian sources.

The Pi^{Z} allele for alpha$_1$-antitrypsin (Table 4.1) does not appear to be present in the Ainu, but it is polymorphic in Aupilagtoq Eskimos (0.02). Six of 151 Aupilagtoq individuals had the Pi^{Z} allele. All of them belonged to the same family (father and 5 children). It is therefore possible that the Pi^{Z} allele has come from Danish admixture. There are few data available for this system, however, and it may have been absent in the original Lapp population. It was only present in the Norwegian Lapps and in the Svettijärvi Skolt Lapps (Fagerhol *et al.*, 1969). It is polymorphic in Caucasians, 0.02 in Norwegians and 0.01 in white Americans (Giblett, 1969) although there are only data from a few populations.

The r^{y} (*CdE*) allele was not found in any of the circumpolar populations (Table 4.1). It is, however, to be noted that in the absence of r (*cde*), r^{y} would be detectable only in homozygotes which are rare. In heterozygotes it would in general be indistinguishable from R^{Z} (*CDE*).

The three HL–A histocompatibility antigens A1, A8 and A12 are common

121

(phenotypic frequencies $\geq$ 20%) in Caucasians. They are absent or uncommon in Mongolians. They have very low frequencies in all of the circumpolar peoples (Table 4.5). The HL–A1 antigen occurs in Caucasians with a frequency of about 30% and is absent in Japanese and other Mongolians.

The HL–A1 antigen is absent in Igloolik Eskimos and it is likely that it is not an Eskimo antigen. It probably entered the MacKenzie Delta and West and South Greenland Eskimos from admixture (Dossetor *et al.*, 1973*a*). The Skolt Lapps had a very low frequency of the HL–A1 antigen and that in the Norwegian and Utsjoki Lapps is less than half that in Caucasians. It could be argued that the Lapp frequencies also came from admixture with Caucasians. But also Finns have a low frequency of the HL–A1 antigen (16.9%), i.e., only about half of that in other Caucasians.

The HL–A8 antigen has a low frequency in Mongolian peoples. It is absent in the Japanese but present in a low frequency in the Ainu. The HL–A8 antigen is present in about 20% of the Caucasians but is absent or rare in the circumpolar populations (Table 4.5). It appears to vary with the degree of admixture.

The HL–A12 antigen has a frequency of over 25% in the majority of Caucasian populations but has a low frequency or is even absent in the circumpolar populations, similar to A8. The Ainu, however, have a considerably higher frequency of A12 than of A8. The higher frequency in Ainu is probably due to admixture with Japanese whose frequency of the HL–A antigen A12 is 13%.

Other HL–A antigens which are absent or have low frequencies in all of the circumpolar populations are A13, A14 and W21. All of these frequencies are considerably lower in the Japanese (for references on histocompatibility testing on circumpolar populations see Dossetor *et al.*, 1973*a*; Kissmeyer-Nielsen *et al.*, 1973; Mittal *et al.*, 1973; Tiilikainen *et al.*, 1973).

The $E_2{}^+$ allele for the C5+ form of cholinesterase had a frequency of about 0.03 in the Ainu, Alaskan Eskimos and Inari Fisher and Mountain Lapps and was 0.06 or more in the Igloolik Eskimos and Skolt Lapps. These were the only populations tested. The frequencies of the $E_2{}^+$ allele in a British population (Robson & Harris, 1966) and a Japanese population (Omoto & Harada, 1972) were about 0.05. The fluctuations in frequencies for this allele may well be due to genetic drift after introduction from outside sources but it seems also possible that the gene existed in all of the original populations.

The frequencies for the *Inv*[1] allele were similar in all of the circumpolar populations except the Utsjoki Lapps (Table 4.1). The frequencies seem to be at a higher level (about 0.20) than the average frequency for Caucasians (0.05–0.10) (Steinberg, 1969) but lower than that of East Asians (0.30) (Steinberg & Kageyama, 1970). The frequency of the Utsjoki Lapps was 0.09 and similar to the Caucasian frequency. The frequency of *Inv*[1] (ranging from

0.09 to 0.28) in the Finnish Lapps is two to five times higher than that found in other north European populations, and it is the highest known among Europeans. In contrast, the linguistically related Finns and Maris have lower Inv^1 frequencies (0.02 to 0.05) than other northern and central European populations so far investigated (Grubb, 1959; Steinberg *et al.*, 1974).

In addition to the above data, the HL–A2 antigen has phenotypic frequencies of about 50% which are similar in the Ainu, Igloolik Eskimos and Skolt Lapps. It has lower frequencies in the Norwegian and Utsjoki Lapps and lowest of all in East Greenland Eskimos. The Japanese and Caucasian frequencies for the HL–A2 antigen are also around 50% (Int. Workshop, 1973).

Genes which were tested in only a few of the populations or were not polymorphic

There were some polymorphisms, e.g. of C3 and Lp lipoprotein, which were only investigated in the Greenland Eskimos and Lapps. The data are summarized in Table 4.1 and more discussion of these may be found in the original papers quoted in the table. Ag^x had as high a frequency in the Ainu (0.66) as in Japanese and other non-Caucasian ethnic groups (Misawa *et al.*, 1971). Ag^x was less frequent in the mixed Thule Eskimos and in those living in Ammassalik than in the West Greenland Eskimos and the Lapp populations. Lp was tested in only the Aupilagtoq Eskimos and Norwegian Lapps with Lp^a being slightly less frequent in the Eskimos (0.16) than in the Lapps (0.19). The Lp^a frequencies found in other northern populations (0.19 in Norwegians and 0.14 in Icelanders) are of the same order of magnitude as those of the Eskimos and Lapps who were tested (Berg & Eriksson, 1973c). The frequency of the $C3^1$ allele was 0.06 in the Aupilagtoq Eskimos and 0.09 in the Svettijärvi Skolt Lapps but only 0.02 in the Inari Mountain Lapps (Berg *et al.*, 1972; Arvilommi, Berg & Eriksson, 1973). Compared to non-Lappish northern populations, e.g. Norwegians and Icelanders, which have $C3^1$ frequencies 0.15 and 0.19, respectively, the frequencies are low in Lapp populations (Berg & Eriksson, 1973b).

The incidence of haemoglobin (Hb) and erythrocyte lactate dehydrogenase (LDH) variants detectable by electrophoresis seems to be low in circumpolar populations. Omoto & Harada (1972) did not find an LDH variant among 125 Ainu. Scott *et al.* (1958) did not find any abnormal haemoglobins in Alaskan Eskimos, Indians or Aleuts. No haemoglobin or LDH variants were observed in blood from 360 Eskimos in Igloolik (McAlpine *et al.*, 1974). In 1576 Lapps from Inari (Finland), 200 Lapps from the region of Kiruna–Gällivare (Sweden), and 153 Eskimos from Aupilagtoq (Greenland) no Hb or LDH variants were observed (Nilsson & Eriksson, 1972). Since this investigation covered about 50% of the Finnish Lapps, it is likely that these lack LDH variants. The frequency of LDH variants (of the LDH–A subunit)

123

in Finns was about 0.1% and in Icelanders 0.4%. The frequency of Hb variants in northern Europe has been estimated at 2 to 5 per thousand (for references *see* Nilsson & Eriksson, 1972).

Studies of the haptoglobin subtypes among the Skolt Lapps showed a high Hp^{1S}/Hp^1 ratio (0.87) as compared with other European populations so far investigated (mainly 0.5–0.7). Also Asiatic peoples seem to have a high frequency of the slow Hp subtype (for references see Ehnholm & Eriksson, 1969).

The frequencies of *Es D^2* (recently observed esterase known as esterase D in red blood cells) (Hopkinson *et al.*, 1973) were similar in 86 Ainu of Iburi (0.335) (Omoto *et al.*, 1974) and in 145 Igloolik Eskimos (0.292) (Cox *et al.*, 1978). The Ainu frequency was similar to that of 1066 Japanese (0.342) (Omoto *et al.*, 1974) and both the Ainu and Eskimos had frequencies that were about three times that in 867 English (0.098) (Hopkinson *et al.*, 1973). Welch & Lee (1974) found that the frequency of *Es D^2* was about twice as high in Finnish Lapps as in the English. The *Es D^2* frequency in Blacks varies between 0.055 to 0.098. The frequencies of the *Es D^2* allele in Asiatic populations (0.227–0.351) are the highest of those populations tested so far (Hopkinson *et al.*, 1973; Welch & Lee, 1974).

There were too few data to discuss for W23–W32 at the first HL–A sublocus and W16, W20 and 407 at the second sublocus in Table 4.5.

Harvald (1976) has suggested that the differences in disease frequencies between Greenland Eskimos and Caucasians could be related to the differences in gene frequencies between the two ethnic populations, particularly in the tissue antigen frequencies. Recently an extraordinarily high correlation between the second segregant series histocompatibility antigen (HL–A), W27, and some well characterized rheumatic diseases has been reported (for references *see* Harvald, 1976). Indeed, more than 90% of patients with ankylosing spondylitis (morbus Bechterew) or Reiter's syndrome are W27 positive (in controls about 8%). These disorders occur frequently in Greenland; and W27 occurs four times as frequently in pure Eskimos as in the Danish population (for further discussion *see* the section on Immunogenetic Aspects, pp. 155–8).

Genetic Distances

Two separate series of distance measures were obtained. The first (Table 4.7) is based on ADA, AK, PGM$_1$, Gc and Hp data from four groups of Lapps (Svettijärvi, Nellim, Fisher and Inari Mountain), the Ainu, Alaskan Eskimo, Canadian Igloolik Eskimo and three Greenlandic Eskimo groups (unmixed Thule, Aupilagtoq and Ammassalik). The second series of genetic distances (Table 4.8) were calculated with ABO, Rh, MN, Kell and P data. In this instance three Lapp groups (Norway, Svettijärvi and Nellim), the

Table 4.7. *Genetic distances among 10 circumpolar populations, based on data from ADA, AK, PGM_1, Gc and Hp systems*

| Population | Location | Skolt Lapps | | Lapps | | Ainu | Eskimos | | | |
		Svettijärvi	Nellim	Fisher	Mountain	Hokkaido	Alaska	Igloolik	Thule	Aupilagtoq
Skolt Lapps	Nellim	0.150								
Lapps	Fisher	0.275	0.385							
	Mountain	0.226	0.299	0.149						
Ainu	Hokkaido	0.281	0.191	0.410	0.299					
Eskimos	Alaska	0.277	0.205	0.456	0.353	0.140				
	Igloolik	0.278	0.229	0.450	0.346	0.195	0.042			
	Thule	0.340	0.254	0.532	0.444	0.220	0.054	0.162		
	Aupilagtoq	0.178	0.181	0.370	0.269	0.248	0.176	0.168	0.263	
	Ammassalik	0.271	0.259	0.459	0.366	0.290	0.144	0.137	0.226	0.134

Table 4.8. *Genetic distances among 11 circumpolar populations, based on data from ABO, Rh, MN, Kell and P systems*

Population	Location	Lapps	Skolt Lapps		Ainu	Eskimos					
		Norway	Svettijärvi	Nellim	Hokkaido	Wainwright	Copper	Ft. Chimo	Thule	Aupilagtoq	Ammassalik
Skolt Lapps	Svettijärvi	0.177									
	Nellim	0.293	0.241								
Ainu	Hokkaido	0.318	0.401	0.415							
Eskimos	Wainwright	0.345	0.326	0.480	0.439						
	Copper	0.375	0.342	0.487	0.441	0.119					
	Fort Chimo	0.263	0.319	0.442	0.403	0.243	0.213				
	Thule	0.481	0.498	0.560	0.436	0.395	0.325	0.298			
	Aupilagtoq	0.272	0.291	0.384	0.370	0.166	0.159	0.175	0.336		
	Ammassalik	0.427	0.346	0.483	0.485	0.190	0.225	0.362	0.447	0.276	
	Scoresbysund	0.289	0.273	0.375	0.451	0.291	0.279	0.235	0.407	0.260	0.366

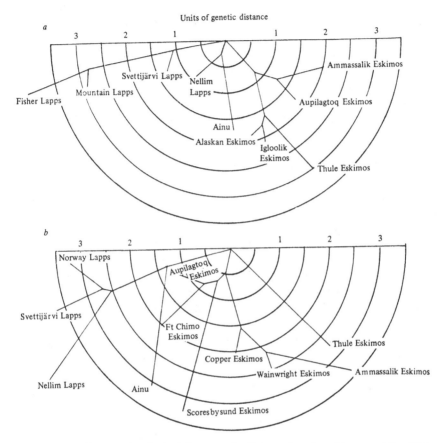

Fig. 4.10. Genetic networks obtained from genetic distances based on (*a*) genes at the *ADA*, *AK*, *PGM*$_1$, *Gc* and *Hp* loci in the Sevettijärvi, Nellim Fisher and Mountain Lapps, the Ainu ,and Alaskan, Igloolik, Aupilagtoq, Thule and Ammassalik Eskimos; (*b*) genes at the *ABO*, *Rh*, *MN*, *Kell* and *P* loci in the Norway, Sevettijärvi and Nellim Lapps, the Ainu, and Wainwright, Copper, Ft Chimo, Thule, Aupilagtoq, Ammassalik and Scoresbysund Eskimos.

Ainu, the Alaskan Wainwright Eskimos, the Canadian Copper (central Arctic) and Fort Chimo (eastern Arctic) Eskimos and the Greenland. Thule (unmixed, northwest Greenland), Aupilagtoq (western Greenland), Ammassalik and Scoresbysund (eastern Greenland) Eskimos were compared. The genetic ('minimum string') networks connecting these circumpolar populations are shown in Fig. 4.10.

Because different populations and different genes were employed in the calculation of the two sets of distance measures, the results are not directly

127

comparable. A permissible observation, however, is that the smallest distances occur within the ethnic aggregates (i.e. within Eskimos, or within Lapps), while the largest distances (with two exceptions) are between these aggregates. There is considerable overlap in the intra-group distances and inter-group distances indicating that some Eskimos are as divergent from each other as are Eskimos and Lapps (and vice versa). The position of the Ainu is intriguing: they appear to occupy an intermediate position between Eskimos and Lapps.

The evolutionary model employed in the calculation of both the genetic distances and the genetic networks excludes (among others) gene flow between populations. Small degrees of admixture that may occur are considered essentially random processes that would augment the action of genetic drift (Cavalli-Sforza & Edwards, 1967). With respect to the circumpolar populations considered in this paper, there has been some gene flow between the different Lapp populations (Skolt Lapps in particular) and between the different Eskimo populations. In addition, however, there is also gene flow from outside into these ethnic aggregates. Thus, the Japanese have mixed with the Ainu and some Scandinavian genes have entered the Lapp and Greenlandic Eskimo populations. The result of these outside sources of gene flow could be both convergence or divergence, depending on what was initially assumed to be the relationship of these circumpolar groups to each other.

Under these circumstances it is safest to consider the networks in Fig. 4.10 as representations of present genetic resemblances (Lalouel, 1974) among the Lapps, Ainu and Eskimos rather than as phylogenetic trees of descent. Such an interpretation leaves unanswered the question of phylogenetic relationship between these groups.

The networks themselves, although based on distances calculated with different genes and with different Lapp and Eskimo populations, show a basic separation of the Eskimos from both the Ainu and the Lapps. The network based on red-cell enzyme and serum-protein data suggests that the Ainu resemble the Nellim Skolt Lapps more than any other group. However, this is not borne out in the network based on blood-group data, in which the Ainu occupy a clearly more intermediate position between the Eskimos and the Lapps.

Summary

It has not been possible to realize fully the objectives set out at the beginning of this chapter. A wealth of data has been collected and yet a cursory glance at the tables makes it clear that more data are needed for the populations already studied to provide the prerequisite of 'the same markers in the same populations' for the distance analysis. In addition, data for more markers in the Siberian populations would be useful for studies of distances and migratory patterns. As pointed out at the beginning of the chapter, the reasons for differences in gene frequencies in such small isolated populations are often random chance or admixture with neighbors or visitors from quite different ethnic groups, and it is difficult to distinguish between the reasons for differences which result from admixture, mutation, genetic drift or natural selection.

The distance analysis tells us that the Ainu tend to occupy an intermediate position in their relation to the Eskimos and Lapps. Furthermore superficial observation of much more of the data than could be used for the distance analysis suggests that more of the allelic frequencies (23; *see* p. 91, genes with frequencies similar in the Ainu and Eskimos but different in Lapps) were similar among the Ainu and the Eskimos than among either the Eskimos and Lapps (13; *see* p. 112, genes with frequencies similar in the Eskimos and Lapps but different in the Ainu) or the Lapps and Ainu (nine; *see* p. 115, genes with frequencies similar in the Lapps and Ainu and different in the Eskimos).

Finally, the frequencies of four alleles formed apparent clines around the Pole (p. 118) and there were 14 alleles or haplotypes whose frequencies were similar or were absent or nearly absent in all of the circumpolar populations (p. 121). The frequencies of only four of these were similar among the three population groups but quite different from all of their neighbors; three HL–A genes, *A12*, *A13* and *W21* were practically absent in the circumpolar populations but polymorphic in both Caucasians and Mongolians and one allele *Inv*[1] had a higher frequency in all of the circumpolar peoples than in both Caucasians and Mongolians even though the *Inv*[1] allele was polymorphic in the latter two populations. From the foregoing one can only speculate that the frequencies for the HL–A genes *A12*, *A13* and *W21* and for the allele *Inv*[1] may have some adaptive significance related to some selective factor which is common to the Ainu, Eskimos and Lapps.

OTHER GENETIC TRAITS

W. LEHMANN, N. E. SIMPSON & A. W. ERIKSSON

Traits other than the monogenetic blood characteristics discussed in the previous section are of interest to population geneticists. This is particularly so if they are: (1) inherited through one or only a few genes, (2) common, (3) of easy determination, (4) variable in frequency in different populations and (5) of constant expression being influenced by environmental factors to a limited extent.

The genetic or morphological characteristics in this section have at least some of these attributes; e.g., the inability to taste phenylthiocarbamide or the presence of dry ear wax are determined by, rather than governing inheritance of autosomal recessive traits. External characteristics such as pigmentation have a complex heredity but they are important in mate selection, in which some phenotypes – males or females – are favored, resulting in assortative mating and the bringing about of changes in gene frequencies and mean phenotype frequencies over a long period of time.

Phenylthiocarbamide Tasting Ability

The frequencies of non-tasters of phenylthiocarbamide (PTC) seem to form a cline; they increase from Japan across Alaska, Canada and Greenland. The PTC non-taster frequency is very low among the Ainu (6.0–6.4%) (Simmons *et al.*, 1953; Omoto, 1970). Alaskan (Allison & Blumberg, 1959) and Canadian (Hughes, 1971) Eskimos have a non-taster frequency of 25.7% and 28.6%, respectively, while the Labrador Eskimos (Sewall, 1939) have a high frequency (40.8%). An extremely high frequency of non-tasters was found by Eriksson *et al.* (1970*b*) among the Aupilagtoq Eskimos (55.1%) in northwestern Greenland. Also Alsbirk & Alsbirk (1972) noted a very high frequency of PTC non-tasters (53.5%) in another western Greenland Eskimo population (Table 4.9).

On the other hand, the frequencies of non-tasters among Lapps are quite variable. During the Scandinavian IBP–HA studies in Finnish Lapland from 1966 to 1970, the PTC tasting ability of various Lapp populations was tested (Eriksson *et al.*, 1970*a*). The frequency of non-tasters (Table 4.9) is quite low among Fisher Lapps (9.1%) in the eastern part of the commune of Inari. This low frequency has also been found among Swedish and Norwegian Lapps, 7% and 6.8% respectively (Allison & Nevanlinna, 1952), although the number of subjects tested was not very large. In comparison, Mellbin (1962) found a frequency of 23% among school children whose parents were nomadic Swedish Lapps. Monn (1969) observed a similar frequency (21%) among Lapp school children in northern Norway.

The frequency of non-tasters is clearly higher among the Skolt Lapps (average 27.3%) than among other Lapps. This high non-taster frequency is similar to the findings of Allison & Nevanlinna for Finns in South Finland. In comparison, Eriksson *et al.* (1970*a*) found a lower frequency (22%) among males in the two northernmost counties of Finland, whereas Kajanoja (1972) found with the solution test a high frequency (35%) among males from six different regions of the country. Also among males in northeastern regions in Finland (Kuusamo, Salla, Savukoski) high frequencies of PTC non-tasters were found.

Several authors have reported a significantly higher taste sensitivity (lower average thresholds) for PTC among women than among men. Other authors have not been able to support this finding. In the Skolt Lapp populations a markedly lower frequency of non-tasters among women was noted. The probable effect of hormones on the PTC tasting ability has been discussed (for references *see* Eriksson *et al.*, 1970*a*).

Table 4.9. *Frequency of PTC non-tasters in circumpolar populations*

Population	Number tested (males and females)	Non-tasters	t-gene frequency	Reference
Ainu (Hidaka)	232	6.0	0.245	Omoto (1970)
Ainu (Hokkaido)	328	6.4	0.253	Simmons *et al.* (1953)
Northern Alaskan Eskimos	68	25.7	0.507	Allison & Blumberg (1959)
Canadian Eskimos (Igloolik)	297	28.6	0.535	Hughes (1971)
Canadian Eskimos (Labrador)	130	40.8	0.639	Sewall (1939)
Greenland Eskimos (Aupilagtoq)	78	55.1	0.742	Eriksson *et al.* (1970*b*)
Greenland Eskimos (Umanaq)	129	53.5	0.731	Alsbirk & Alsbirk (1972)
Iceland (Husavik)	332	38.6	0.621	Lehmann *et al.* (unpublished observation)
Norwegian Lapps	78	6.8	0.261	Allison & Nevanlinna (1952)
Norwegian Lapps	255	17.6	0.420	Monn (1969)
Swedish Lapps	62	7.0	0.265	Allison & Nevanlinna (1952)
Finnish Lapps Inari				
Sevettijärvi Skolt Lapps	310	27.7	0.527	Eriksson *et al.* (unpublished observation)
Nellim Skolt Lapps	145	26.2	0.512	Eriksson *et al.* (unpublished observation)
Fisher Lapps	209	9.1	0.301	Eriksson *et al.* (unpublished observation)
Mountain Lapps	106	15.1	0.389	Eriksson *et al.* (unpublished observation)
Offspring of Fisher and Mountain Lapps	61	8.2	0.286	Eriksson *et al.* (unpublished observation)
Offspring of Skolt Lapps and other Lapps	22	9.1	0.302	Eriksson *et al.* (unpublished observation)
Offspring of Skolt Lapps and Finns	98	14.3	0.378	Eriksson *et al.* (unpublished observation)
Offspring of other Lapps and Finns	78	24.4	0.494	Eriksson *et al.* (unpublished observations)
Utsjoki Lapps	124	19.4	0.440	Eriksson *et al.*
Finns (Northeastern Finland)	331	38.4	0.619	(unpublished observations)
Finns (Northern Finland)	761	22.1	0.470	Eriksson *et al.* (1970*a*)
Finns (Helsinki)	202	29.2	0.540	Allison & Nevanlinna (1952)
Finland (males only)	1537	35.0	0.592	Kajanoja (1972)
Maris (USSR) (males only)	321	26.5	0.515	Kajanoja (personal communication)

The human biology of circumpolar populations

Several authors have demonstrated an impairment of PTC taste sensitivity with increase in age. Eriksson *et al.* (1970*a*) observed a tendency in this direction among male Skolts, but there was no obvious trend among women. The variations in PTC tasting ability between different populations, however, cannot be based on the different age structures of the populations alone nor on the fact that different testing methods have been used (cf. Kajanoja, 1972). The question has also been posed as to whether tobacco smoking influences PTC tasting ability. It may be that age and tobacco smoking have a cumulative effect. This would explain the increase in frequencies in non-tasters among the ageing male Skolts; most of them are habitual smokers.

Conclusion

The frequency of PTC non-tasters is low among the Ainu and the Lapps from the Lappish core area. The Skolt Lapps have frequencies similar to those of other Europeans. In sharp contrast to other Mongolian peoples Eskimos have very high frequencies of PTC non-tasters. The Greenland Eskimos have the highest frequencies of PTC non-tasters so far noted.

Cerumen dimorphism

There are two different main types of ear wax: the wet, yellow and soft type (genotype WW or Ww), and the dry, grey and brittle (genotype ww). Table 4.10 shows that the frequency of dry cerumen was much higher in the Ainu of Hokkaido (49.7–53.7%) than in Eskimos (23.2–41.1%) and 'pure' Lapps (3.0–12.6%). In the Ainu who were thought to be relatively pure, i.e., with little admixture from the neighboring Japanese, the frequency was slightly lower than that in the Ainu who were considered to have more Japanese admixture. The difference between the frequencies in the two Ainu populations is, however, not significant. The Japanese neighbors of the Ainu in Chiba Prefecture have a high frequency of dry cerumen (80.1%) (Omoto, 1970), a frequency which is in agreement with others reported for Japanese (Nakajima & Hirano, 1968). According to a review by Matsunaga (1962) the frequency of dry cerumen is higher than 90% in some parts of Japan and in Chinese, Tungus and other Mongolians. The dry cerumen is generally regarded as a Mongolian marker. It is worth emphasizing that the frequency of dry cerumen is considerably lower in the Ainu than in the Japanese. Omoto (1970) reports that there is considerable Japanese genetic admixture in the Ainu and among the about 16 000 Ainu in Hokkaido, there are relatively few pure Ainu. Albeit, Omoto (1970) suggests that the frequency of dry ear wax in the original Ainu was much higher than in Caucasians in the past. In his opinion there is no reason to assume a close relationship between the Ainu and Caucasian populations.

In comparison with other circumpolar populations the frequency of dry

Table 4.10. *Cerumen dimorphism in circumpolar people and some other populations*

Population	Number tested	Per cent dry	Frequency of gene w	Reference
Hidaka Ainu (Hokkaido)	322	53.7	0.733	Omoto (1970)
Hidaka Ainu (Hokkaido) ('pure')	185	49.7	0.705	Omoto (1970)
Japanese (Chiba Prefecture)	1071	80.1	0.895	Nakajima & Hirano (1968)
Alaskan Eskimos (Wainwright)	90	41.1	0.641	Pawson & Milan (1974)
Alaskan Eskimos (Anchorage Hospital)	67	31.3	0.560	Pawson & Milan (1974)
American Indians				
Navaho, full-blooded Indians	162	69.7	0.835	Petrakis et al. (1967)
All South West Indians	251	66.9	0.818	Petrakis et al. (1967)
Sioux, full-blooded Indians	113	46.9	0.685	Petrakis et al. (1967)
Choctaw Indians	432	21.0	0.458	Martin & Jackson (1969)
Iceland (Husavik)	322	1.2	0.111	Lehmann et al. (unpublished observation)
Germans (Münster)	514	3.1	0.176	Matsunaga & Ebbing (1956)
Caucasians (USA)	368	1.3	0.114	Petrakis (1969)
Finnish Lapps				
Inari				
Sevettijärvi Skolt Lapps	254	11.4	0.338	Eriksson (1973)
Nellim Skolt Lapps	111	12.6	0.355	Eriksson (1973)
Fisher Lapps	146	5.5	0.234	Eriksson (1973)
Mountain Lapps	101	3.0	0.172	Eriksson (1973)
Offspring of Fisher and Mountain Lapps	63	6.4	0.252	Eriksson (1973)
Offspring of Skolt Lapps and other Lapps	26	0.0	0.000	Eriksson (1973)
Offspring of Finns and Skolt Lapps	26	0.0	0.000	Eriksson (1973)
Offspring of Finns and other Lapps	73	4.1	0.203	Eriksson (1973)
Utsjoki Lapps	120	14.2	0.376	Eriksson et al. (unpublished observation)
Finns (Northeastern Finland)	323	2.5	0.157	Eriksson et al. (unpublished observation)

ear wax was relatively low in Lapps. The frequency of dry cerumen in the Skolt Lapps and the Utsjoki Lapps was considerably higher than in other Lapps and Finns. The frequencies for dry ear wax in the Fisher and Mountain Lapps in Inari are similar to those found in Icelanders, Germans and white Americans.

133

The human biology of circumpolar populations

The frequencies of dry cerumen vary from one Arctic or sub-Arctic indigenous population to another with noticeable differences. The low frequency in the Inari Fisher and Mountain Lapps (Eriksson, 1973b) may have been partly the result of gene flow from the neighboring Finns; a hypothesis which can only be supported by the low frequency of dry cerumen in northeastern Finns, around 1% (Eriksson *et al.*, unpublished).

The Aupilagtoq Eskimos have a frequency for dry ear-wax type which is intermediate between Caucasian and Mongolian frequencies. Caucasian admixture may be responsible for the lower frequency of dry cerumen in these Eskimos than in Mongolian populations. It is of interest that the dermatoglyphic studies on the Aupilagtoq Eskimos (Simon, 1972) and those on the Eskimos in Kraulshavn (north of Aupilagtoq) showed that the dermal patterns of these populations were definitely different from those of Mongolians. This supports the theory that the Mongolians separated as Proto-Mongolians and were isolated; this leads to different dermatoglyphic patterns in the different subsequent races. Omoto (1970) suggested a similar theory for the cerumen dimorphism in the Ainu.

The reason for the variability of the frequencies of dry and wet types of cerumen is unknown. McCullough & Giles (1970) have suggested that the frequencies vary with humidity and that selective differences are related to the frequency of infections of the middle ear. The authors recognized, however, that there are not sufficient data to support such a hypothesis. Studies in middle and southeastern Asia, by Petrakis *et al.* (1971), did not support this hypothesis either. The answer will have to await more data. Biochemical studies of cerumen as have been done in the Japanese (Matsunaga, 1962) have also not elucidated the reason for cerumen variability in different populations.

Conclusion

The Ainu have much higher frequencies of dry cerumen than Eskimos (23.2–41.1%) and Lapps (3.0–12.6%). However, the frequency of dry cerumen in the Ainu (49.7–53.7%) is notably lower than in other Asiatic populations studied so far (> 80%). The frequency in Skolt Lapps (11.8%) is considerably higher than in Fisher Lapps (5.5%), Mountain Lapps (3.0%) and Finns in northeastern Finland (1.0%).

Isoniazid inactivation

The inactivation of isoniazid (INH), one of the most effective tuberculostatic drugs to be used on northern peoples, is genetically controlled and is polymorphic. The difference between rapid and slow inactivators of INH lies in their respective different activities of acetyl transferase (for references *see* Tiitinen, Mattila & Eriksson, 1967, 1968, 1973).

134

The frequency of rapid inactivators of INH is high in Eskimos (95%) (Armstrong & Peart, 1960), Koreans (89%), the Japanese (88%) and the Ainu (88%) (Sunahara, Urano & Ogawa, 1961). The frequency is more than 70% among American Indians, but is below 45% among the majority of white populations (Harris, Knight & Selin, 1958; Kalow, 1962; Hannegren, Borgå & Sjöquist, 1970).

The Lapp populations in the parish of Inari in northern Finland have considerably higher frequencies of rapid INH activators than other Europeans. The Fisher Lapps in the village of Nellim who until recently were extremely isolated, had a frequency of 80% ($N = 15$), the other Fisher Lapps 68% ($N = 25$), i.e. a mean of 72.5% in Fisher Lapps. The Nellim Skolt Lapps had a frequency of 50% ($N = 26$), the Sevettijärvi Skolt Lapps 59% ($N = 22$), and the Mountain Lapps 50% ($N = 16$). The 29 offspring of Fisher and Mountain Lapps had 76% of rapid INH inactivators. The number of investigated Lapps is not high, but the subjects were carefully sampled. All were 'pure' Lapps and not closely related, cousins of first degree at the utmost (Tiitinen *et al.*, 1967, 1968, 1973). These results are in agreement with the findings that in the world as a whole the frequency of rapid INH inactivators seems to show an increase in a northern direction (Sunahara *et al.*, 1961; Hannegren *et al.*, 1970). However, a possible selective mechanism as reason for this gradient is still unknown and the low frequencies of rapid inactivators of INH in Finns (36% and 39%) (Tiitinen *et al.*, 1967) and Swedes (32%) (Hannegren *et al.*, 1970) are in sharp contrast to the high frequencies in Lapps.

Conclusion

The frequencies of rapid INH inactivators are in Eskimos and Ainu among the highest noted. The Lapps in Finland, and particularly the Fisher Lapps have considerably higher frequencies than other Europeans.

Pigmentation: color of hair, iris and skin

One of the most conspicuous hereditary traits is the pigmentation of skin, hair and eyes.

Hair color

A screening of the hair color was made in different Lapp populations by the same investigator with the same 'Fischer-Saller Haarfarbentafel' with 30 different hair sample hues.

Fig. 4.11 shows that the darkest hair colors (U–Y) have a rather high frequency (about 75%) in the Skolt Lapps. Also the Mountain and Fisher

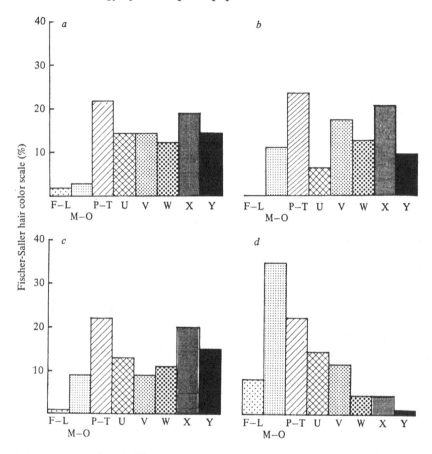

Fig. 4.11. Frequencies of different hair hues in females aged 20–50 yr, according to the Fischer–Saller hair-color scale. *a*, Skolt Lapps (*N* = 106); *b*, Fisher Lapps (*N* = 64); *c*, Mountain Lapps (*N* = 100); *d*, Swedish-speaking Åland Main Islanders (*N* = 321). F–L, Blond; M–O, Dunkel-Blond; P–T, Braun; U–Y, Braun-Schwarz.

Lapps are very dark-haired compared to Finns, the corresponding percentages of the dark-haired being about 70%, 65% and 38%, respectively.

In Swedish Lapp females at the age of 25–50 yr the frequency of 'dark-haired' was 78% (Dahlberg & Wahlund, 1941).

Also other hirsute characteristics among Lapps are more European than Asiatic. The hair of Lapps is relatively soft and sometimes wavy, and not course and wiry, as among Mongolians.

Iris color

The eye colors are very dark among the Ainu and unmixed Eskimos (for

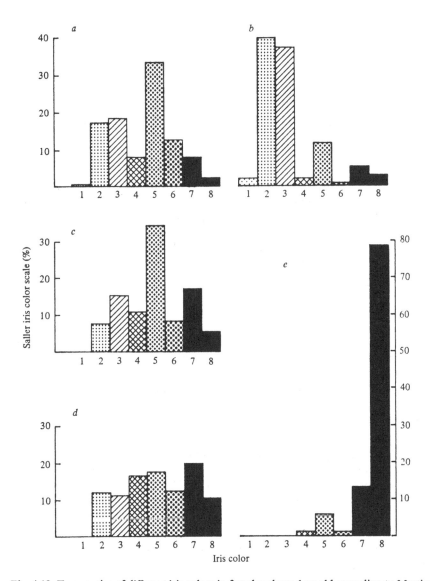

Fig. 4.12. Frequencies of different iris colors in females above 1 yr old according to Martin Saller's iris-color scale. *a*, Skolt Lapps (*N* = 239); *b*, Oulu Finns (*N* = 316); *c*, Mountain Lapps (*N* = 116); *d*, Fisher Lapps (*N* = 97); *e*, Northwest Greenlandic Eskimos (*N* = 84). Abscissa scale 1–8, light blue–dark brown.

further details *see* Chapter 6, Ophthalmology by H. Forsius). The iris colors among Lapps are darkest among Fisher and Mountain Lapps and lightest among Skolt Lapps (Fig. 4.12). In comparison with other northwestern European populations the Lapps have rather strongly pigmented irises. However, about 30–50% of the Lapps in Finland have blue eyes. The darkest iris colors (nos 7–8 in Martin-Saller's scale) appear in a frequency of about 30% among Fisher Lapps, 20% among Mountain Lapps, and lower than 10% in Skolt Lapps. In contrast, the Finns have the fairest eye colors which have been noted. Lappish females have more strongly pigmented irises than Lappish males (Forsius *et al.*, 1970).

Skolt Lapps have considerably lighter irises than the Fisher and Mountain Lapps. However, the Skolt Lapps have a relatively high frequency of dark hair colors. The Skolt Lapps seem to have been rather mixed with neighboring populations, e.g. Finns (Karelians) and Russians. Many other hereditary characteristics also differ in the Skolt Lapps from those in other Lapps from the Lappish core area.

Also in the Swedish Lapps the frequency of light irises seems to be rather high. Dahlberg & Wahlund (1941) calculated an average frequency of 'dark-eyed' among females to be 68% and among males 52%.

Skin color

Lundborg & Wahlund (1932) state about the skin pigmentation in Swedish and Norwegian Lapps: 'The color of the skin is light grey to brownish, mostly with a faint yellowish tint. Bryn points out that this color of the skin must be designated as a genuine Lappic character.' Preliminary studies with an EEL-skin reflectometer indicate, however, that the skin color among the majority of the Finnish Lapps does not deviate markedly from the surrounding fair Nordic populations (Eriksson *et al.*, unpublished observations).

In July 1972, Guy Pawson (personal communication) examined skin color in 96 Eskimos at Wainwright using the Photovolt Model 97 Reflectance Meter and amber, blue and green tristimulus filters. Readings were made at glabella on the forehead and on the inner upper arm. In the green wavelength range (550 nm) inner arm measurements of Wainwright Eskimos averaged 33% reflectance as compared to 36% for a sample of US college students, 21% for urban Quechua in Peru and 8% for Negroes from previously published information. In the Eskimos it can be assumed that measurements on the inner arm would give the closest approximation to genetically controlled skin color since this site is rarely exposed to the sun. The measurements showed the Eskimo subjects from North Alaska to be surprisingly light in color despite their high tanning capacity.

Recent studies suggest that cold injury may have been a contributing factor in the evolution of 'white' skin (Post, Daniels & Binford, 1975). The antirachitic selection for light types has also been stressed. Variation in solar ultraviolet at different latitudes may have caused racial differentiation in man. At least the Eskimos with their medium dark skin are an exception from the noted correlation between skin pigmentation and equatorial latitudes (Loomis, 1967).

Mongolian spot

The blue Mongolian spot, caused by accumulation of corium pigment in the sacral region, is common (about 80%) among newborn children of Asian people but is rare among Finnish Lapps (Jaana Koutonen, personal communication, 1975).

Conclusion

In comparison with the majority of northwestern European populations, Scandinavians, Finns and northern Slavs, who are the fairest people in the world, the Lapps are a great deal more pigmented. However, in comparison with peoples from southern and eastern Europe – or Eskimos and Ainu – the Lapps have a rather light complexion. Reflectometer studies have shown both North Alaskan Eskimos and Lapps to be surprisingly light in skin color despite their high tanning capacity.

Middle phalangeal hair

Eskimos*

Anthroposcopic observations including frequency counts of the presence or absence of middle phalangeal hair (MPH) were obtained on adult Eskimos living in northwestern Alaska during the US IBP study of these populations. The subjects lived in four coastal communities: Wainwright, Point Hope, Barrow and Kaktovik (Barter Island), and one inland village: Anaktuvuk Pass. The total sample numbers 277 individuals, 122 males and 155 females. A further subdivision consisted of 24 males and 37 females of mixed origin and 98 male and 118 female pure Eskimos. These designations of mixed ('hybrid') and pure ('non-hybrid') status are based on genealogical data of F. A. Milan and P. L. Jamison and census data from the US Bureau of Indian Affairs. All of known mixed origin are recorded as such, so if errors of assignment occur they are more likely to be designated as mixed and not as pure, rather than the reverse.

The mixed are generally second, third or fourth generation offspring from

* The authors are most grateful to Prof. P. L. Jamison, PhD, Department of Anthropology, Indiana University, Bloomington, USA, who wrote the section on Eskimos.

Table 4.11. *Percentage distribution of middle phalangeal hair among north Alaskan Eskimos (Jamison, unpublished)*

	Males				Females		
		Middle phalangeal hair				Middle phalangeal hair	
Digit	Number tested	Present	Absent	χ^{2a}	Number tested	Present	Absent
Right hand:							
Second	122	4.1	95.9	2.18	155	1.3	98.7
Third	122	19.7	80.3	0.79	154	15.6	84.4
Fourth	122	33.6	66.4	4.71*	155	21.9	78.1
Fifth	122	13.9	86.1	7.65†	155	4.5	95.5
Left hand:							
Second	122	1.6	98.4	0.62	154	0.6	99.4
Third	122	21.3	78.7	4.02*	154	12.3	87.7
Fourth	121	32.2	67.8	5.23*	154	20.1	79.9
Fifth	121	10.7	89.3	2.96	154	5.2	94.8

^{a} χ^2 values based on frequency counts rather than percentages.
* $P < 0.05$.
† $P < 0.01$.

matings between natives and members of nineteenth-century whaling crews in the Arctic. Admixture includes genes from US Whites and Blacks and, rarely, South Pacific islanders, with a predominance of the first category. Because of the small numbers involved, the mixed Eskimos will be treated as a group and contrasted with pure Eskimos.

Table 4.11 lists frequencies for the presence and absence of MPH in male and female Eskimos. Percentages for digits two through five are given, for both left and right hands. Chi square (χ^2) values are also listed as a test of the significance of the differences. Table 4.11 demonstrates that MPH is most frequently found on the third and fourth digits of both right and left hands, males and females.

Table 4.11 shows that in every case the frequency appears to be higher among males than among females. There are, however, only four instances of significant differences: on the right hand, males have significantly greater incidence on their fourth and fifth digits; on the left hand, the significantly greater male incidence occurs on the third and fourth digits. Over both sexes, MPH occurs on less than 5% of second digits, 15–21% of third digits, 20–34% of fourth digits and 5–14% of fifth digits.

Table 4.12 lists the presence or absence of MPH by sex and also by mixed and pure origin of the Eskimos. The χ^2 values in the middle of the mixed/pure columns test for differences between the admixed and non-admixed individuals of each sex. The χ^2 values down the middle of Table 4.12 itself test for

Table 4.12. *Percentage distribution of middle phalangeal hair among north Alaskan Eskimos (Jamison, unpublished)*

	Males								Females						
		Pure hair				Mixed hair				Pure hair				Mixed hair	
Digit	Number tested	present	absent	χ^{2a}	Number tested	present	absent	χ^{2a}	Number tested	present	absent	χ^{2a}	Number tested	present	absent
Right hand:															
Second	98	4.1	95.9	0.00	24	4.2	95.8	2.48	118	0.8	99.2	0.76	37	2.7	97.3
Third	98	15.3	84.7	6.01*	24	37.5	62.5	0.30	118	12.7	87.3	3.17	36	25.0	75.0
Fourth	98	27.6	72.4	8.19†	24	58.3	41.7	2.95	118	17.8	82.2	4.94*	37	35.1	64.9
Fifth	98	10.2	89.8	5.78*	24	29.2	70.8	5.56	118	2.5	97.5	4.47*	37	10.8	89.2
Left hand:															
Second	98	1.0	99.0	1.18	24	4.2	95.8	1.20	117	0.0	100.0	3.18	37	2.7	97.3
Third	98	16.3	83.7	7.38†	24	41.7	58.3	3.04	117	8.5	91.5	6.47*	37	24.3	75.7
Fourth	97	24.7	75.3	12.56†	24	62.5	37.5	1.90	117	17.1	82.9	2.79	37	29.7	86.5
Fifth	97	8.2	91.8	3.18	24	20.8	79.2	3.51	117	2.6	97.4	6.84*	37	13.5	86.5

ᵃ χ^2 = Chi square values based on frequency counts rather than percentages.
* = $P < 0.05$.
† = $P < 0.01$.

141

significant differences between the sexes in non-admixed individuals. Interestingly enough, with the exception of the fifth digit on the right hand, all of the significant differences between the sexes seen in Table 4.11 disappear when pure Eskimos are compared without the influence of the admixed individuals.

All the cases of mixed origin appear to have a higher frequency of MPH than the cases of pure origin. Among males the differences, however, are significant only for digits 3 through 5 on the right hand and digits 3 and 4 on the left. The females of mixed origin have significantly greater incidence on digits 4 and 5 on the right hand and digits 3 and 5 on the left.

When comparing overall frequencies in mixed and pure Eskimos disregarding sex, on the second digits mixed have an incidence of 3–4%, pure have 1–4%; for the third digit the incidences are 24–42% in mixed and 8–16% among pure; fourth digits display MPH in 30–62% of all mixed and 17–28% of the pure; and, finally, 11–29% of the mixed and 2–10% of the pure have this hair on their fifth digits.

Lapps

The frequency of MPH taking all of the fingers together was highest in the Nellim Skolt Lapps (89.2%) of all the Finnish Lapps who were studied. The next highest frequency was in the Svettijärvi Skolts, the next in the Inari Fisher and then in the Inari Mountain Lapps (Table 4.13). The differences between the frequencies in the last three populations were small. The frequency of MPH among the offspring from Inari Fisher and Mountain Lapps was 86.9%, which was, strikingly, almost as high as that for the Nellim Skolt Lapps. The frequency in the mixed offspring from both kinds of Skolt Lapps and Finns was somewhat less (73.2%) than in the Skolts or the offspring of Fisher and Mountain Lapps in Inari.

Comparison

When the frequencies of the MPH of Eskimos from throughout the circumpolar region are compared, considerable differences are seen among them (Table 4.13). Eskimos living in Labrador, Canada, have the lowest frequency according to Sewall (1939). About the 146 investigated pure Eskimos he concludes: 'Only 1.3% showed hair on the second digital phalanx.' Of the 55 mixed subjects 3.6% had hair on the second phalanx. Low frequencies are found also among northwestern Greenland Eskimos from Upernavik (27.8%) (Jørgensen, personal communication, 1974) and Canadian Eskimos living in Igloolik (28.7%) (Hughes, 1971) with a slightly higher frequency (32.1%) for northwest Alaskan Eskimos (Jamison, personal communication, 1974). The frequency for northwestern Greenland Eskimos from Aupilagtoq

Table 4.13. *Frequencies of subjects with one or more fingers with middle phalangeal hair*

Population	Number tested[a]	Per cent	Reference
Eskimos			
Eskimos (Northwest Alaska)	277	32.1	Jamison (unpubl.)
Eskimos (Igloolik, Canada)	143	28.6	Hughes (1971)
Eskimos (Aupilagtoq, North-west Greenland)	143	51.1	Jörgensen (unpubl.)
Eskimos (Upernavik)	126	27.8	Jörgensen (unpubl.)
Lapps (Inari, Finland)			
Sevettijärvi Skolt Lapps	271	80.8	Eriksson *et al.* (unpubl.)
Nellim Skolt Lapps	111	89.2	Eriksson *et al.* (unpubl.)
Fisher Lapps	145	78.6	Eriksson *et al.* (unpubl.)
Mountain Lapps	104	77.9	Eriksson *et al.* (unpubl.)
Offspring of Fisher and Mountain Lapps	61	86.9	Eriksson *et al.* (unpubl.)
Offspring of Skolt Lapps and other Lapps	21	66.7	Eriksson *et al.* (unpubl.)
Offspring of Skolt Lapps and Finns	71	73.2	Eriksson *et al.* (unpubl.)

[a] Males and females are pooled, at least in the samples of Aupilagtoq Eskimos and Finnish Lapps. Children were also included. In particular the populations of Svettijärvi Skolt Lapps and the offspring of Skolt Lapps with Finns and with other Lapps included many children.

(51.1%) was intermediate in value (Jørgensen, personal communication, 1974). If MPH of Finnish Lapps is compared to the above, the Lapp frequencies are higher than all of the Eskimo frequencies.

Analysis of the frequency of MPH by age in Aupilagtoq Eskimos showed that although there was an increase in the frequency with age in both sexes, the increase did not occur between the ages of 10 and 60 yr. No sex differences were found. When the data for the Aupilagtoq Eskimos were observed finger by finger as Jamison did for the data from Alaskan Eskimos (Table 4.11), MPH most frequently occurred on the fourth, then the third, fifth and second fingers. Jamison observed a similar pattern of frequencies.

Conclusion

There were higher frequencies of middle phalangeal hair (MPH) among Eskimo males than Eskimo females. However, this sexual distinction practically disappears when only non-admixed Eskimos are compared. Mixed Eskimos show considerably greater frequencies of MPH than do non-mixed Eskimos. The frequencies of subjects with MPH on one or more fingers vary considerably. The Eskimo populations have relatively low frequencies of

MPH, which is in good agreement with frequencies found in other Mongoloid populations. The Lapps, like other Europeans, have rather high frequencies of subjects with MPH.

Dermatoglyphics

Eskimos*

No attempt will be made to review Eskimo dermatoglyphics for all of the variables found upon the digital and palmar skin surfaces. Space limitations permit only the briefest mention of digital pattern distributions consisting of arch, loop and whorl frequencies. It was also decided to combine data from both sexes to facilitate comparing a maximum number of groups. This action perhaps was not very consequential since, for the most part, the sample sex ratios were nearly balanced except for a few instances where sample composition showed an excess of males.

The first reported use of dermatoglyphics in researching Eskimo biology appeared in a brief paragraph within the Report of the Canadian Arctic Expedition 1913–18, Volume XIII, which dealt with the physical characteristics of Copper Eskimos of the Coronation Gulf area (Jenness, 1923). No further studies of Eskimo dermatoglyphics appeared until Midlo & Cummins (1931) described digital and palmar prints obtained from 60 St Lawrence and four King Island Eskimos of Alaska. Shortly thereafter, Abel (1934) and Cummins (1935) published dermal pattern analyses of Eskimos from Barrow, Alaska and Scoresbysund, Greenland, respectively.

From these few pioneer works the number of Eskimo groups with dermatoglyphic traits investigated has rapidly expanded to include nearly 20 localities. A fairly representative sampling of Eskimo inhabitants of Alaska, Canada and Greenland is now available. Perhaps at this time it would be useful to survey the existing results in order, firstly to indicate what the general picture looks like at present, and secondly to provide an aid in planning new research efforts in Eskimo dermatoglyphics.

Table 4.14 lists 18 Eskimo localities from which digital pattern distributions are currently available. These samples include a total of some 2100 Eskimos of whom somewhat less than one-half were from Alaska, and slightly more than one-quarter each were from Canada and Greenland. There is noticeable variability between the samples in terms of their arch-loop-whorl frequencies. To point out the extremes in this variation an East Greenland series from Scoresbysund is noteworthy for having a very high whorl pattern percentage (72.2%), while Wainwright Eskimos had the lowest whorl percentage (23.6%). Accordingly in these two series, loop pattern frequencies were very nearly reversed to those of whorls.

* The authors are most grateful to Professor Robert J. Meier, PhD, Department of Anthropology, Indiana University, Bloomington, USA, who wrote the section on Eskimos.

Table 4.14. *Percentage arch–loop–whorl pattern frequencies in circumpolar populations, arranged according to decreasing whorl frequency (Meier, 1974)*

Population	Number tested	Arches	Loops	Whorls	Reference
Alaska					
St Lawrence Island	59	4.4	48.8	46.8	Midlo & Cummins, 1931
Barrow	30	2.3	51.3	45.5	Meier, 1974
Point Hope	218	4.5	50.0	45.5	Meier, 1974
Kodiak Island	96	6.5	51.9	41.6	Meier, 1966
Barrow	223	3.4	60.5	36.2	Meier, 1974
Barter Island	39	1.5	64.4	34.1	Meier, 1974
Anaktuvuk Pass	71	1.3	66.2	32.5	Meier, 1974
Wainwright	239	4.3	72.1	23.6	Meier, 1974
Canada					
Coronation Gulf	24	3.8	41.4	54.8	Jenness, 1923
Southampton Island	614	4.6	52.5	43.0	Popham, 1953
East Central Arctic (Baffin Island)	452	2.9	62.7	34.4	Auer, 1950
Greenland					
Scoresbysund	68	0.8	26.9	72.2	Abel, 1934
Scoresbysund	38	3.0	38.8	58.2	Ducros & Ducros, 1972
Ammassalik	36	1.6	40.9	57.4	Gessain, 1959
Julianehåb	140	3.5	54.1	42.5	Cummins & Fabricius-Hansen 1946
Scoresbysund: Dane-mixed Eskimo	14	5.0	60.7	34.2	Abel, 1934
Aupilagtoq	147	7.5	62.0	30.7	Simon, 1972
Kraulshavn	122	8.1	63.9	28.0	Frehse, 1974
Finland					
Inari Lapps (Fisher and Mountain)	222	7.1	50.3	42.2	Saatmann, 1971
Skolt Lapps	235	10.4	55.8	33.8	Lehmann et al., 1970
Offspring of Inari Lapps and Finns	43	5.0	65.0	30.0	Flitz, 1971
Offspring of Skolt Lapps and Finns	74	9.8	61.7	28.5	Lehmann et al., 1970
Finns	395	7.1	65.9	27.0	Lehmann, et al., 1970
Finns (males only)	1141	8.3	64.8	26.9	Chit, 1972 (males only)
USSR Lapps (Kola Peninsula) (males only)	84	14.9	57.2	27.9	Chit, 1972 (males only)

There is also good agreement between Alaska, Canada and Greenland in that all three countries have both high and low whorl frequencies in their various Eskimo groups. In order to illustrate this more clearly and also to incorporate some additional information into the comparison, it is convenient to utilize a constructed dermatoglyphic variable called pattern intensity (PI) (Cummins & Midlo, 1961; Holt, 1968), which can be defined as the number of triradii found in the pattern of an individual's ten fingers. However, in some cases for this survey it was necessary to calculate PI as the number of loops added to twice the number of whorls, and the sum was then divided by the total number of individuals. This calculated PI would serve as a close approximation of individual PI values because loops usually have only one triradius and whorls generally have two triradii. Of course, the calculated PI does not permit dispersion statistics or tests of significance of observed differences between Eskimo groups.

PI values for each of the Eskimo groups by country of origin are given in Table 4.15, along with the year in which the dermal prints were collected. These PI figures not only demonstrate a consistent variability within each of the three countries, but they also clearly suggest that the earlier the sample was collected, the higher the PI value would be. To illustrate this relationship, Eskimos from the earliest samples of each of the countries were grouped together (*see* bottom of Table 4.15) and it was found that their weighted mean PI (15.5) was approximately two full points above Eskimos in general (13.4). Two major alternatives emerge to explain the apparent marked decrease in PI and in essence a sharp decrease in whorl frequency, in Eskimos over the past three to four decades. These alternatives are that admixture from non-Eskimo sources (mainly Caucasian but some Negroid) have contributed lower PI values over successive generations, or that sample size has become larger and possibly better representative of larger dermatoglyphic areas, and hence, was less affected by local family (or founder effect) peculiarities.

Unfortunately, these alternatives become inseparably intertwined because of generally small effective population sizes in Eskimos. As a consequence, the more successful the attempt is in removing persons with non-Eskimo ancestry from the sample, the more likely it is that the sample will then consist of a small set of highly interrelated individuals. This problem appears to apply to the study of Abel (1934) with Scoresbysund Eskimos, and to a lesser degree to Kodiak Island (Meier, 1966).

In general, it can be surmised that Alaskan and probably West Greenland coast samples of all Eskimo groups have been influenced the most by gene flow from non-Eskimo sources acting upon aboriginal dermatoglyphic distributions. This matter has been brought out for consideration by Cummins & Fabricius-Hansen (1946), Simon (1972) and Meier (1973, 1974).

Small effective population size and relative breeding isolation also quite probably played substantive roles in bringing about dermatoglyphic diversity in Eskimos. Given the operation of random evolutionary processes (including random drift and founder effect), some of the small, mobile, yet separated Eskimo hunting bands could have become fairly fixed with high frequencies of any of the three dermal pattern types. Examples of this situation would include Kodiak Island (with a high percentage of arches), Wainwright (with a high percentage of loops), and Point Hope (with a high percentage of whorls).

146

Scoresbysund (with the highest whorl frequency) and Aupilagtoq (with a high frequency of arches) might also fit into the same picture. Of course, gene flow could have been operating at the same time, and since Eskimo populations undergoing admixture were small in number, gene flow could have had a much greater effect than if the populations were large.

The remarkable dermatoglyphic similarity between Eskimos grouped by country of origin (the weighted mean PI values of Greenland, Canada and Alaska differ by less than 0.5 or a meaningless one half of a triradius) is achieved when the effects of admixture and/or random evolutionary processes are confounded by the grouping procedure. This finding might suggest that Eskimo groups of the different countries have experienced somewhat similar histories affecting their dermatoglyphic characteristics. That is to say, within a given country some groups experienced greater or lesser amounts of admixture, and some groups were more or less isolated into small hunting bands over extended time periods.

This brief survey has perhaps brought out a problem which would exist if one wanted to define what a typical or aboriginal set of dermatoglyphic traits might have been for Eskimos in general. Quite probably there never was strict uniformity in these traits. It would seem likely that substantial variability in dermatoglyphics would inevitably develop given the size, structuring and mobility of Eskimo populations. Would this variability in dermatoglyphics be any different from that apportioned to anthropometric or serological traits? The results of Eskimo research conducted by scientists from many nations over recent years might now be processed toward answering this question. Only through an integrated study of several biological, and non-biological, parameters will it be possible to understand better the operation of evolutionary processes in Eskimos in particular, and in our species as a whole.

Lapps

In 1966–9 studies of dermatoglyphics were made in the Finnish Lapps in the Inari parish. Table 4.14 shows that there was considerable variation in the frequencies of pattern between the Lappish groups just as there was between the Eskimo groups. For example, the frequency of whorls in the offspring of Inari Lapps and Finns as well as in offspring of Skolt Lapps and Finns is less than in the non-half-breed Inari and Skolt Lapps. In fact, the frequencies of the half-breed Lapps seem to be closer to the Finnish frequencies although the sample sizes for the half-breed Lapps were small. The frequency of arches was highest in Skolt Lapps and lowest in the half-breed Inari Lapps. When the data were divided by right and left and by sex there were some differences which were significant using a χ^2 test. For example, the male Skolt Lapps had more whorls and fewer loops and arches than females from the same population. There were also more whorls on the fingers of the right than on those of

147

Table 4.15. *Eighteen Eskimo and three Lapp groups arranged according to country of origin and decreasing magnitude of pattern intensity*

Series	Year of collection	Pattern intensity
Alaska		
Barrow	1932	14.4
St Lawrence Island	1930	14.2
Point Hope	1970	14.1
Kodiak Island	1962	13.5
Barrow	1971	13.3
Barter Island	1970	13.3
Anaktuvuk Pass	1970	13.1
Wainwright	1969	11.9
Weighted Mean of Alaskan Groups		13.3
Canada		
Coronation Gulf	1913–16[a]	15.1
Southampton Island	1950	13.8
East-Central Arctic (Baffin Island)	1950?	13.2
Weighted Mean of Canadian Groups		13.4
Greenland		
Scoresbysund	1929–31[b]	17.3
Ammassalik	1936	15.6
Scoresbysund	1971	15.5
Julianehåb	1939	14.0
Aupilagtoq	1968	12.9
Kraulshavn	1968	12.0
Weighted Mean of Greenlandic groups		13.7
Weighted Mean of all groups		13.4
Weighted Mean of Barrow (1932, St Lawrence Island)	(1930)	
Coronation Gulf (1913–16), Scoresbysund and Ammassalik (1936)	(1929–31)	15.5
Lapland (Finland)		
Inari Lapps	1969	13.4
Offspring of Inari Lapps and Finns	1969	12.5
Skolt Lapps	1966, 1969	12.3
Weighted Mean of Lapp Groups (Inari and Skolt Lapps)		12.7

[a] The results for the Eskimo populations are compiled by Meier (1974), those for the Lapps by Lehmann *et al.* (unpublished observations).
[b] The original reports gave the years of organized expeditions to Canada, Greenland and Lapland.

Table 4.16. *Frequencies of single flexion crease ('Vierfingerfurche') in Finnish Lapps*

	Right hand		Left hand	
Population	Number tested	%	Number tested	%
Pure Skolt Lapps	256	2.0	264	3.4
Half breed Skolts	35	3.6	34	1.8
Inari Lapps (Mountain and Fisher Lapps)	215	3.3	214	2.4

the left hand. The Inari Lapps had the highest frequency for whorls taken together of all of the Lappish populations studied and the frequency of Skolts was intermediate to that of the Inari Lapps and that of the Finns.

When the frequencies of the whorls in the Lapps were compared with those of the Eskimos (Table 4.14), those of the Inari Lapps were similar to those of the Eskimos from Kodiak Island in Alaska, Southampton Island, Canada and Julianehåb, Greenland and those of the Skolt Lapps were similar to a few other Alaskan and Canadian Eskimo populations.

The frequencies of arches in the USSR Lapps, Skolt Lapps and offspring of Skolt Lapps and Finns were higher than in all of the Eskimo populations and the frequency in Inari Lapps was similar to those of the Greenlandic Eskimos of Aupilagtoq and Kraulshavn (Simon, 1972; Frehse, 1974). The frequency of arches was particularly variable among the Eskimo and Lapp populations.

The frequencies of arches in the USSR Lapps, Skolt Lapps and offspring Lapps than in Skolt Lapps and in the offspring of Inari Lapps and Finns. The index was very similar in the two latter populations. From Table 4.15, in general, the weighted mean for PI was lowest in Lapps (12.7), highest in the Greenlandic Eskimos (13.7), next highest in the Canadian Eskimos (13.4) and lowest among the Alaskan Eskimos (13.3) but the latter PI was still not as low as that of the Lapps. The Skolt Lapps and the Eskimos in Aupilagtoq and Kraulshavn had similar PI.

A single transverse crease ('Vierfingerfurche', ape hand, simian crease) in place of the usual two creases was found in Finnish Lapps to be the same as among other Europeans (2–4%) (Table 4.16).

Conclusion

In the 18 Eskimo populations studied there is considerable variation in the arch–loop–whorl pattern frequencies. The east Greenlandic Eskimos have high frequencies (57–72%) of whorls and low frequencies of arches (1–3%). The Eskimos in Wainwright (Alaska) had a whorl percentage of only 24%.

The pattern intensity (PI) values are higher in the earlier collected Eskimo samples. This may be a consequence of non-Eskimo gene flow in the last generations or of differences in sampling. When compared with Eskimos the Lapps, as other Europeans, had on the average a lower frequency of whorls (30–42%) and a higher frequency of arches, of which the USSR–Lapps and the Skolt Lapps had the highest frequency (over 10%).

Some hereditary pathological traits

So far as known the Ainu do not have high frequencies of any particular hereditary pathological traits. Among Canadian Eskimos the rare Laurence–Moon–Bardet–Biedl syndrome is not uncommon. Among the Lapps the congenital dislocation of the hip is very much in evidence (*see* pp. 151–5).

Many diseases occur in Greenlandic Eskimos with other frequencies and other clinical characteristics than in Danes. Diabetes, ischemic heart disease, asthma, duodenal ulcer and multiple sclerosis are rare in Greenland, whereas other disorders, e.g. congenital malformations, have higher frequencies than in Denmark. Evaluation of the genetic determination of these ethnic traits is going on (Harvald, 1976).

The Eskimos in Greenland have lower lipid values than Scandinavian populations, but Greenlandic Eskimos living in Denmark have rather high values (Bang & Dyerberg, 1972).

Even if few data so far have been presented, it appears that in comparison with other northwestern European populations Lapps have a low frequency of obesity, hypertension, cholecystopathy, juvenile diabetes mellitus, varicsoe veins, myopia and color sense disturbances (*see* Chapter 5).

In the 355 Svettijärvi Skolt Lapps only one female with diabetes mellitus was found. These Lapps are still living in rather natural conditions with reindeer breeding as the most important source of income. However, in the 154 semi-urbanized Nellim Skolt Lapps there are four females and one male with diabetes mellitus. One of the females got insulin-dependent diabetes at the age of about 20 yr. All the three other Nellim Skolts were over 50 yr of age when they got diabetes.

The blood pressure of 331 Inari Lapps and 221 Skolt Lapps over the age of 20 yr was studied by Sundberg *et al.* (1975). The systolic pressure was found to rise more strongly with age in females than in males. Neither sex displayed any distinct effect of age on diastolic pressure. Comparison with Finns and Åland Islanders revealed fairly equal systolic blood-pressure values in females, while definitely lower values were found in males of Inari Lapps and Skolt Lapps. The Inari Lapps and Skolt Lapps displayed no such clear age dependence of the diastolic blood pressure as occurs in Finns. These findings are in agreement with the hypothesis according to which the setting of resting blood-pressure level is more influenced by different kinds of load associated with technological development and with the degree of urbanized mode of life than by genetic factors.

Coronary heart disease mortality figures for Finns, and particularly for eastern Finns, are among the highest known (Aromaa *et al.*, 1975). According to various medical teams (Lewin & Eriksson, 1970) working in Finnish and Norwegian Lapp populations during the IBP expeditions, there seems to be no evidence of high incidence of coronary heart disease among Lapps.

Studies on lipids in different populations in Finland (Björkstén *et al.*, 1975) show that the Inari Lapps (including the Skolt Lapps) and rural Finns had nearly identical serum lipid levels in spite of the differences in the genetic background and also in way of living. Age-adjusted mean log triglyceride values were in all age groups considerably lower among Lapps than among Finns. The prevalence of hyperlipidemias was also examined. The type IIa was rather prevalent among Lapps.

In the Finns there is an enrichment of some rare hereditary diseases. Some of these diseases, particularly cornea plana congenita recessiva but also aspartylglucosaminuria and dystrophia retinae pigmentosa et dysacusis syndrome, have a relatively high frequency in northern Finland. However, according to genealogical investigations and studies on Lapps there is no evidence that some of the rare genetic diseases in Finns should be emerging from the Lapps (for references and further discussions *see* Eriksson, 1973*a* and Norio *et al.*, 1973).

Conclusion

Diabetes, ischemic heart disease, bronchial asthma, duodenal ulcer and multiple sclerosis are rare in Greenland Eskimos, whereas congenital malformations have higher frequencies than in Scandinavians. Eskimos in Greenland have considerably lower lipid values than Lapps and other northwestern European populations. Among Lapps juvenile diabetes, cholecystopathy, obesitas, varicose veins and myopia appear to be relatively rare. In contrast to Finns, Lapps do not seem to have a high incidence of coronary heart disease. The Finnish Lapps have relatively high serum-cholesterol levels, but the triglyceride values are considerably lower among Lapps than among Finns. The type IIa hyperlipidemia is rather prevalent among Finnish Lapps.

Congenital dislocation of the hip

The prevalence of congenital dislocation of the hip (CDH) in Europeans is 1 to 2 per 1000 (Wynne-Davies, 1970). In some Amerindian tribes high frequencies of CDH have been reported, but in Eskimos CDH seems to have a frequency similar to that in Whites.

In both ancient Lapp skeletons and present day Lapps frequencies of CDH

151

Table 4.17. *Frequency of congenital dislocation of the hip in Lapps, Finns, Norwegians and Swedes*

Population	Frequency per 1000	Sex ratio (male: female)	Reference
Lapps (Norway)	50 ± 8	1:6	Wessel (1918)
Lapps (Norway)	34 ± 12		Schreiner (1935)
Lapps (Finland)	52 ± 31		Näätänen (1936)
Lapps (Sweden)	25 ± 6	1:20	Mellbin (1962)
Norway, Finnmark	2.4 ± 0.2	1:3	Getz (1955)
Norway	1 ± 0.1	1:5	Getz (1955)
Finland	10.6 ± 0.1	1:6	Laurent (1953)
Sweden	2.2 ± 0.2		Palmén (1961)

There was bilateral congenital dislocation in the hip in 30–40% of cases studied.

between 25 and 50 per 1000 are reported (for references *see* Table 4.17), i.e. the highest noted in the world. Wessel (1918) was the first to state that the Skolt Lapps did not have as high a frequency of CDH as other Lapps. In the 637 Skolt Lapps (including offspring of Skolt Lapps and Finns) investigated during 1966–70 only one female with CDH was found (Table 4.18). The frequencies of CDH in the other Lapp populations in Inari are high. Even if the Skolt Lapps are included the frequency of CDH is high (2 of 634 males and 20 of 654 females) and in rather good agreement with the frequencies in Table 4.17 (A. W. Eriksson, T. Lewin & P. Luukka, unpublished observations).

Family studies are consistent with a multifactorial etiology of CDH. Susceptibility to CDH varies with the degree of acetabular dysplasia (which appears to be polygenically inherited) and with the degree of laxity of the joint capsule, frequently found to be autosomal dominant (Wynne-Davies, 1970).

In CDH a number of associations with environmental factors has also been identified, e.g. excess in firstborn and in breech births. The frequency of CDH is highest among peoples who keep their babies' legs adducted and extended, e.g. Amerindians and Lapps. The Eskimo women carry their babies with their legs abducted around the women's backs inside their parkas.

CDH occurs more often in babies born in the winter than in the summer months, perhaps because tight swaddling may elicit frank disease in a predisposed baby. In the past, Lapp mothers used to swaddle their babies very tightly in a cradle, so that babies could not move their legs. This cradle, called 'komsa', or 'gietka', was a very practical nomad cradle, hollowed from a log, covered with leather and provided with a protective hood; the cradle was carried on a band across the shoulder or it was attached to the pommel of the reindeer's pack saddle. Nowadays these cradles are not used because the Lapps are not nomads any longer and because the health nurses have for-

Table 4.18. *Frequencies of congenital dislocation of the hip in Finnish Lapp populations (Inari)*

| | | Age groups (yr) | | | | | | | | Total number studied | |
| | | 0–19 | | 20–39 | | 40–59 | | 60– | | | |
Population		CDH	N	CDH	N	CDH	N	CDH	N	CDH	N
Sevettijärvi Skolt Lapps	males	0	87	0	48	0	26	0	15	0	176
	females	0	76	0	45	0	33	0	25	0	179
Nellim Skolt Lapps	males	0	16	0	22	0	21	0	13	0	72
	females	0	25	0	21	1	21	0	15	1	82
Fisher Lapps	males	0	38	0	34	0	37	1	22	1	131
	females	1	31	0	32	5	36	6	32	12	131
Mountain Lapps	males	0	15	0	18	0	9	1	17	1	59
	females	0	20	1	20	1	27	2	11	4	78
Offspring of Mountain Lapps and Fisher Lapps	males	0	13	0	15	0	7	0	3	0	38
	females	1	24	0	8	0	8	0	8	1	48
Offspring of Fisher and Mountain Lapps and Skolt Lapps	males	0	16	0	2	0	0	0	0	0	18
	females	0	16	0	1	0	0	0	1	0	18
Offspring of Skolt Lapps and Finns	males	0	62	0	5	0	1	0	0	0	68
	females	0	54	0	5	0	1	0	0	0	60
Offspring of other Lapps and Finns	males	0	44	0	20	0	4	0	0	0	72
	females	0	27	0	18	1	13	0	0	1	58
Total	males	0	291	0	164	0	105	2	74	2	634
	females	2	273	1	150	8	139	8	92	19	645

bidden their use. It seems that during the last generation the frequency of CDH has decreased considerably, although it is still high in many Lapp populations. Below the age of 30 yr no cases among 382 males and only two among 356 females had CDH. But in the ages over 50 yr two among 124 males and 14 among 158 females had CDH (Table 4.18).

Lapps, and especially Lapp women, have a particular gait. This and the high frequency of CDH have been supposed to have connections to the particular morphology of the hip joint in Lapps, i.e. an extreme femoral antetorsion and a shallow acetabulum (Getz, 1955). CDH is by far the most frequent malformation of the locomotor apparatus. This is what is to be expected in traits which are recent from an evolutionary point of view. Also the frequency of spondylosis is high among Lapps, about 14%, i.e. about three times as high as in other Scandinavians (Schreiner, 1935; T. Lewin & J. Edgren, unpublished observations). The high frequency of orthopaedic ailments of the lower back and extremities may be due to developmental patterns which are still not sufficiently stabilized in the genotypes of Lapps (Getz, 1955).

Andrén (1962) has stressed that the cause of CDH is elongation of the joint capsule. This elongation is largely ascribed to maternal hormones (oestrogens and particularly relaxin) which also produce temporary instability of the pubic symphysis in the infant. This enables the femoral head to slip out of the acetabulum. Andrén (1962) found that there is a decreased breakdown of oestrogens in newborn infants with CDH. A disorder of oestrogen metabolism is probably a causal factor of CDH. This hypothesis is supported by the fact that CDH is about six times as common among females as among males in whom androgens abolish the effect of oestrogens.

It is a paradoxical phenomenon that Lapps, who live in desolate areas and have always been forced to use their legs, have a frequency of CDH that is 20–30 times higher than that in other northern European populations. Lappish skeletons, and particularly the lower extremities, are characterized by certain traits which are different from those found in other Scandinavian skeletal material (Getz, 1955, 1957). Lapps have higher frequencies of curved lower legs, enlarged epiphyses and flat foot than other Scandinavians (Mellbin, 1962; A. W. Eriksson & P. Luukka, unpublished observations). The curved legs seem to be specific for Lapps and thus not only the result of rickets in the first years of life. However, X-ray studies have not been done. The comparatively high frequency of enlarged epiphyses and enamel hypoplasia of the permanent teeth seems, however, mainly to be a result of the lack of vitamin D (Mellbin, 1962).

Until recent times, before the vitamin D-prophylaxis was introduced, Lapps are known to have had a high frequency of rachitis (Ekvall, 1940; Mellbin, 1962; for further references *see* Lewin & Hedegård, 1971). This may be a consequence of the fact that the rather pigmented Lapps cannot synthesize vitamin D as efficiently as the blond Finns and Scandinavians. White skin seems to be an adaptation to northern latitudes. The one exception to the correlation between latitude and skin color is the Eskimo. But his diet contains several times the minimum preventive dose of vitamin D (Loomis, 1967).

As a consequence of the rachitis many Lapp women may have had a narrow pelvis. In such women the selective significance is obvious. However, the oestrogens and particularly relaxin impart an extra elasticity to the pelvis in preparation for the physical demands of childbirth. This may have been of particular importance in women with a narrow pelvis. Thus it would be of interest to know whether Lapp women have prenatally a high level of these hormones or whether the newborn of Lapps have a delayed oestrogen metabolism as babies with CDH have (Andrén, 1962).

Daiger, Schanfield & Cavalli-Sforza (1974) showed that the vitamin-D binding of α-globulin was identical with group specific component (Gc) protein. In the light of this finding the world distribution of the Gc alleles was studied by Mourant, Domaniewska-Sobzak & Tills (1975). With some exceptions (including the Lapps) it was found that the Gc^2 allele has its highest frequencies in areas of low sunshine incidence.

It is not yet known whether differential binding of vitamin D may be a possible mechanism affecting the Gc polymorphism. The Lapps are exceptional in having one of the lowest Gc^2 gene frequencies noted in spite of the

Genetic studies

fact that the neighboring majority populations have Gc^2 frequencies higher than 0.20 (*see* the first part of this chapter and Table 4.2).

Conclusion

Lapps, with the exception of the Skolt Lapps, have a frequency of congenital dislocation of the hip (CDH) that is more than 20 times higher than in other populations, including the Ainu and Eskimos. The probable role of rickets in causing narrow pelvis and labor difficulties is discussed in contrast to maternal hormones, which may facilitate the labor in women with narrow pelvis, but increase the risk of CDH, particularly in female babies. It is not yet known whether the low Gc^2 allele frequencies in Lapps can be explained by differential binding capacity of vitamin D. In Arctic populations with insufficient dietary intake of vitamin D the ultraviolet-deprivation rickets may have been an important selective factor.

Immunogenetic aspects

During the IBP–HA field expeditions of 1969–70 in Inari, Finnish Lapland, studies of immunological patterns and occurrence of certain parasites were conducted. Very low frequency of antibodies to toxoplasma and echinococcus was found among Finnish Lapps (Huldt, 1975). On the other hand anti-trichinella antibodies were found in the same frequency as in other parts of Scandinavia (I. Ljungström, unpublished).

The antibody pattern to a number of viral agents differed in many respects from that of other Scandinavian populations. For instance lower frequencies of antibodies to Coxsackie A7 and A9 were found while on the other hand antibodies to B1 were more commonly seen. The same is true for ECHO-virus type 7 while antibodies to ECHO 6, 9, 11 and 25 were rare. Notably many Skolt Lapps had antibodies to mumps virus, adenovirus and herpes simplex. High frequency of antibodies was found to measles virus in the age groups 0–29 yr while only 60% in the age groups 30–80 yr had antibodies (M. Lagercrants, in preparation).

The existence of the mosquito-borne arbovirus in the far north has been verified by finding antibodies against Inkoo virus among Lapps. Inkoo virus is a Finnish representative of the California encephalitis virus group. The increase in the prevalence of antibodies with age suggests endemic arbovirus activity. The antibody percentages in the different age groups were: 1–4 yr 14%, 5–9 yr 41%, 10–14 yr 73%, 15–19 yr 88% and in adults 99% (85% if only titres of $\geq$ 10 are included). The prevalence was over four times higher in Inari than in southern Finland and is among the highest in the world recorded for the California group. In Lapland the virus vectors, mosquitos, are a plague. The high antibody frequency seems to be a consequence of numerous mosquito attacks, because the Lapps live in close contact with nature.

155

Some members of the California virus are pathogenic for children, and it seems to be worth studying the pathogenicity of Inkoo virus (Brummer-Korvenkontio, 1973, 1974).

The levels of anti-bacterial antibodies in Finnish Lapps did not differ significantly from those of neighbouring populations. Also the levels of total IgM, IgA and IgG were normal throughout (L. Nilsson & Ö. Ouchterlony, personal communication, 1975).

A low incidence of autoantibodies was found (M. Werner, unpublished). The significance of these observations in Lapps is not yet clear, but they may support the suggestion of a genetic tendency towards the production of auto-antibodies. The low incidence of autoantibodies is in agreement with the clinical findings (A. W. Eriksson, W. Lehmann, T. Lewin & P. Luukka, un-published observations) that autoimmune diseases (e.g. rheumatoid arthritis) seem to be rare among Finnish Lapps. Also Ekvall (1940) did not find acute or chronic rheumatic polyarthritis in any nomadic Lapp in Västerbotten (Sweden).

During the IBP–HA studies on 1288 subjects belonging to the Lapp families in Inari no patients with Downs syndrome or other apparent major chromo-somal aberrations have been found. Also one of the general practitioners in the district of the parishes of Inari and Utsjoki, Dr L. Kirjarinta, is of the opinion that autoimmune disorders and chromosomal aberrations seem to be rare among Lapps. These findings are in agreement with the suggestion that the autoantibody may be a factor which increases the probability of nondisjunc-tion during gametogenesis (*see* review by Fialkow, 1969). However, the Lapp series are small and further studies have to be conducted (L. Kirjarinta, in preparation).

In the pathogenesis of diabetes mellitus the hypothesis of an autoimmune component has been discussed. In fact, juvenile diabetes is very rare in the Lapps and has probably not been observed in the Eskimos. Nerup *et al.* (1974) have shown that individuals with HL–A8 and/or W15 (particularly when they have both) have an increased susceptibility to the juvenile onset type of diabetes over those without the specific HL–A types. Thomsen *et al.* (1975) have shown that, apart from the HL–A8 and W15 increase among juvenile diabetes, there is an even greater increase of LD–8a (the lymphocyte-dependent gene in the major histocompatibility complex) than of HL–A8 but not of LD–W15a. They point out that there is a marked association between autoantibodies and LD–8a and suggest that LD–8a may be responsible for organ specific autoimmune diseases in general.

From Table 4.5 in this chapter it can be seen that the frequency of HL–A8 is extremely low in Eskimos as well as in Lapps. This may explain the lack of autoimmune diseases in general and diabetes in particular in the Eskimos and Lapps. Thomsen *et al.* (1975) have also suggested that W15 may have a special influence on the pancreatic beta cell. This gene is common, however, in many of the populations studied (Table 4.5) which probably suggests that there are unknown factors also influencing the development of diabetes (*see* also Gamble & Taylor, 1973). The rarity of juvenile diabetes in the Lapps and Eskimos might be due to the very low frequency of HL–A8 in these populations. On the other hand, the low frequency of HL–A8 might be due to a selection against those with the HL–A type if many of them developed juvenile diabetes and died before reproducing as they would have done until quite recently.

Evidence has been provided that the neutralist theory cannot alone account for the diversity of the HL–A system. In other words, HL–A is subject to natural selection. So far HL–A and disease association studies have yielded several important hints as to the pathogenesis of various diseases, but few in relation to the concept of natural selection, mainly because the association concerns diseases which are rather rare or influence the reproduction so little that they cannot have any great significance from the point of view of natural selection. Hitherto, there is no convincing evidence that HL–A plays a major role in resistance to infections, such as tuberculosis, which, in Eskimos and Lapps particularly, killed many young people before reproductive age.

It is open to speculation whether the micro-organisms have to a large extent been responsible for the present distribution of HL–A antigens in various populations and whether we are now paying the price for having become resistant to many micro-organisms by being more susceptible to a variety of autoimmune and hyperimmune conditions (Svejgaard *et al.*, 1975). Against this background it is of interest to note that in both Lapps and Eskimos there is a very low frequency of juvenile diabetes and a low frequency of HL–A antigen 1 and 8 which have associations to juvenile diabetes. On the other hand some well characterized rheumatic conditions, e.g. morbus Bechterew and Reiters syndrome, are very frequent in Greenlandic Eskimos (Harvald, 1976), but seem to be rare in Lapps, in spite of the fact that both of these Arctic populations have a high frequency of the antigen HL–A–W27 which seems to be strongly associated with these rheumatic diseases.

An association between Gm type and antibody has been suggested as a likely selective force. High mean IgG3 concentrations have been noted to be associated with sera homozygous and heterozygous for Gm (3, 5, 13, 14) (for further details *see* Steinberg *et al.*, 1973 in the first part of this chapter).

In Alaskan Eskimos (Paul *et al.*, 1951) and Swedish Lapps (Mellbin, 1962) low incidence of poliomyelitis and mumps antibodies have been reported. Whether the low incidence of some antibodies among the Arctic, non-cosmopolitan populations is only due to the isolation of these peoples and their dispersion over relatively wide areas or whether there are also genetic factors which may be of importance for forming antibodies against pathogens, cannot yet be answered.

The human biology of circumpolar populations

Conclusion

Finnish Lapps have a low frequency of antibodies to toxoplasma and echinococcus. Frequency patterns of antibodies to certain viral agents (e.g. Coxsachie A7 and A9, ECHO-virus type) differed in many respects from those of neighboring majority populations. The levels of antibacterial antibodies, IgM, IgA and IgG were similar to findings in other Fennoscandian populations. Lapps have a low incidence of autoanitbodies and also autoimmune diseases, e.g. rheumatoid arthritis, seem to be rare. The prevalence of some but not all of the autoimmune diseases in Lapps and Eskimos could be a result of different frequencies of various HL–A types.

Acknowledgements

The authors gratefully acknowledge the typing of the manuscript by Mrs Anneke Brussel, Institute of Human Genetics, Free University of Amsterdam, and Mrs Shirley Friend, Department of Paediatrics, Queen's University and the drawing of the figures by Mr S. B. R. Morton of the Instructional Communications Unit, Queen's University, Kingston, Ontario, Canada. We are most grateful to all of those investigators who allowed us to include their unpublished data. We also express our thanks to Dr A. E. Mourant, London, for his valuable comments on the manuscript, to Professor A. G. Steinberg for his careful reading of the sections on gammaglobulin allotypes and to Miss Marja-Riitta Eskola, BSc., Helsinki, who compiled the Lapp and other Scandinavian genetic data and checked the references. The expeditions to Lapland (Finland) and Greenland have been supported by financial assistance in the form of grants from the World Health Organization, the Deutsche Forschungsgemeinschaft, the Nordiska Kulturfonden, the Danish Ministry for Greenland, the Finnish Council for Medical Sciences, and the Sigrid Jusélius Foundation.

References
Genetic markers in blood

Ageheim, H. & Bergström, M. (1972). Adenosine deaminase polymorphism in a Swedish population. *Acta Genet. med. Roma*, **21**, 135–8.
Allison, A. C., Broman, B., Mourant, A. E. & Ryttinger, L. (1956). The blood groups of the Swedish Lapps. *J. roy. anthrop. Inst.*, **86**, 87–94.
Allison, A. C., Hartmann, O., Brendemoen, O. J. & Mourant, A. E. (1952). The blood groups of the Norwegian Lapps. *Acta path. microbiol. Scand.*, **31**, 334–8.
Arvilommi, H., Berg, K. & Eriksson, A. W. (1973). C3 types and their inheritance in Finnish Lapps, Maris (Cheremisses) and Greenland Eskimos. *Humangenetik*, **18**, 253–9.
Beckman, G., Beckman, L. & Cedergren, B. (1971). Population studies in northern Sweden. II. Red cell enzyme polymorphism in the Swedish Lapps. *Hereditas*, **69**, 243–8.

Beckman, G. & Pakarinen, A. (1973). Superoxide dismutase. A population study. *Hum. Hered.*, **23**, 346–51.

Beckman, L. (1959). A contribution to the physical anthropology and population genetics of Sweden. Variations of the ABO, Rh, MN, and P blood groups. *Hereditas*, **45**, 1–189.

Beckman, L., Broman, B., Jonsson, B. & Mellbin, T. (1959). Further data on the blood groups of the Swedish Lapps. *Acta genet. Basel*, **9**, 1–8.

Beckman, L. & Holmgren, G. (1961). Transferrin variants in Lapps and Swedes. *Acta genet. Basel*, **11**, 106–10.

Beckman, L. & Holmgren, G. (1963). On the genetics of the human transferrin variants B_1, B_2 and D_1. *Acta genet. Basel*, **13**, 361–5.

Beckman, L. & Mellbin, T. (1959). Haptoglobin types in the Swedish Lapps. *Acta genet. Basel*, **9**, 306–9.

Berg, K. (1969). Genetic studies of the adenylate kinase (AK) polymorphism. *Hum. Hered.*, **19**, 239–48.

Berg, K. (1973). Studies of polymorphic traits for the characterization of populations. The populations of Scandinavia. *Israel J. med. Sci.*, **9**, 1147–55.

Berg, K., Arvilommi, H. & Eriksson, A. W. (1972). Genetic marker systems in arctic populations. II. Polymorphism of C3 in Finnish Lapps. *Hum. Hered.*, **22**, 481–7.

Berg, K. & Eriksson, A. W. (1971). Genetic marker systems in arctic populations. I. Lp and Ag data on the Greenland Eskimos. *Hum. Hered.*, **21**, 129–33.

Berg, K. & Eriksson, A. W. (1973*a*). Genetic marker systems in arctic populations. V. The interited Ag(x) serum lipoprotein antigen in Finnish Lapps. *Hum. Hered.*, **23**, 241–6.

Berg, K. & Eriksson, A. W. (1973*b*). Genetic marker systems in arctic populations. VI. Polymorphism of C3 in Icelanders. *Hum. Hered.*, **23**, 247–50.

Berg, K. & Eriksson, A. W. (1973*c*). Genetic marker systems in arctic populations. VII. Genetic variation in serum lipoproteins in Icelanders. *Hum. Hered.*, **23**, 251–6.

Berggård, I. & Bearn, A. G. (1962). Excretion of haptoglobin in normal urine. *Nature, Lond.*, **195**, 1311–12.

Bjarnason, O., Bjarnason, V., Edwards, J. H., Fridriksson, S., Magnusson, M., Mourant, A. E. & Tills, D. (1973). The blood groups of Icelanders. *Ann. hum. Genet.*, **36**, 425–58.

Bodmer, J. & Bodmer, W. F. (1973). Population genetics of the HL-A system. A summary of data from the Fifth Int. Histocompatibility Testing Workshop. *Israel J. med. Sci.*, **9**, 1257–68.

Braend, M., Efremov, G., Fagerhol, M. K. & Hartmann, O. (1965). Albumin and transferrin variants in Norwegians. *Hereditas*, **53**, 137–42.

Brewer, G. J. (1967). Achromatic regions of tetrazolium stained starch gels: inherited electrophoretic variation. *Amer. J. hum. Genet.*, **19**, 674–80.

Brönnestam, R., Beckman, L. & Cedergren, B. (1971). Genetic polymorphism of the complement component C3 in Swedish Lapps. *Hum. Hered.*, **21**, 267–71.

Camoens, H., Monn, E. & Berg, K. (1972). Genetic marker systems in arctic populations. III. Polymorphism of red cell adenosine deaminase (ADA) in Norwegian Lapps. *Hum. Hered.*, **22**, 561–5.

Camoens, H., Monn, E. & Berg, K. (1973). Genetic marker systems in arctic populations. IV. Polymorphism of red cell acid phosphatase in Norwegian Lapps. *Hum. Hered.*, **22**, 69–71.

Cavalli-Sforza, L. L. & Edwards, A. W. F. (1967). Phylogenetic analysis: models and estimation procedures. *Amer. J. hum. Genet.*, **19**, 233–57.

The human biology of circumpolar populations

Chen, S.-H., Giblett, E. R., Anderson, J. E. & Fossum, B. L. G. (1972). Genetics of glutamic–pyruvic transaminase: its inheritance, common and rare variants, population distribution, and differences in catalytic activity. *Ann. hum. Genet.*, **35**, 401–9.

Chown, B. & Lewis, M. (1959). The blood group genes of the Copper Eskimo. *Amer. J. phys. Anthrop.*, **17**, 13–18.

Cleve, H. (1973). The variants of the group-specific component. A review of their distribution in human populations. *Israel J. med. Sci.*, **9**, 1133–46.

Cox, D. W., Simpson, N. E. & Jantti, R. (1978). Group-specific component, alpha$_1$-antitrypsin and esterase D in Canadian Eskimos. *Hum. Hered.*, **28**, 341–50.

Dossetor, J. B., Howson, W. T., Schlaut, J., McConnachie, P. R., Alton, J. D. M., Lockwood, B. & Olson, L. (1973*a*). Study of the HL–A system in two Canadian Eskimo populations. In *Histocompatibility Testing 1972*, ed. J. Dausset & J. Colombani, pp. 325–32. Baltimore: Williams & Wilkins.

Dossetor, J. B., McConnachie, P. R., Stiller, C. R., Alton, J. D. M., Olson, L. & Howson, W. T. (1973*b*). The major histocompatibility complex in Eskimos. *Transplant. Proceedings*, **5**, 209–13.

Duncan, I. W., Scott, E. M. & Wright, R. C. (1974). Gene frequencies of erythrocytic enzymes of Alaskan Eskimos and Athabaskan Indians. *Amer. J. hum. Genet.*, **26**, 244–6.

Ehnholm, C. & Eriksson, A. W. (1969). Haptoglobin subtypes among Finnish Skolt Lapps. *Ann. Med. exp. Fenn.*, **47**, 52–4.

Eriksson, A. W. (1973). Genetic polymorphisms in Finno-Ugrian populations. Finns, Lapps and Maris. *Israel J. med. Sci.*, **9**, 1156–70.

Eriksson, A. W., Fellman, J., Forsius, H., Gustafsson, B., Lindström, C., Kirjarinta, L., Kirjarinta, M., Lehmann, W., Benöhr, H. & Waller, H. D. (1967). Red cell glutathione reductase deficiency among Skolt Lapps. *Bull. Europ. Soc. hum. Genet.*, **1**, 73–6.

Eriksson, A. W., Fellman, J., Forsius, H., Lehmann, W., Lewin, T. & Luukka, P. (1976). The origin of the Lapps in the light of recent genetic studies. *Proc. 3rd int. Symp.*, *Yellowknife*, NWT, Circumpolar Health, 1974, ed. R. J. Shepard & S. Itoh; pp. 169–182. Toronto & Buffalo: Univ. Toronto Press.

Eriksson, A. W., Fellman, J., Kirjarinta, M., Eskola, M.-R., Singh, S., Benkmann, H.-G., Goedde, H. W., Mourant, A. E., Tills, D. & Lehmann, W. (1971*a*). Adenylate kinase polymorphism in populations in Finland (Swedes, Finns, Lapps), in Maris, and in Greenland Eskimos. *Humangenetik*, **12**, 123–30.

Eriksson, A. W., Kirjarinta, M., Fellman, J., Eskola, M.-R. & Lehmann, W. (1971*b*). Adenosine deaminase polymorphism in Finland (Swedes, Finns, and Lapps), the Mari Republic (Cheremisses), and Greenland (Eskimos). *Amer. J. hum. Genet.*, **23**, 568–77.

Eriksson, A. W., Kirjarinta, M., Lehtosalo, T., Kajanoja, P., Lehmann, W., Mourant, A. E., Tills, D., Singh, S., Benkmann, H.-G., Hirth, L. & Goedde, H. W. (1971*c*). Red cell phosphoglucomutase polymorphism in Finland – Swedes, Finns, Finnish Lapps, Maris (Cheremisses) and Greenland Eskimos, and segregation studies of PGM_1 types in Lapp families. *Hum. Hered.*, **21**, 140–53.

Eriksson, A. W., Kirjarinta, M. & Gürtler, H. (1970). Population genetic characteristics of a Greenland Eskimo population. *Arctic Anthrop.*, **7**, 3–5.

Eskola, M.-R., Ehnholm, C., Fellman, J. & Eriksson, A. W. (1971). Studies on plasma protein polymorphism among Finnish Lapps. *Proc. 2nd int. Symp. circumpolar Health, Oulu, Finland, 21–4 June, 1971,* (*abstract, p.*) 26.

160

Genetic studies

Eskola, M.-R., Kirjarinta, M., Nijenhuis, L. E., van den Berg-Loonen, E., van Loghem, E., Isokoski, M., Sahi, T. & Eriksson, A. W. (1976). Genetic polymorphisms in Utsjoki Lapps. *Proc. 3rd int. Symp., Yellowknife, NWT, Circumpolar Health,* 1974, ed. R. J. Shephard & S. Itoh, pp. 188–193. Toronto & Buffalo: Univ. Toronto Press.

Fagerhol, M. K., Eriksson, A. W. & Monn, E. (1969). Serum Pi types in some Lappish and Finnish populations. *Hum. Hered.,* **19**, 360–4.

Fernet, P., Langaney, A. & Robbe, P. (1971*a*). Résultats sérologiques de la mission de 1969 à Ammassalik. Cahier du CRA no. 11–12, in *Bull. et Mém. de la Soc. d'Anthrop. de Paris,* **8**, pp. 173–85, *Anthropologie Biologique et sociale des Ammassalimiut, Enquêtes programmées par Robert Gessain, no. 8.*

Fernet, P., Mortensen, W. S., Langaney, A. & Robert, J. (1971*b*). Hémotypologie du Scoresbysund (Est Groënland). Cahier du CRA no. 11–12, in *Bull. et Mém. de la Soc. d'Anthrop. de Paris,* **8**, pp. 177–85, *Anthropologie Biologique et Sociale des Ammassalimiut Enquêtes Programmées par Robert Gessain, no. 9.*

Giblett, E. R. (1969). *Genetic markers in human blood.* Oxford and Edinburgh: Blackwell Scientific Publications.

Gilberg, Å. & Persson, I. (1967). Serum protein types in polar Eskimos. *Acta genet. Basel,* **17**, 422–32.

Grubb, R. (1959). Hereditary gamma globulin groups in man in *CIBA Foundation Symposium on Biochemistry of Human Genetics,* pp. 264–78. London: Churchill.

Gustafsson, B., Gustafsson, C., Kirjarinta, M., Eriksson, A. W., Sjöblom, L. & Lehmann, W. (1971). Some aspects on the red cell glutathione reductase deficiency among Skolt Lapps. *Proc. 2nd int. Symp. circumpolar Health, Oulu, Finland, 21–4 June, 1971, (abstract),* p. 26.

Gutsche, B. B., Scott, E. M. & Wright, R. C. (1967). Hereditary deficiency of pseudocholinesterase in Eskimos. *Nature, London,* **215**, 322–3.

Harris, H. (1970). *The principles of human biochemical genetics.* Amsterdam: Elsevier-North Holland Publishing Co.

Harris, H. & Hopkinson, D. A. (1972). Average heterozygosity per locus in man: an estimate based on the incidence of enzyme polymorphisms. *Ann. hum. Genet.,* **36**, 9–20.

Harvald, B. (1975). Current genetical trends in the Greenlandic population. *Proc. 3rd int. Symp., Yellowknife, NWT, Circumpolar Health, 1974,* ed. R. J. Shepard & S. Itoh, pp. 165–169. Toronto & Buffalo: Univ. Toronto Press.

Heiken, A. (1962). The frequency of the Duffy blood group factor Fy[a] in Sweden. *Acta path. microbiol. Scand.,* **55**, 384.

Heiken, A. (1965). A genetic study of the MNSs blood group system. *Hereditas,* **53**, 187–211.

Heiken, A. & Rasmuson, M. (1966). Genetical studies on the Rh blood group system. *Hereditas,* **55**, 192–212.

Hirschfeld, J. & Beckman, L. (1961). Distribution of the Gc-serum groups in northern and central Sweden. *Acta genet. Basel,* **11**, 185–95.

Hopkinson, D. A., Mestriner, M. A., Cortner, J. & Harris, H. (1973). Esterase D: a new human polymorphism. *Ann. hum. Genet.,* **37**, 119–37.

Int. Workshop. (1973). Appendix A. The Code for the analysis of the data of the ed. J. Dausset & J. Colombani, pp. 679–719. Baltimore: Williams & Wilkins. fifth int. Histocompatibility Workshop. In Histocompatibility Testing 1972,

Kimura, M. (1968). Genetic variability maintained in a finite population due to mutational production of neutral and nearly neutral isoalleles. *Genet. Res.,* **11**, 247–69.

The human biology of circumpolar populations

Kirjarinta, M. (1976). Correlation between glutathione reductase activity and acid phosphatase phenotypes. A biochemical population genetic study in Skolts and other Lapps. *Proc. 3rd int. Symp., Yellowknife*, NWT, Circumpolar Health, *1974*, ed. R. J. Shephard & S. Itoh, pp. 194–197. Toronto & Buffalo: Univ. Toronto Press.

Kirjarinta, M., Fellman, J., Gustafsson, C., Keisala, E. & Eriksson, A. W. (1969). Two rare electrophoretic variants of erythrocyte enzyme in Finland. *Scand. J. clin. lab. Invest.*, **23**, Suppl. 108, 46.

Kissmeyer-Nielsen, F., Kjerbye, K. E., Lamm, L. U., Jørgensen, J., Bruun Petersen, G. & Gürtler, H. (1973). Study of the H–LA system in Eskimos. In *Histocompatibility Testing 1972*, ed. J. Dausset & J. Colombani, pp. 317–24. Baltimore: Williams & Wilkins.

Kornstad, L. (1959). The frequency of the Rh antigen C^w in 2750 Oslo blood donors. *Vox Sang.*, **4**, 225–30.

Kornstad, L. (1972). Distribution of the blood groups of the Norwegian Lapps. *Amer. J. phys. Anthrop.*, **36**, 257–66.

Kurczynski, T. W. & Steinberg, A. G. (1967). A general program for maximum likelihood estimation of gene frequencies. *Amer. J. hum. Genet.*, **19**, 178–9.

Lalouel, J.-M. (1974). Controversial issues in human population genetics. *Amer. J. hum. Genet.*, **26**, 262–4.

Levin, M. G. (1958). Blood groups among the Chuckchi and Eskimo. Reprinted from *Sovietskaja Etnografija*, **5**, 113–16 by *Arct. Anthrop.*, **1**, 87.

Levin, M. G. (1959). New material on the blood groups among the Eskimo and Lamuts. Reprinted from *Sovietskaja Etnografija*, 3, 98–9 by *Arct. Anthrop.*, 1, 91.

Matsumoto, H. & Miyazaki, T. (1972). Gm and Inv allotypes of the Ainu in Hidaka area, Hokkaido. *Jap. J. hum. Genet.*, **17**, 20–6.

McAlpine, P. J., Chen, S.-H., Cox, D. W., Dossetor, J. B., Giblett, E., Steinberg, A. G. & Simpson, N. E. (1974). Genetic markers in blood in a Canadian Eskimo population with a comparison of allele frequencies in circumpolar populations. *Hum. Hered.*, **24**, 114–42.

Milan, F. A. (1970). The demography of an Alaskan Eskimo village. *Arct. Anthrop.*, **7**, 26–43.

Misawa, S. & Hayashida, Y. (1968). On the blood groups among the Ainu in Shizunai, Hokkaido. *Proc. Jap. Acad.*, **44**, 83–8.

Misawa, S. & Hayashida, Y. (1970). On the blood groups among the Ainu in Niikappu, Hokkaido, *J. anthrop. Soc. Nippon*, **78**, 177–86.

Misawa, S., Hayashida, Y. & Okochi, K. (1971). Distribution of Ag(x) and Ag(y) antigens in the Ainu. *Jap. J. hum. Genet.*, **16**, 30–4.

Mittal, K. K., Hasegawa, T., Ting, A., Mickey, M. R. & Terasaki, P. I. (1973). Genetic variation in the HL–A system between Ainus, Japanese, and Caucasians. In *Histocompatibility Testing 1972*, ed. J. Dausset & J. Colombani, pp. 187–95. Baltimore: Williams & Wilkins.

Monn, E. (1969). Red cell phosphoglucomutase (PGM) types of Norwegian Lapps. *Hum. Hered.*, **19**, 264–73.

Monn, E., Berg, K., Reinskou, T. & Teisberg, P. (1971). Serum protein polymorphisms among Norwegian Lapps. Studies on the Lp, Ag, Gc and transferrin systems. *Hum. Hered.*, **21**, 134–9.

Mourant, A. E. (1954). *The distribution of the human blood groups*. Oxford: Blackwell Scientific Publications.

Mourant, A. E., Kopec, A. C. & Domaniewska-Sobczak, K. (1975). *The distribution of human blood groups and other biochemical polymorphisms*, 2nd edn. Oxford University Press.

Genetic studies

Nevanlinna, H. R. (1947). The distribution of the Rh groups among Finnish blood donors. *Ann. Med. exp. Fenn.*, **25**, 146.

Nevanlinna, H. R. (1972). The Finnish population structure. Agenetic and genealogical study. *Hereditas*, **71**, 195–236.

Nielsen, J. C., Mårtensson, L., Gürtler, H., Gilberg, Å. & Tingsgård, P. (1971). Gm types of Greenland Eskimos. *Hum. Hered.*, **21**, 405–19.

Nilsson, L.-O. & Eriksson, A. W. (1972). Screening for haemoglobin and lactate dehydrogenase variants in the Icelandic, Swedish, Finnish, Lappish, Mari and Greenland Eskimo Populations. *Hum. Hered.*, **22**, 372–9.

Olaisen, B. & Teisberg, P. (1972). Erythrocyte alanine aminotransferase polymorphism in Norwegian Lapps. *Hum. Hered.*, **22**, 380–6.

Omoto, K. (1972). Polymorphisms and genetic affinities of the Ainu of Hokkaido. *Hum. Biol. Oceania*, **1**, 278–88.

Omoto, K. (1973). Polymorphic traits in peoples of Eastern Asia and the Pacific. *Israel J. med. Sci.*, **9**, 1195–215.

Omoto, K., Aoki, K. & Harada, S. (1975). Polymorphism of esterase D in some population groups in Japan. *Hum. Hered.*, **25**, 378–81.

Omoto, K. & Harada, S. (1972). The distribution of polymorphic traits in the Hidaka Ainu. II. Red cell enzyme and serum protein groups. *J. Faculty of Science, The University of Tokyo, Sec. V*, **4**, 171–211.

Omoto, K. & Misawa, S. (1974). The genetic relations of the Ainu. In: *The Biological Origins of the Australians*, ed. R. L. Kirk & A. G. Thorne. Canberra: Inst. Aboriginal Studies.

Omoto, K. & Misawa, S. (1974). The genetic relations of the Ainu. In: The Biological Origins of the Australians, ed. R. L. Kirk & A. G. Thorne. Canberra: Inst. Aboriginal Studies.

Persson, I. (1968). The distribution of serum types in West Greenland Eskimos. *Acta genet. Basel*, **18**, 261–70.

Persson, I. (1970). Anthropological investigations of the population of Greenland. *Meddel. Grønland, Udgivne af kommissionen for videnskabelige undersøgelser i Grønland.*, **180**, p. 78. København: CA Reitzels Forlag.

Persson, I., Rivat, L., Rousseau, P. Y. & Ropartz, C. (1972). Ten Gm factors and the Inv system in Eskimos in Greenland. *Hum. Hered.*, **22**, 519–28.

Persson, I. & Swan, T. (1971). Serum β-lipoprotein polymorphism in Greenlanders. Frequency of the Ag(x) factor. *Hum. Hered.*, **21**, 384–7.

Persson, I. & Tingsgård, P. (1965). A deviating Gc type. *Acta genet. Basel.*, **15**, 51–6.

Persson, I. & Tingsgård, P. (1966). Serum types in East Greenland Eskimos. *Acta genet. Basel*, **16**, 84–8.

Persson, I. & Tingsgård, P. (1968). Serum protein types in East Greenland Eskimos. *Acta genet. Basel*, **18**, 61–9.

Race, R. R. & Sanger, R. (1968). *Blood groups in man. 5th edn.* Oxford & Edinburgh: Blackwell Scientific Publications.

Reed, T. E. & Schull, W. J. (1968). A general maximum likelihood estimation program. *Amer. J. hum. Genet.*, **20**, 579–80.

Reinskou, T. & Kornstad, L. (1965). The Gc types of the Norwegian Lapps. *Acta genet. Basel*, **15**, 126–33.

Robson, E. B. & Harris, H. (1966). Further data on the incidence and genetics of the serum cholinesterase phenotype C_5+. *Ann. hum. Genet.*, **29**, 403–8.

Scott, E. M., Duncan, I. W., Ekstrand, V. & Wright, R. C. (1966). Frequency of polymorphic types of red cell enzymes and serum factors in Alaskan Eskimos and Indians. *Amer. J. hum. Genet.*, **18**, 408–11.

Scott, E. M., Griffith, I. V., Hoskins, D. D. & Schneider, R. G. (1958). Lack of

163

abnormal hemoglobins in Alaskan Eskimos, Indians, and Aleuts. *Science, Washington*, **129**, 719–20.

Scott, E. M. & Powers, R. F. (1974). Properties of the C_5 variant form of human serum cholinesterase. *Amer. J. hum. Genet.*, **26**, 189–94.

Scott, E. M., Weaver, D. D. & Wright, R. C. (1970). Discrimination of phenotypes in human serum cholinesterase deficiency. *Amer. J. hum. Genet.*, **22**, 363–9.

Singh, S., Jensen, M., Goedde, H. W., Lehmann, W., Pyörälä, K. & Eriksson, A. W. (1971). Pseudocholinesterase polymorphism among Lapp populations in Finland. *Humangenetik*, **12**, 131–5.

Singh, S., Saternus, K., Münsch, H., Altland, K., Goedde, H. W. & Eriksson, A. W. (1974). Pseudocholinesterase polymorphism among Ålanders (Finno-Swedes), Maris (Cheremisses, USSR) and Greenland Eskimos, and the segregation of some E_1 and E_2 locus types in Finnish Lapp families. *Hum. Hered.*, **24**, 352–62.

Spencer, N., Hopkinson, D. A. & Harris, H. (1964). Quantitative differences and gene dosage in the human red cell acid phosphatase polymorphism. *Nature, London*, **201**, 299–300.

Steinberg, A. G. (1969). Globulin polymorphisms in man. *Ann. Rev. Genet.*, **3**, 25–52.

Steinberg, A. G. & Kageyama, S. (1970). Further data on the Gm and Inv allotypes of the Ainu: Confirmation of the presence of a $Gm^{2, 17, 21}$ phenogroup. *Amer. J. hum. Genet.*, **22**, 319–25.

Steinberg, A. G. & Matsumoto, H. (1964). Studies on the Gm, Inv, Hp and Tf serum factors of Japanese population and families. *Hum. Biol.*, **36**, 77–85.

Steinberg, A. G., Stauffer, R., Blumberg, B. S. & Fudenberg, H. H. (1961). Gm phenotypes and genotypes in SU whites and negroes; in American Indians and Eskimos; in Africans; and in Micronesians. *Amer. J. hum. Genet.*, **13**, 205–13.

Steinberg, A. G., Tiilikainen, A., Eskola, M.-R. & Eriksson, A. W. (1974). Gammaglobulin allotypes in Finnish Lapps, Finns, Åland Islanders, Maris (Cheremis), and Greenland Eskimos. *Amer. J. hum. Genet.*, **26**, 223–43.

Stern, C. (1973). *The principles of human genetics*, 3rd edn. San Francisco: W. H. Freeman & Company.

Szathmáry, E. J. E. (1974). Genetic relationships of arctic and subarctic populations: evolutionary implications. *Anthropological Series no. 16*. University of Toronto: Department of Anthropology.

Teisberg, P. (1971). C3 types of Norwegian Lapps. *Hum. Hered.*, **21**, 162–7.

Thorsby, E., Bratlie, A. & Teisberg, P. (1971). HL–A polymorphism of Norwegian Lapps. *Tiss. Antigens*, **1**, 137–46.

Tiilikainen, A., Eriksson, A. W., MacQueen, J. M. & Amos, D. B. (1973). The HL–A system in the Skolt population. In *Histocompatibility Testing 1972*, ed. J. Dausset & J. Colombani, pp. 85–92. Baltimore: Williams & Wilkins.

Virtaranta, K., Kirjarinta, M., Eriksson, W. A., Sahi, T. & Isokoski, M. (1976). Erythrocyte alanine aminotransferase polymorphism in a Lappish population *Proc. 3rd int. Symp., Yellowknife*, NWT, Circumpolar Health, 1974, ed. R. J. Shephard & S. Itoh, pp. 197–200. Toronto & Buffalo: Univ. Toronto Press.

Waller, H. D., Benöhr, H. C., Lehmann, W. & Eriksson, A. W. (1972). Aktivierung der Glutathionreduktase roter Blutzellen mit Flavin-Adenin-Dinucleotid (FAD). *Klin. Wschr.*, **50**, 462–6.

Welch, S. & Lee, J. (1974). The population distribution of genetic variants of human esterase D. *Humangenetik*, **24**, 329–31.

Yamazaki, T. & Maruyama, T. (1974). Evidence that enzyme polymorphisms are selectively neutral, but blood groups are not. *Science, Washington*, **183**, 1091–2.

Zolotareva, I. M. (1966). Blood group distribution of the peoples of northern Siberia. *7th int. Congr. Anthrop. Ethol. Sci.*, pp. 502–8. Moscow: Nauka Publishing House.

Other genetic traits

Abel, W. (1934). Finger- und Handlinienmuster Ostgrönländischer Eskimos. *Wissenschaftliche Ergebnisse der Deutschen Grönland-Expedition Alfred Wegener 1929 und 1930–1*, **6**, 43–65.

Allison, A. C. & Blumberg, B. S. (1959). Ability to taste phenylthiocarbamide among Alaskan Eskimos and other populations. *Hum. Biol.*, **31**, 352–9.

Allison, A. C. & Nevanlinna, H. R. (1952). Taste-deficiency in Lappish and Finnish populations. *Ann. Eugen. Lond.*, **17**, 113–14.

Alsbirk, K. E. & Alsbirk, P. H. (1972). PTC taste sensitivity in Greenland Eskimos from Umanaq. Distribution and correlation to ocular anterior chamber depth. *Hum. Hered.*, **22**, 445–52.

Andrén, I. (1962). Pelvic instability in new borns. With special reference to congenital dislocation of the hip and hormonal factors. *Acta radiol.* Suppl. 212, 76 pp.

Armstrong, A. R. & Peart, H. A. (1960). A comparison between the behaviour of Eskimos and non-Eskimos to the administration of isoniazid. *Amer. Rev. Tuberc.*, **81**, 588–94.

Aromaa, A., Björkstén, F., Eriksson, A. W., Maatela, J., Kirjarinta, M., Fellman, J. & Tamminen, M. (1975). Serum cholesterol and triglyceride concentrations of Finns and Finnish Lapps: basic data. *Acta med. Scand.*, 198, 13–22.

Auer, J. (1950). Fingerprints in Eskimos of the Northwest territories. *Amer. J. phys. Anthrop.*, **8**, 485–8.

Bang, H. O. & Dyerberg, J. (1972). Plasma lipids and lipoproteins in Greenlandic West Coast Eskimos. *Acta med. Scand.*, **192**, 85–94.

Björkstén, F., Aromaa, A., Eriksson, A. W., Maatela, J., Kirjarinta, M., Fellman, J. & Tamminen, M. (1975). Serum cholesterol and triglyceride concentrations of Finns and Finnish Lapps: Interpopulation comparisons and occurrence of hyperlipidemia. *Acta med. Scand.*, 198, 23–33.

Brummer-Korvenkontio, M. (1973). Arboviruses in Finland. V. Serological survey of antibodies against Inkoo virus (California group) in human, cow, reindeer, and wildlife sera. *Amer. J. Trop. Med. Hyg.*, **22**, 654–61.

Brummer-Korvenkontio, M. (1974). Bunyamwera arbovirus supergroup in Finland. A study on Inkoo and Batai viruses. *Commentationes Biologicae (Helsinki)*, **76**, 1–52.

Chit, H. L. (1972). Über das Hautleistensystem der Bevölkerung Finnlands. *Ann. Acad. Sci. Fenn. A. V. Med.*, **151**, 1–26.

Cummins, H. (1935). Dermatoglyphics in Eskimos from Point Barrow. *Amer. J. phys. Anthrop.*, **20**, 13–17.

Cummins, H. & Fabricius-Hansen, V. (1946). Dermatoglyphics in Eskimos of West Greenland, *Amer. J. phys. Anthrop.*, **4**, 395–402.

Cummins, H. & Midlo, C. (1961). *Finger prints, palms and soles: an introduction to dermatoglyphics.* New York: Dover Publications Inc.

Dahlberg, G. & Wahlund, S. (1941). *The race biology of the Swedish Lapps. Part II. Anthropometrical survey.* Uppsala: Almqvist & Wiksells Boktryckeri, A.B.

Daiger, S. P., Schanfield, M. S. & Cavalli-Sforza, L. L. (1974). Group specific component (Gc) proteins bind vitamin D. *Amer. J. hum. Genet.*, **26**, 24 A.

Ducros, J. & Ducros, A. (1972). Nombre des Crêtes et Dessines des Doigts d' Ammassalimiut (Scoresby Sound, Groenland Oriental) et des Populations Eskimo et Asiatiques. *L'Anthropologie*, **76**, 711–26.

Ekvall, S. (1940). On the history and conditions of life of the West Bothnian nomad Lapps, their food and health conditions. *Acta med. Scand.*, **105**, 329–59.

The human biology of circumpolar populations

Eriksson, A. W. (1973a). Genetic polymorphisms in Finno-Ugrian populations: Finns, Lapps and Maris. *Israel J. med. Sci.*, 9, 1156–70.

Eriksson, A. W. (1973b). Über die genetische Struktur bei Lappen. In *Studies in the anthropology of the Finno-Ugrian peoples. Stencil no. 7*, pp. 109–27. Helsinki: Archaeological Institute of the University of Helsinki.

Eriksson, A. W., Fellman, J., Forsius, H. & Lehmann, W. (1970a). Phenylthiocarbamide tasting ability among Lapps and Finns. *Hum. Hered.*, 20, 623–30.

Eriksson, A. W., Kirjarinta, M. & Gürtler, H. (1970b). Population genetic characteristics of a Greenland Eskimo population. *Arctic Anthrop.*, 7, 3–5.

Fialkow, P. J. (1969). Genetic aspects of autoimmunity. *Progr. med. Genet.*, 6, 117–67.

Flitz, M. (1971). Über das Hautleistensystem der Finger, Hände und Füsse von Finnlappen. Kiel: Inaugural dissertation.

Forsius, H., Luukka, H., Lehmann, W., Fellman, J. & Eriksson, A. W. (1970). Irisfärg, korneabrytningsförmåga och korneatjocklek bland skoltsamer och finnar (Ophthalmogenetical studies on Skolt Lapps and Finns: corneal thickness, corneal refraction and iris pigmentation). *Nord. Med.*, 84, 1559–61.

Frehse, W. (1974). Über das Hautleistensystem der Kraulshavn-Eskimos. Kiel: Inaugural Dissertation.

Gamble, D. R. & Taylor, K. W. (1973). Coxsackie B virus and diabetes. *Brit. med. J.*, 1, 289–90.

Gessain, R. (1959). Dermatoglyphes Digitaux et Palmaires des Eskimos D'Angmassalik. *Bull. Soc. d'Anthropologie*, 10, Xth Series, 233–50.

Getz, B. (1955). The hip joint in Lapps and its bearing on the problem of congenital dislocation. *Acta orthop. Scand.*, 24, Suppl. 18, 81 pp.

Getz, B. (1957). The proximal tarsus in the light of a biometrical investigation in Lappic skeletons. *Acta morph. Neerl.-Scand.*, 1, 188–201.

Hannegren, A., Borgå, O. & Sjöquist, F. (1970). Inactivation of isoniazid (INH) in Swedish tuberculous patients before and during treatment with para-aminosalicylic acid (PAS). *Scand. J. resp. Dis.*, 51, 61–9.

Harris, H. W., Knight, R. A. & Selin, M. (1958). Comparison of isoniazid concentrations in the blood of people of Japanese and European descent. *Amer. Rev. Tuberc.*, 78, 944–8.

Harvald, B. (1975). Current trends in medical research in Greenland. *Proc. 3rd int. Symp., Yellowknife*, NWT, Circumpolar Health, *1974*, ed. R. J. Shephard & S. Itoh, pp. 12–15. Toronto & Buffalo: Univ. Toronto Press.

Holt, S. B. (1968). *The genetics of dermal ridges.* Springfield, Illinois: Charles C. Thomas.

Hughes, D. H. (1971). Genmarkierer bei den Eskimos in Igloolik in der kanadischen Arktis. In W. Lehmann (Hrsg.): Der Mensch in der Arktis. *Anthrop. Anz.*, 33, 125–6.

Huldt, G. (1974). Prevalence of parasitic infections in northern Scandinavia. *Proc. int. Ecolog. Symp., Luleå*, 1971. ed. E. Bylund, H. Linderholm & O. Rune, pp. 176–180. Luleå: Luleå Alltryck Ab.

Jenness, D. (1923). Physical characteristics of the Copper Eskimos. *Rep. Canad. Arct. Exp., 1913–18*, 7, 45–6B.

Kajanoja, P. (1972). A contribution to the physical anthropology of the Finns. *Ann. Acad. Sci. fenn. A. V. Med.*, 153, 1–12.

Kalow, W. (1962). Pharmacogenetics. In *Heredity and the response to drugs*, pp. 93–104. Philadelphia: W. B. Saunders Co.

Laurent, L. E. (1953). Congenital dislocation of the hip. *Acta chir. scand.*, Suppl. 179, 133pp.

166

Lehmann, W., Jürgens, H. W., Forsius, H., Eriksson, A. W., Haack, M., Junkelmann, B., Bahlmann, E. & Pape, G. (1970). Über das Hautleistensystem der Skoltlappen. *Z. Morph. Anthrop.*, **62**, 61–99.

Lewin, T. & Eriksson, A. W. (1970). The Scandinavian International Biological Program, Section for Human Adaptability, IBP/HA. Scandinavian IBP/HA Investigations in 1967–9. *Arctic Anthrop.*, **7**, 631–69.

Lewin, T. & Hedegård, B. (1971). Human biological studies among Skolt Lapps and other Lapps. *Proc. Finnish dental Soc.*, **67**, Suppl. 1, 63–70.

Loomis, W. F. (1967). Skin-pigment regulation of vitamin-D biosynthesis in man. *Science, Washington*, **157**, 501–6.

Lundberg, H. & Wahlund, S. (1932). The race biology of the Swedish Lapps. Part I. General Survey. Prehistory. Demography. Future of the Lapps. Uppsala: Almqvist & Wiksells Boktryckeri A.B.

Martin, L. M. & Jackson, J. F. (1969). Cerumen types in Chocktaw Indians. *Science, Washington*, **163**, 677–8.

Matsunaga, E. (1962). The dimorphism in human normal cerumen. *Ann. hum. Genet., Lond.*, **25**, 273–86.

Matsunaga, E. & Ebbing, H. C. (1956). Über Ohrschmalztypen bei Deutschen und Japanern. *Z. menschl. Vererb.-u. Konstit-Lehre*, **33**, 404–8.

McCullough, J. M. & Giles, E. (1970). Human cerumen types in Mexico and New Guinea: a humidity-related polymorphism in 'Mongoloid' peoples. *Nature, London*, **226**, 460–2.

Meier, R. J. (1966). Fingerprint patterns from Karluk Village, Kodiak Island. *Arctic Anthrop.*, **3**, 206–10.

Meier, R. J. (1973). Dermatoglyphic variation in five Eskimo groups from Northwestern Alaska. *IXth Intern. Congr. Anthropological and Ethnological Sciences, Inc. Chicago*.

Meier, R. J. (1974). Evolutionary processes in Eskimo dermatoglyphics. *Arctic Anthrop.*, **11**, 20–8.

Mellbin, T. (1962). The children of Swedish nomad Lapps. A study of their health, growth and development. *Acta paediat., Uppsala*, **51**, Suppl. 131, 97 pp.

Midlo, C. & Cummins, H. (1931). Dermatoglyphics in Eskimos. *Amer. J. phys. Anthrop.*, **16**, 41–9.

Monn, E. (1969). Further data on the genetics of the ABO–, MN– and PTC-systems of the Norwegian Lapps. *Hum. Hered.*, **19**, 678–83.

Mourant, A., Domaniewska-Sobczak, K. & Tills, D. (1975). *Sunshine and the geographical distribution of the Gc alleles.* Abstract of Communication, 18 Apri 1975. London: Society for the Study of human Biology.

Näätänen, E. K. (1936). Über die Anthropologie der Lappen in Suomi. *Ann. Acad Sci. Fenn. Med. A.*, **47**, 2.

Nakajima, A. & Hirano, I. (1968). Distribution and inheritance of ear-wax types A study on inhabitants in Awa-District, Chiba Prefecture. *Jap. J. hum. Genet.* **13**, 201–7.

Nerup, J., Platz, P., Andersen, O. O., Christy, M., Lyngsøe, J. & Poulsen, J. E. (1974). HL–A antigens and diabetes mellitus. *Lancet*, **2**, 864–6.

Norio, R., Nevanlinna, H. R. & Perheentupa, J. (1973). Hereditary diseases in Finland; rare flora in rare soil. *Ann. clin. Res.*, **5**, 109–41.

Omoto, K. (1970). The distribution of polymorphic traits in the Hidaka Ainu. I. Defective colour vision, PTC taste sensitivity and cerumen dimorphism. *J. Fac. Sci. Univ. Tokyo, Sec. V*, **3**, 337–55.

Palmén, K. (1961). Preluxation of the hip joint. Diagnosis and treatment in the new-

167

born and the diagnosis of congenital dislocation of the hip joint in Sweden during the years 1948–60. *Acta paediat., Uppsala,* **50**, Suppl. 129, 71pp.

Paul, J. R., Riordan, J. T. & Kraft, L. M. (1951). Serological epidemiology: Antibody patterns in North Alaska Eskimos. *J. Immunol.,* **66**, 695–713.

Pawson, S. & Milan, F. A. (1974). Cerumen types in two Eskimo communities. *Amer. phys. Anthrop.,* **41**, 431–2.

Petrakis, N. L. (1969). Dry cerumen – a prevalent genetic trait among American Indians. *Nature, London,* **222**, 1080–1.

Petrakis, N. L., Molohon, K. T. & Tapper, D. J. (1967). Cerumen in American Indians: Genetic implications of sticky and dry types. *Science, Washington,* **158**, 1192–3.

Petrakis, N. L., Pingle, U., Petrakis, S. J. & Petrakis, S. L. (1971). Evidence for a genetic cline in earwax types in the Middle East and Southeast Asia. *Amer. J. phys. Anthrop.,* **35**, 141–4.

Popham, R. E. (1953). A comparative analysis of the digital patterns of Eskimos from Southampton Island. *Amer. J. phys. Anthrop.,* **11**, 203–13.

Post, P. W., Daniels, Jr. F. & Binford, Jr. R. (1975). Cold injury and the evolution of 'white' skin. *Hum. Biol.,* **47**, 65–80.

Saatmann, J. (1971). Über das Hautleistensystem der Finger und Hände von Lappen. Kiel: Inaugural dissertation.

Schreiner, K. E. (1935). *Zur Osteologie der Lappen I & II.* Oslo: A. W. Brøgger.

Severin, E. (1956). The incidence of congenital dislocation of the hip and pes equinovarus in Sweden. *Nord. Med.,* **55**, 221–3.

Sewall, K. W. (1939). Blood, taste, digital hair and color of eyes in Eastern Eskimos. *Amer. J. phys. Anthrop.,* **25**, 93.

Simmons, R. T., Graydon, J. J., Semple, N. M. & Kodana, S. (1953). A collaborative genetical survey in Ainu: Hidaka, Island of Hokkaido. *Amer. J. phys. Anthrop.,* **11**, 47.

Simon, U. (1972). Über das Hautleistensystem von Grönland-Eskimos aus Aupilagtoq. Kiel: Inaugural dissertation.

Sunahara, S., Urano, M. & Ogawa, M. (1961). Genetical and geographic studies on isoniazid inactivation. *Science, Washington,* **134**, 1530–1.

Sundberg, S., Luukka, P., Lange Andersen, K., Eriksson, A. W. & Siltanen, P. (1975). Blood pressure in adult Lapps and Skolts. *Ann. clin. Res.,* **7**, 17–22.

Svejgaard, A., Platz, P., Rijder, L. P., Nielsen, L. S. & Thomsen, M. (1975). HL–A and disease associations – a survey. *Transplant. Rev.,* **22**, 3–43.

Thomsen, M., Platz, P., Andersen, O. O., Christy, M., Lyngsøe, J., Nerup, J., Rasmussen, K., Ryder, L. P., Nielsen, L. S. & Svejgaard, A. (1975). MLC typing in juvenile diabetes mellitus and idiopathic Addison's disease. *Transplant. Rev.,* **22**, 125–47.

Tiitinen, H., Mattila, M. J. & Eriksson, A. W. (1967). Isoniazid inactivation in Finns and Lapps. *Bull. Europ. Soc. hum. Genet.,* **1**, 77–8.

Tiitinen, H., Mattila, M. J. & Eriksson, A. W. (1968). Comparison of the isoniazid inactivation in Finns and Lapps. *Ann. Med. intern. Fenn.,* **57**, 161–6.

Tiitinen, H., Mattila, M. J. & Eriksson, A. W. (1973). Isoniazid inactivation in Finns and Lapps as demonstrated by various methods. *Amer. Rev. resp. Dis.,* **108**, 375–8.

Wessel, A. B. (1918). Laaghalte slegter in Finmarken. *Tidsskr. N. Laegefor.,* **38**, 337–68.

Wynne-Davies, R. (1970). A family study of neonatal and late-diagnosis congenital dislocation of the hip. *J. med. Genet.,* **7**, 315–33.

5. Craniofacial studies

ALBERT A. DAHLBERG

There have been seven craniofacial studies in IBP and they are representative of all the populations in the circumpolar regions. The groups included in the studies were the Ainu of Hokkaido, Japan (Hanihara, 1968), Eskimo of Igloolik, Canada (Mayhall, 1970a; Colby & Cleall, 1972, 1974); Eskimo of the Northern Foxe Basin, Canada (Colby & Cleall, 1972, 1974); Eskimo of Greenland (Jakobsen, 1970); Eskimo of Wainwright, Alaska (Dahlberg, 1968, 1969, 1970; Mayhall, 1970a, b; Hylander, 1972; Merbs, in Cederquist, 1975); Alaskan Eskimo (Bang, 1971); Lapps of Finland (Hedegard, 1971; Kirveskari, 1971, 1972, 1973); and Lapps and other northern population groups of the Soviet Union (Zoubov, 1971). Additional reports by members of these groups are many and include those by Alexandersen (1970, 1971), Bang (1972), Carlsson et al., (1971) Helkimo et al. (1971), Lewin (1971), Masuda, Tamada & Tanaka (in Hanihara et al., 1975).

At the November 1967 session of the Working Party Conference for the IBP Study of Circumpolar Populations, certain recommendations were made relating to studies, equipment and procedure. Not all the study groups have been able to follow these guidelines for various reasons. Some lacked the necessary equipment, personnel or consent for some studies, particularly those involving X-rays.

The resulting reports were far-reaching but many could not be used for comparisons. To bring the materials of the investigations together, the reports are presented in this chapter partially by direct quotations and partially by paraphrasing. Some comments by the present writer are added.

The data and discussions in the studies of the dental and craniofacial regions of the north are extensive. They are adequate for use in considering possible adaptive significance in the severe demands of the circumpolar climate and environmental conditions. Injected into all of this are the effects of foreign cultural aspects of nutrition and life which alter the selective directions the human organisms were taking. These in themselves have served to test the significances of the traits and structures studied.

The studies reflect the fact that the circumpolar peoples have their roots in two main groups, namely Mongoloid and Caucasoid, with a sharp separation of the two with the demarcation between Greenland and Scandinavia. The Lapps do of course demonstrate some traits derived from Mongoloid sources, but have more from the so-called Caucasoid. An increasing frequency of certain Mongoloid traits is described from west to east in the Soviet Union northland by Zoubov. The Ainu again are individualistic in many ways, but are preponderately Mongoloid.

Differences in frequencies of trait details vary widely between some population groups and are seemingly more significant for some groups in such characteristics as palate size and shape and the ratio of tooth dimension to the supporting boney base.

It was unfortunate that comparisons could not be made between many of the populations. Cephalometric roentgenograms were available only for Wainwright Alaskan, Northern Foxe Basin, and Igloolik Canadian Eskimos and for some skeletal precontact Eskimos from a northwest Hudson's Bay area and some from Point Hope, Alaska in the Smithsonian Institution collections. For other studies, though, enough material and data have been collected to give a useful comprehensive view of odontology in the Arctic regions. The materials of Zoubov had the advantage of a wide geographic and population range treated by a single base of trait classification.

Ainu (Japan)

Professor Kazuro Hanihara with T. Masuda, T. Tanaka and M. Tamada conducted the odontological studies of the Ainu. In their report (Hanihara *et al.*, 1975) they stated that 'although the Ainu are not necessarily a population which acquired high adaptability to cold, it is generally accepted that the Arctic Mongoloids acquired their highly adaptive characteristics to cold during the maximum stage of the last glaciation. This means that the other Mongoloid populations who lived in a mild climate might have retained some archaic characteristics to a greater or lesser extent. In this respect, it is quite probable that some of the characteristics of the Ainu might have been derived from the ancestral groups of the modern Mongoloids, and not from other racial stocks. If this hypothesis is accepted, the characteristics which are found in common among the Ainu and other Mongoloid populations may be reasonably explained.' They felt that much more additional evidence is necessary to prove this assumption. Based on the present findings, however, theories which attribute the origin of the Ainu to the Caucasoids or to the Australoids have to be rechecked from the viewpoint of the adaptability of this population. The large amount of data obtained through recent activity of the IBP may play an important role in unveiling the racial history of the Ainu.

Hanihara *et al.* based their studies on plaster dental casts collected from some 600 individuals in southern Hokkaido. They were mostly full-blooded Ainus and some hybrid Wajins (Japanese living in Hokkaido). Of these 105 individuals were selected whose rate of admixture was calculated to be 0.5 or less on the basis of pedigree analysis. The average rate of admixture of this group was estimated to be 0.23.

170

Selected dental traits of the Ainu and hybrids were compared with those of Australian Aborigines, Pima Indians, Eskimos and Caucasoids. The Ainu were found to be quite similar to the Wajin and Eskimos, but are very different from the Caucasoids. The biological distance between the Ainu and the Australian Aborigines is not large, but they are quite different from each other in frequency and development of the fovea anterior appearing in the lower second molar. In addition, the Ainu seem to differ quite a bit from the Pima Indians, but we find few striking discrepancies between these two populations as regards the frequency distributions of their crown characteristics.

On the basis of these studies it seems that the Ainu might have been derived from Mongoloid stock and have a common ancestor with neighboring populations such as the Wajin, Eskimos and American Indians. The same results were obtained from investigations of finger and palm-print patterns (Kimura, 1961), red-cell-enzyme systems and serum-protein groups (Omoto, 1972), and blood groups (Misawa & Hayashida, 1972).

In contrast to this, differences between the Ainu and the Aborigines, although not so large as those between the Ainu and the Caucasoids, are supported by osteological studies (Yamaguchi, 1967) and polymorphic characteristics (Omoto, 1972), so that the theory which attributes the origin of the Ainu to the Australoid stock seems to have little support.

The differences between the Ainu and the Wajin have attracted the attention of many anthropologists and anatomists. The similarities of the Ainu to the Caucasoids or the Australoids were seen in characteristics such as hairiness, long headedness, a simple pattern of the cranial sutures, well-developed glabella, deeply depressed nose root, widely projected cheek bones, relatively massive mandible, edge-to-edge bite and the flatness of the extremity bones of the Ainu.

The investigators felt that 'most of these characteristics are not necessarily indicators of a certain racial stock but are generally regarded as archaic characteristics'. They said that the relatively scant body hair and beard in the Mongoloids is likely to be one characteristic acquired in the course of adaptation to cold climate, though the Ainu are not necessarily a population that acquired high adaptability to cold.

In a personal communication Hanihara stated: 'Although the final conclusion of my work has not yet been obtained, multivariate analysis showed that morphology of the dentition of the Ainu was very close to those of the Japanese, relatively similar to those of the Australian Aborigines, Amerindians and Eskimos, and very far from those of the Caucasians and American Negroes. This trend is quite parallel to the result obtained from analyses of the blood groups, serum-protein types, dermatoglyphics, etc., and seems to give us important basic knowledge to solve the problem on the origin of the Ainu.'

The human biology of circumpolar populations

Eskimo of Canada

The efforts of the Canadian group of the IBP have been directed towards the collection and analysis of morphological, epidemiological, taste sensitivity and craniofacial growth data by John Mayhall, University of Toronto and J. F. Cleall, University of Manitoba. These data-collecting efforts have been mainly at Igloolik and Hall Beach with some previously collected material housed at the University of Manitoba relating to craniofacial growth by J. F. Cleall and W. B. Colby.

In formal annual reports No. 3 and No. 4 of the Canadian IBP and in an additional separate report in 1973, Mayhall summarizes his findings on epidemiological, morphological and taste sensitivity data.

Mayhall found that most discrete dental traits studied fell within normal limits for Eskimoid populations previously studied. The metric analysis of tooth size also showed the same trends. Studies continue on tooth-loss patterns and tooth-emergence timing.

The males from Igloolik and Hall Beach were reported as having a lower percentage of anterior teeth lost in the mandible than did the ancient Thule culture groups. However, the number of missing third molars was reversed. This seemingly larger third molar loss may be explained by the inability in earlier studies to ascertain accurately which teeth were congenitally missing and thus eliminate them from the sample. The higher rate of caries in recent populations might also be responsible. Tables XIII–XV of the Canadian IBP Progress Report show the changes in the dental caries rates for the four-year period 1969–73. During this time professional dental care was minimal and preventive measures were sporadic at best. The caries rate of 8.25 at Hall Beach is a 53% increase over the incidence of 1969, while the increase at Igloolik is 71.12%.

Comparisons of the total mean caries experience for Hall Beach and Igloolik reveal that while Igloolik still has a lower caries rate it is rapidly approaching the levels seen in Hall Beach. This may reflect the patterns of acculturation and the availability of southern food. Four years ago, Hall Beach had more accessibility to caries-producing food through the DEW-line site, while Igloolik was somewhat more isolated in this regard. There was also a greater number of Eskimos employed in Hall Beach. In the intervening years Igloolik has had greater amounts of food shipped in through the Hudson's Bay Company and the Cooperative, and a much higher percentage of men are employed, at least part-time.

Taste sensitivity

Mayhall (1970a) found a highly significant statistical difference between the Igloolik Eskimos and the University of Toronto students in their ability to taste sucrose ($\chi^2 = 80.7297$, d.f. $= 4$).

172

Craniofacial studies

Colby & Cleall (1974) have analyzed the cephalometric X-rays of the Eskimos of the Northern Foxe Basin living in the settlement of Igloolik.

They used linear and angular measurements to evaluate and compare the craniofacial development of this relatively isolated population. Development of the craniofacial region appeared to follow normal trends as for other populations. Many morphological differences were observed in the craniofacial complex of this group when compared to a Caucasian control population. The most notable of these were:

1. Nearly all linear measurements of the craniofacial complex were larger in the Eskimo population.
2. The greater overall size of the Eskimo craniofacial complex was established before three years of age and in all probability was present at birth.
3. Greater cranial depth was found to be due to expansion of the occipital region in a posterior direction rather than to concomitant increases in anterior and posterior cranial depth.
4. Linear measurements of anterior cranial depth, anterior cranial base, and mid-facial depth were all smaller in the Eskimo population.
5. Mandibular corpus length was greater in the Eskimo group, but the anterioposterior position of the mandible in relation to the anterior cranial base was not prognathic.
6. The chin point, pogonion, was more recessive in relation to the NB plane in the Eskimo population.
7. The maxillary and mandibular incisors were significantly more procumbent in the Eskimo group, and together with the anteriorly positioned maxilla produced a marked bimaxillary protrusion.

Eskimo of Greenland

Jan Jakobsen of the Royal Dental College of Copenhagen is pursuing studies in odontology including morphology, microbiology, epidemiology and dental health programs. He has not obtained cephalometric data but is initiating a course of action and inquiry in an effort to obtain such materials.

Morphology

Trait analysis of expression of morphological units (Carabelli's cusp) is being pursued by Jakobsen.

Dental casts were made on 49 Thule Eskimos of K'anak' in 1962 by A. and T. Dahlberg and supplemented by others made by Sorenson in 1964. Unpublished data and reports by Dahlberg, Kobayashi and Dahlberg include the following findings:

The shovel-shaped characteristic of the central and lateral upper incisors was present in a great degree and extent. However, the lingual lateral rims are not as prominent and large as are many of the Indian populations of the

American continent generally. 16% (8) showed no shovelling which may be accounted for by admixture. The shovel-shaped feature was measured by relative depth of the sulcus on the lingual surface as compared to the lateral rims on the mesial and distal aspect of the same surface.

In the lower molar occlusal surface patterns the inhabitants of K'anak' showed the traditional high frequency of five cusped first molars with the distribution trending toward the four cusps in the second molars and back to an intermediate frequency distribution of the two forms in the third molars. In both the metacone and the hypocone reductions, the trend was toward greater reduction in the second and third molars than is seen in most populations.

Carabelli's cusp was notably missing in all but one individual, he having a high proportion of Danish ancestry. There was a general distribution of the smaller phenomena of pits, lines, grooves and eminences in the area but not of cusps. This conforms with the findings of P. O. Pedersen and others relating to the extremely low incidence of Carabelli's cusp in other groups in southwest Greenland.

Only two individuals showed any suggestion of the protostylid phenomena and these were of the small category. Many showed a large pit at the base of the buccal groove but in 70% or more, both male and female, individuals showed no protostylid structures whatsoever.

Similarly to what is seen in certain other related groups, no instances of torus palatinus were observed in the K'anak' sample. The torus mandibularis was evident in a very marked form in only one male, in moderate manifestation in one female and two males and in a slight form in one female. The mandibular torus was missing completely in 87.8% of the sample investigated.

Epidemiology

Dental caries is in high frequency among most primitive populations that have come in contact with the foods and ways of civilization. The dentitions of the north Greenland, Thule, Eskimo have been well cared for by the dental service of the Greenland Ministry Public Health authorities. This is strongly in evidence and perhaps additionally, the retention of some of the ancient habits of food and culture may have been instrumental in holding the DMF rate considerably lower than what it was observed to be in any of the corresponding age groups in the Kodiak and Alaskan native populations.

In the last 40 to 50 yr the incidence of dental caries in the Greenland population has increased at an incredible speed and has become a serious public health problem. Jakobsen & Hansen reported (1974) that the prevalence of caries in first and seventh grade children in 14 communities was 1.5 to 2 times higher than in Denmark. Untreated caries was 4 to 5 times more frequent in Greenland than in Denmark. Isolated communities, previously almost without caries, now showed the same high prevalence as in other parts of Greenland. They stated that there was no doubt but that the increased frequency of caries was due to a steep increase in consumption of sugar and sweets. The study revealed that women had more caries than men, a difference already apparent at the age of 14 to 15 yr. In areas where the drinking water contained 0.5–0.6 ppm fluoride, the caries prevalence was lower when compared with the low fluoride areas.

According to Jakobsen & Hansen (1974), all people in Greenland receive free dental treatment. Today there are 18 dental clinics with 31 units, 20 full-time dentists, and three dentists working only in the summer period. There are 36 chairside assistants and five laboratory technicians. In Greenland all dentists and auxiliaries are employed by the Ministry for Greenland. Mobile equipment for boats is available. The population of Greenland is 48 000 (40 000 born in Greenland and 8000 Danes). The ratio is one dentist to 2300 inhabitants. In south Denmark there is one dentist to 1470 inhabitants. The costs for dental care in Greenland were 6 000 000 Danish Kroner in 1973, which is 7.5% of all the expenses for health in that part of Denmark. The equipment has been standardized and is serviced by experts from Denmark.

Every second year the Ministry for Greenland arranges courses for the dentist by bringing faculty members from Denmark. Courses have also been arranged for the chairside assistants. In 1974 great emphasis was placed on preventive measures. Progress has been made in recent years by expanding facilities for dental treatment in Greenland.

Eskimo of Wainwright, Alaska

The odontological study group of the United States working in Wainwright, Barrow, and Point Hope, Alaska pursued investigations of craniofacial growth and morphology (cephalometric X-rays), measurements, oral epidemiology, tooth eruption sequence and timing, dental morphology (dental casts), and typing of salivary samples (Wainwright only).

Craniofacial growth and morphology

The first 1968 cephalometric X-rays of the Wainwright population were made by a unit under the direction of Dr Dotter of the University of Oregon, Portland. In subsequent years the cephalometric X-rays under the direction of A. Dahlberg of the University of Chicago with J. T. Mayhall, T. Dahlberg, P. Walker and C. F. Merbs (1969–72) used the techniques of special head-film orienting devices with precise target-sagittal plane-film distance of 60 inches (150 cm) recommended by orthodontic craniometric workers. These were useful for comparisons and studies of craniofacial growth and description. The data study has recently been done by Robert Cederquist in the form of a Master's thesis (Cederquist, 1975).

175

The human biology of circumpolar populations

The roentgenographic cephalometric investigation revealed:

1. Great sex and age similarities in shape of the craniofacial complex.
2. Differences in size of the craniofacial area, in that all linear measurements were greater in males than in females, resulting in a larger male face both vertically and in depth.
3. That females tend to have more dental protrusion.
4. That anterio-posterior differences in maxillary and mandibular apical base relation are greater in males in the younger age groups. In the older groups the tendency is reversed, which is however not significant.
5. That there is a tendency for males to show more midfacial flatness than females.
6. That the degree of total facial flatness is similar in both sexes.

The investigation also revealed that the most notable changes that occur with age are:

1. Increase in all linear dimensions for both females and males.
2. The posterior facial height increases relatively more than the anterior facial height causing an anterior rotation of the mandible. This phenomenon is more pronounced in males than in females.
3. Mandibular basal prognathism and mandibular base position are increasing in both males and females.
4. Reduction in apical base relation (ANB-angle) in both sexes.
5. Both males and females show reduction in horizontal overbite (overjet).
6. The gonial angle becomes less obtuse in both sexes.
7. Decrease in the axial inclination of maxillary and mandibular incisors in males.
8. In both females and males the degree of both midfacial flatness and total facial flatness is increasing.

The craniofacial morphology of the adults in the Wainwright sample was also compared with craniofacial data from two roentgenographic cephalometric studies on adult Eskimo skeletal material. One of these comprised pre-contact material from northwest Hudson's Bay areas (Hylander, 1972) and the other consisted of skulls from northwest Alaska dated A.D. 1860 (Dahlberg, unpublished observations).

The Wainwright Eskimos and the skeletal material showed great similarities in craniofacial morphology, e.g. degree of maxillary prognathism, degree of midfacial flatness, most linear dimensions. However, some distinguishing features were also observed:

1. The contemporary Wainwright adult Eskimos have less obtuse cranial base, and the Wainwright males have a longer posterior cranial base than do males in the skeletal material.
2. There is a positional difference of the mandible, being more retrognathic in the skeletal material.
3. The difference in maxillary and mandibular apical base relation is less in the Wainwright Eskimos.
4. The Wainwright group show more labial inclination of maxillary and mandibular incisors than do the Alaskan skulls.
5. The anterio-posterior dimension of the frontal sinus is greater in the Wainwright group than in the Hudson Bay material.

176

Measurements

For all teeth except the maxillary second premolars, the mesiodistal crown diameters of the Wainwright Eskimo males are larger than those of the Wainwright females. But in the maxilla, except for the first molars, no significant differences are indicated. Significant differences exist between the sexes in the first molars, both in the maxilla and the mandible.

The Alaskan Wainwright Eskimos have larger mesiodistal crown diameters in each tooth than occurs in the American Whites. The differences between the mean mesiodistal crown diameters of the Wainwright Eskimos and the Aleuts were calculated. These comparisons were made in separate groups of males and females. There are large differences between the mean values of the Alaskan Wainwright Eskimos. Their teeth are larger than are those of the Aleuts in both the male and female groups.

It is recognized that there is a functional advantage of large teeth over small ones, selective within individual dentitions for one or more tooth regions or groups. Wainwright Eskimo teeth range much larger than Aleuts and some others. Similarly, greater efficiency can be ascribed to thicker, sturdier enamel and teeth with more crenulations. Such structural entities though may have limited advantages, differing with the amount and character of wear to which the teeth are subjected. Differential wear increases the number and nature of the cutting edges on the occlusal enamel. The extent of meat and fish eating in the Arctic results in rounded edges of the teeth rather than the sharp ones produced by abrasive loaded foods. All the groups of the circumpolar regions have been subjected to heavy tooth wear for cultural reasons and because of the natures of their foods.

Certain morphological features such as the shovel shape in combination with large tooth size and bone support for the teeth are significant. More shovel shape exists in the large teeth of the Eskimo and Ainu than in the Lapps. The moderately worn shovel-shaped tooth has a decided advantage in cutting edges and shearing capability than does the non-shoveled tooth.

The Alaskan Wainwright Eskimos have a larger mesiodistal crown diameter in both sexes than for example do those of the Japanese.

Oral Epidemiology

During July of 1968 all available residents of Wainwright were examined by John Mayhall for evidence of oral pathology and oral hygiene. The high dental caries rates summary for this group was published earlier (Mayhall, Dahlberg & Owen, 1970*a*). Another paper presented the results of the examination for torus mandibularis (Mayhall, Dahlberg & Owen, 1970*b*).

Table 5.1. *Caries experience in Wainwright, Alaska, Eskimos (decayed, missing and filled teeth per individual)*

Age (yr)	Males				Females			
	No.	Decayed[a]	Missing	Filled[a]	No.	Decayed[a]	Missing	Filled[a]
0–10	37	3.27	0.11	1.59	44	3.73	0.16	1.84
11–20	22	1.59	1.41	7.82	28	1.32	1.75	10.53
21–30	8	6.38	11.00	8.25	9	3.67	15.00	6.11
31–40	13	2.62	14.85	1.62	15	2.60	19.40	2.13
41–50	10	1.10	16.70	1.00	3	5.67	14.33	2.67
51+	16	2.19	20.12	0.94	13	0.31	25.46	0.62
Total	106	2.71	7.59	3.24	112	2.63	7.64	4.28

[a] Decayed and filled permanent and deciduous teeth per individual.

178

Dental Caries

The examination was made with a mirror and explorer aided by a battery-operated spotlight. No radiographs were available. In order to summarize the total caries experience of the population, decayed and filled permanent and deciduous teeth were combined. Only permanent teeth are included in the category of missing teeth. Inspection of Table 5.1 reveals the efficacy of the US Public Health Service's priorities of dental treatment. The USPHS has concentrated upon school age children for the last two to fifteen years. This is reflected in the relatively low rates of decayed teeth in the under twenty years of age groups. It can also be seen that the number of filled teeth is large in the 11–20 yr and 21–30 yr groups, indicating a large amount of dental treatment in the past. However, this latter group (21–30 yr) now exhibits a high number of decayed teeth, indicating that preventive measures employed earlier have lost their effectiveness.

The number of missing teeth reaches appalling proportions at an early age. The low periodontal rates would indicate that the vast majority of the missing teeth are the result of caries. There is no doubt that dental caries is the most prevalent disease in this group. The oldest male who is caries free is only six years of age, while the oldest female is only three. Of those over the age of thirty, twenty-three are edentulous.

Mayhall states that 'In both the males and the females the peak of caries involvement occurs in the 21–30 age group with a leveling off in later years in the females and a drop in the DMFT in the males. This latter drop is indicative of dietary changes, with the younger population consuming high proportions of cariogenic food and the older people demonstrating less exposure to this type of food and, instead, subsisting primarily on "native" food.

The dietary effects on the rates of dental caries in Eskimos have been demonstrated by Russell and his co-workers (1961), Bang & Kristoffersen (1972), Mayhall (1970), and Möller and co-workers (1972). The first two of these studies were conducted in Alaska and showed conclusively that a change from a diet of locally available food to a diet of commercially processed food results in an increase in the caries rate. The same general trends have been seen in the Northwest Territories of Canada (Mayhall) and Greenland (Möller *et al.*).'

Periodontal Disease

Mayhall reported that periodontal disease is extremely rare in all age groups in Wainwright (Table 5.2). The great majority of patients evidenced only a slight gingival inflammation around single teeth. These results are surprising in light of the results from other Alaskan Eskimos reported by Kristoffersen & Bang (1973). They reported much higher level of the disease after reporting

Table 5.2. *Periodontal indices of Wainwright, Alaska, Eskimos*

Age (yr)	No.	Male Indices	No.	Female Indices
0–10	21	0.02	34	0.01
11–20	21	0.06	28	0.03
21–30	8	0.02	8	0.08
31–40	11	0.40	13	0.10
41–50	7	0.97	3	0.15
51+	10	0.54	4	0.02

a low level in 1957. Russell and co-workers (1961) also noted higher levels which may indicate that the criteria utilized in this study were different from the other studies. However, a comparable study of Canadian Eskimos (Mayhall, 1976) demonstrated higher levels in the culturally less advanced Foxe Basin people. This study was accomplished by the same individual who conducted the Wainwright oral examinations. There was 'no significant increase in disease levels in the four-year interval of the study in Canada, although there had been a tremendous shift in diet toward processed food and away from the native foods. A cross-sectional study of the same Foxe Basin groups earlier in which the subjects were divided into groups according to their dietary intake failed to reveal significant differences in periodontal disease' (Mayhall, 1970).

Oral Hygiene

Mayhall used Greene & Vermillion's (1960) simple system for quantifying the amount of soft debris and supragingival calculus on the teeth (Table 5.3). This system records the number of thirds of the tooth covered by these two contaminants. Mayhall reported that in Wainwright there was three to four times more debris on the teeth while the calculus deposition was comparable to the Canadian Eskimo values.

'The Eskimos' primary problem with their teeth is the onslaught of dental caries', according to Mayhall. 'A disease that was unknown in the Eskimos only a few decades ago now is the most prevalent disease in Arctic populations. While it is impossible to ascribe an exact causation, there can be no doubt that the great change in diet of the Eskimo has been a major element. Studies from throughout the Arctic over the past twenty years have conclusively shown that the dietary of the southern resident is ill-suited to the Eskimo dentition. This could be partially the result of the unique shape of their teeth (Mayhall, 1972) which contain a number of built-in food traps.

While the Wainwright Eskimos do not as yet show the high levels of periodontal disease, it will probably be only a short time before they reach the

Table 5.3. *Oral hygiene of Wainwright, Alaska, Eskimos*

Age (yr)	Males			Females		
	No.	Calculus index	Debris index	No.	Calculus index	Debris index
0–10	22	0.0	1.00	34	0.0	0.64
11–20	21	0.01	0.71	28	0.0	0.75
21–30	8	0.36	0.69	8	0.12	0.73
31–40	11	0.82	1.31	13	0.55	0.86
41–50	7	0.87	1.30	3	0.45	0.66
51+	10	0.74	0.90	4	0.88	0.82

levels seen in some other Alaskan Eskimos. The beginning of this may be occurring now with the deterioration of the oral hygiene of these people.

The Eskimo of a few years ago with his legendary good, strong teeth is now being replaced by one with large numbers of missing or diseased teeth, certainly not the trend we would like to see continue.'

Intermaxillary relationships

A very high percentage of the Wainwright population have a mandibular prognathism. This becomes a problem in the construction of dentures. In many of the children as young as two and three years of age an edge-to-edge bite (Class II, malocclusion) existed, suggesting a genetic rather than a developmental, functional or environmental entity.

Morphological traits of the dentition

The morphology of the teeth of the Wainwright Eskimos is similar to other Mongoloid groups such as the American Indian and Asiatics, but differs markedly from white and Negroes. Shovel-shaped incisors are in great evidence.

The protosylid is found in various categories of expression and particularly the undulating surface type. They are practically all bilateral.

Several instances of bilateral, barrel-shaped, upper lateral incisors are to be found in the population. There are also two instances of bilateral occurrence.

Carabelli's cusps are quite general but only in the very small expression type, frequently represented only by the furrows.

Two individuals showed the enamel pearls bilaterally in the upper second premolars.

The human biology of circumpolar populations

Saliva analysis for secretion of blood group substances

Analyses were done on 221 individuals of the Wainwright Alaskan Eskimo population by D. Owen. Anti-sera were tested by trial runs on saliva from secretor and nonsecretor individuals of known ABO blood groups.

All 221 specimens were from secretor individuals; 55.6% belonged to group A, 31.7% to group O, 7.7% to group B, and 5.0% to group AB. All but six cases agreed with field blood typing data.

Eskimo of Alaska

G. Bang of the University of Bergen, Norway, has reported on morphological and epidemiological studies of the Alaskan Eskimo (Bang & Hasund, 1971, 1972, 1973).

The object of his studies was to investigate: (1) the effects of various conditions on the development of dental decay and periodontal disease, (2) the possible effect of a high intake of seafoods on the incidence of dental caries, (3) whether the low caries incidence in primitive-like Eskimos could be attributed to a superior developmental microstructure of the teeth, (4) the morphology of the teeth.

Bang reported that 'During 2 years of a 10-year study period of 280 Eskimos, representative families were subjected to a detailed dietary study and samples of their food and water analyzed. Ground sections of teeth from contemporary and ancient Eskimo skulls were prepared for microscopy. Tooth morphology was studied on casts made from impressions taken in alginate. The study confirmed that there is a close relationship between the length of time the Eskimos have been exposed to western civilization and the caries incidence. Marked changes in caries susceptibility can occur in a relatively short period of time. A high intake of salt water fish had no appreciable effect on the incidence. All the teeth examined showed faulty microstructure.'

He concluded that the coastal and inland Alaskan Eskimos possess a faintly developed Carabelli's cusp in 42.7%. 'No sex difference in the occurrence of Carabelli's cusp was evident and no family showed any difference in the distribution of the frequencies when each family was compared to the rest of the population.'

He suggested that the general tendency toward a higher prevalence of Carabelli's cusp in the coastal Eskimos may be due to more admixture with white people. 'Alaskan Eskimos have a significantly higher frequency of Carabelli's cusp than do Aleuts. Statistical evaluation revealed that in the Alaskan dentition the formation of Carabelli's cusp is independent of the size of the molars and the suppression of the third molars.'

Of the first molars, 97% had four cusps, only 39.6% of the second molars and 15.2% of the third molars, the reduction occurring by elimination of the hypocone. No statistically significant sex difference in the trend towards reduction in the cusp numbers was found.

Bang felt that the higher occurrence of four cusps in upper second molars and certain other features in the coastal group was due to the 'pronounced racial admixture of white people.'

He stated that 'Alaskan Eskimos have a tendency towards a lower frequency of four cusps on all three maxillary molars than Aleuts. Only the second molar exhibited a statistically significant difference in this respect.' He claimed by a statistical evaluation that the reduction of cusps in the first and third molars is independent of the size and form of the molars. For the second molar, however, 'the teeth with four well-developed cusps showed a significantly larger buccolingual diameter'.

Lapps of Finland

The Scandinavian group of odontological scientists was headed by Bjorn Hedegård and Thord Lewin with a staff including Kirveskari, Hansson, Karlsson, Herkimo, Carlsson, and others. Their studies were concerned with morphological features, oral diseases, function of the masticatory system, bite forces, alveolar bone changes, chewing efficiency, growth and nutrition.

Field studies and dental cast collections of the Finnish Lapps were the subject of the academic dissertation on morphological studies of Pentti Kirveskari (1974). He pointed out that the Skolt population has varied extensively during the last 300 to 400 years, from near extinction to 515 in 1967. Hybrid Skolts had numbered fewer than 30 until the 1950s, but in 1967 had increased to 115. The number of hybrids was in fact greater than the census figure from corrections based on serological findings.

The loss of teeth due to dental decay followed the pattern of other studies in the change from aboriginal to 'modern' culture.

Kirveskari proposed the idea that 'the population structure of Skolts may have a bearing on the occurrence and variability of the dental traits'. He thought that genetic isolation of Skolts had continued long enough to permit genetic drift and selection to bring about considerable changes in gene frequencies. He attributed this to the small number of founders and the several population bottlenecks. He pointed out that phenotypic variability of the whole population may even increase, as shown for instance in the Tristanites (Bailit, 1966).

He made a point of the considerable variability in tooth measurements between Norwegian Lapp populations (Selmer-Olsen, 1949). He also discussed the sub-population status of groups of Skolts of northern Finland and the differences of dermal systems as well as of dental traits between them.

He suggested that the Carabelli trait, cusp number of mandibular M1, and the sixth cusp vary significantly and could serve as useful discriminators between related groups (Rosenzweig & Silberman, 1967; Smith, 1973). In cases of admixture or less close relationship, he thought along with other authors that the Carabelli trait, cusp number of molar teeth, groove pattern on mandibular M1, sixth cusp, protostylid, and shovel shape also appear to have discriminating power (Turner, 1967, 1969; Alexandersen, 1970; Bang & Hasund, 1971, 1973).

Lapps, who are very much on the borderline as regards the presence or absence of the shovel and Carabelli's traits were considered by him to be a problem in repeatability of class identification. Detailed classification trait analysis increases the discriminating power considerably (Bang & Hasund, 1971; Turner & Scott, 1973). In his Lapp study, both subjective and multi-classed evaluation of the shovel shape and the metric depth of the fossa were presented as having discriminating power. Considering the trace shovel as non-affected gave the best discrimination with the present–absent classification. It is remarkable that the distinction was better between Ne- and Se-Skolts than between Skolts and Hybrids. The depth of the lingual fossa on maxillary lateral incisors was different between Se- and Ne-Skolts but not between Skolts and Hybrids. The situation was the opposite for the maxillary central incisor. Apparently, differences in shoveling, if any, are to be found on maxillary lateral incisors and on mandibular incisors and canines when closely related populations are compared. Admixture is likely to bring about changes on the stable maxillary central incisors.

Kirveskari reported that Skolt Lapps showed a combination of dental traits that is typical neither of the Caucasoid nor of the Mongoloid race. If the origin of Skolt Lapps can be traced to either of these races, it is more likely to be found in the Caucasoid race. The apparently Mongoloid elements of their tooth morphology may well be the result of drift during the long isolation, pleiotropic effects of other traits associated, for instance, with cold adaptation, early admixture, or any combination of these. The basic pattern of their anterior as well as posterior teeth, however, clearly resembled that of the Caucasoid race.

'The Hybrid population showed frequencies of dental traits that in all likelihood are between those of the parent populations. Other Lapps differ considerably.'

The large frequency of dental caries reduced the number of teeth available for effective and significant study of the three sub-populations. Also as a rule the number of morphological classes used had to be reduced in order to make them large enough for meaningful comparisons.

'The shovel-shape and the depth of the lingual fossa of anterior teeth showed discriminating power; the two Skolt populations differed with regard to the maxillary lateral, and mandibular canine, lateral and central incisor. None of these teeth showed significant differences when all Skolts were compared to Hybrids but, instead, the maxillary central incisor differed moderately between these groups.

The labial surface morphology of maxillary incisors did not differ between the Skolt populations. However, the central incisor of Hybrids was clearly different from that of Skolts.'

The occurrence of lingual tubercles on canines and incisors was generally similar in the subpopulations, except for the lateral incisor, which differed significantly between Se-Skolts and Ne-Skolts.

'The degree of hypocone reduction on maxillary second molars showed a statistically significant difference between Skolts and Hybrids. It was not possible to draw conclusions from the distributions of other molar traits.

From a racial viewpoint, the frequencies of Carabelli's cusp and the deflecting wrinkle of the metaconid conformed with the Mongoloid dental complex. On the other hand, the reduction of the cusp number of mandibular molars and the high degree of modification of the fissure pattern represented typically Caucasoid features in the dentition of Skolts.

The shovel shape of anterior teeth was absent or only vaguely expressed in the majority of Skolts. Semi-shovel and shovel forms were infrequent. The depth of the lingual fossa was slightly greater than in Caucasoids but remarkably less than in Mongoloids.

The relatively low frequency of the sixth cusp and the absence of the distal trigonid crest in Skolts were other features reminiscent of Caucasoid dentitions.

No racial conclusions could be drawn from the occurrence of lingual tubercles on anterior teeth, from the labial surface morphology of maxillary incisors, from the reduction of maxillary molar cusps, and from the occurrence of the seventh cusp, protostylid, and occlusal tubercles of premolars.

On the whole, Skolt Lapps were found to be distinctly different from both Caucasoids and Mongoloids. They possess a highly individual dental trait complex. If Lapp origins are traceable to the Mongoloid or the Caucasoid racial stock, the dental traits of Skolt Lapps suggest the latter alternative.'

Peoples of the north (Soviet Union)

A. A. Zoubov (1971) reported on eight morphological traits of the dentition of six populations in the north of the Soviet Union (Table 5.4). His materials were the results of laboratory studies of casts made from 'accurate dental wax prints' and determination of 'some features with the dental mirror'.

His results indicate a considerable variability of the dental features in the Soviet Arctic populations.

He combined the high and middle degrees of the shovel shape of incisors (semi- and full-shovel categories) and presented percentages for only the central or first permanent upper incisor. The lowest percentage, 3.6%, is

Table 5.4. *Frequencies (in percentage and radians) of dental characters in several populations of circumpolar area*

	Scolt Lapps (132)	Northern Lapps (49)	Kola Lapps (124)	Kola Russians (44)	Kola Komi (55)	Nenets (50)
Shovel-shaped first upper incisor	18.8 0.89	31.3 1.19	23.4 1.03	3.6 0.38	11.5 0.69	56.0 1.69
Carabelli's cusp on the first upper incisor (all degrees together)	12.7 0.73	13.3 0.75	28.3 1.12	34.6 1.26	34.0 1.25	16.0 0.82
Types 3 and 3+ of the second upper molar	71.3 2.01	68.5 1.95	54.9 1.67	56.3 1.70	73.3 2.06	56.5 1.70
6-cusped first lower molar	7.8 0.57	12.0 0.71	7.1 0.54	0.0 0.00	7.1 0.54	8.3 0.58
4-cusped first lower molar	16.2 0.83	16.0 0.82	7.0 0.54	16.0 0.82	0.0 0.00	12.0 0.71
4-cusped second lower molar	77.6 2.16	74.2 2.10	73.5 2.06	94.1 2.65	74.1 2.10	39.1 1.35
Uninterrupted distal trigonid crest on the first lower molar	3.2 0.36	6.2 0.50	7.1 0.54	4.16 0.41	4.5 0.43	25.0 1.05
Deflecting wrinkle of the metaconid on the first lower molar	32.5 1.21	30.0 1.16	32.7 1.22	25.0 1.05	13.9 0.76	29.0 1.38

seen in the Kola Russians with three Lapp groups intermediate and the highest, 56%, for the Nenets of Siberia.

The frequencies of the Carabelli's cusp range nonspecifically between 12.7% and 34.6% with all categories of this trait bulked together.

The distal trigonid crest which separates the two trigonid cusps from the three posterior talonid cusps on lower first permanent molars is of considerable interest. A frequency gradient for this trait runs from west to east in the Soviet Union with the lowest, 3.2%, in the Skolt Lapps and the highest, 25%, in the Nenets.

Four cusped second lower molars are seen in 39% of the Siberian Nenets and in 94% of the Kola Russians.

Zoubov used a graphic method (Fig. 5.1) of illustrating his ideas of 'eastern versus western tendencies' (Mongoloid versus Caucasoid he does not use). This method pitted his interpretation of eastern complex traits (shovel shaped incisors, distal trigonid crest, deflecting wrinkle) against the western (Carabelli's cusp, four cusped first and four cusped second lower molars, and reduced hypocones in the second upper molars). He points out the fact that the 'percentages of the eastern component is very different in the Circumpolar area, being lowest in Russians, a little higher in Komi, still higher in

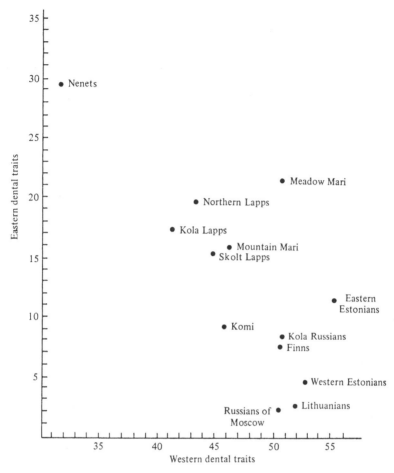

Fig. 5.1. The average percentage of eastern and western dental traits in circumpolar groups compared with some other populations (Zoubov, 1971).

Lapps and highest in Nenets'. He goes on to say that 'there is closeness between the groups which are alike also in their general racial type, which depend on the combination of the peculiarities of two great racial stocks. On the graph (Fig. 5.1) three main groupings occur. The ones having the lowest percentage of Mongoloid admixture in racial type also have the lowest percentage of eastern traits in dental type. In the middle of the graph are situated the groups with the higher level of eastern admixture. Here are all three groups of Lapps and two groups of Mari. It is known that these populations were found to be close in their racial type and have practically the same percentage of Mongoloid admixture. They are alike also in their dental type.

187

The human biology of circumpolar populations

Finally, the Nenets have the most pronounced Mongoloid racial type and also have the highest percentage of eastern traits in the dental complex. So the dental type is associated with the general racial type and depends upon the origin hybridization, and other genetic factors much more than upon the similar environmental conditions of the Circumpolar area.' He states that 'the dental features can be used as good additional criteria for interpopulation comparisons' and contends that 'many other examples might be given of use of dental morphology for study of interpopulation differences on the basis of data obtained in other regions of our and other countries'.

Summary

The studies described in this report included morphological, physiological and developmental descriptions and the epidemiological evaluations of the circumpolar regions concerned in IBP. These entities followed the same general paths of development and specialization in structural advantage witnessed by their gene root sources in the world populations. Local small group advantages by inbreeding, results of founder's effects, and nutritional and environmental results are to be found but are not of special relevance to cold adaptation. None of the group reports on craniofacial and dental studies made any suggestion or claim of adaptive significance of traits to survival in the cold north.

The often proposed idea that the protrusive flat face of the Eskimo is adapted to withstand cold climate does not seem to be borne out by investigations. Laboratory experiments of Steegman (1967, 1970) rejected this hypothesis. His experiments showed increased malar surface temperature as the face width increased. Fatty tissue padding may protect deeper structures but permits surface temperature to fall.

The Wainwright and the Northern Foxe Basin samples both have a similar cranial base flexion and maxillary prognathism. Colby & Cleall (1974) found the mandible to be moderately retrognathic in their study. The Wainwright sample, on the other hand, exhibit a well-developed and forward positioned lower jaw, a finding which is consistent with anthropometric descriptions of Eskimo mandibular morphology (Hrdlicka, 1928; Oschinsky, 1974). In both the Wainwright and the Northern Foxe Basin samples the gonial angle was observed to become less obtuse with increasing age. The size of the gonial angle was in both samples well within the range of what had earlier been reported for Eskimo skeletal material from Alaska, both precontact and from the middle of the last century (Cameron, 1923; Hrdlicka, 1940; Dahlberg, unpublished observations; Dahlberg *et al.*, 1978).

Despite some contrasting features, great similarities exist in the craniofacial area between ancient skeletal materials of Point Hope and Hudson's Bay and the modern populations. The main differences between female

188

groups can be summarized as increased cranial bass flexion and larger frontal sinus in the Wainwright sample. In males, the Wainwright material also showed more cranial base flexion, longer posterior cranial base, larger frontal sinus and a more prognathic mandible than the Canadian Eskimo skulls did.

The lifetime functional potential of the dentition is less in groups having congenital or otherwise missing teeth. Small interbred groups of Eskimos did have such a problem. It is not unusual though to see older individuals doing a very good job of mastication with only one or two badly worn opposing teeth in the mouth. Dental caries was not the problem in the past that it is today, but excessive wear was.

The morphology, growth and other nutritional and developmental studies of the craniofacial components of the circumpolar peoples contribute to understanding tissue responses and occlusal functional relationships in their broad environmental backgrounds. However, they do not seem to give any proof or indication that they were significant in the adaptation of these populations for survival in the cold north.

References

Alexandersen, V. (1970). Tandmorfologisk variation hos eskimoer og andre mongoloide populationer. *Tandlaegebladet*, **74**, 587–602.

Alexandersen, V. (1971). Odontometry: biological variations in Arctic populations. *Proc. 2nd int. Symp. circumpolar Health, Oulu, Finland, 21–4 June 1971*.

Bailit, H. L. (1966). Tooth size variability, inbreeding, and evolution. *Ann. N. Y. Acad. Sci.*, **134**, 616–23.

Bang, G. (1971). Dental disease and diet, tooth structure and morphology in the Alaskan Eskimo. *Proc. 2nd int. Symp. circumpolar Health, Oulu, Finland, 21–4 June 1971*.

Bang, G. & Hasund, A. (1971). Morphological characteristics of the Alaskan Eskimo dentition. I. Shovel-shape of incisors. *Amer. J. phys. Anthrop.*, **35**, 43–7.

Bang, G. & Hasund, A. (1972). Morphological characteristics of the Alaskan Eskimo dentition. II. Carabelli's cusp. *Amer. J. phys. Anthrop.*, **37**, 35–9.

Bang, G. & Hasund A. (1973). Morphological characteristics of the Alaskan Eskimo dentition. III. Number of cusps on the upper permanent molars. *Amer. J. phys. Anthrop.*, **38**, 721–5.

Bang, G. & Kristoffersen, T. (1972). Dental caries and diet in an Alaskan Eskimo population. *Scand. J. dent. Res.*, **80**, 440–4.

Cameron, J. (1923). Osteology of the western and central Eskimo. The Copper Eskimos. In *Report of the Canadian Arctic Expedition 1913–18*, **12**, part C. Ottawa.

Carlsson, G. E., Helkimo, E. & Helkimo, M. (1971). Bite force in a Lapp population of Northern Finland. *Proc. 2nd int. Symp. circumpolar Health, Oulu, Finland, 21–4 June 1971*.

Cederquist, R. (1975). Craniofacial description of Wainwright Alaskan Eskimos. MA thesis, Department of Anthropology, University of Chicago.

Colby, W. B. & Cleall, J. F. (1972). The craniofacial complex of the Northern Foxe

Basin Eskimo: a cross-sectional study. Paper presented at International Association of Dental Research, annual meeting, Las Vegas, March, 1972.

Colby, W. B. & Cleall, J. F. (1974). Radiographic cephalometric analysis of the craniofacial region of the Northern Foxe Basin Eskimo. *Amer. J. phys. Anthrop.*, **40**, 159–70.

Dahlberg, A. A., Cederquist, R., Mayhall, J. & Owen, D. (1978). Eskimo craniofacial studies. In *Eskimos of Northwestern Alaska, a biological perspective*, ed. P. Jamison, S. L. Zegura & F. A. Milan. Stroudsburg, Pennsylvania: Dowden, Hutchinson & Ross.

Dahlberg, A. A. & Matsumoto, M. (1909). Cephalometric studies of Wainwright Alaskan Eskimos. Report paper, annual meeting of the Amer. Assoc. for the Advancement of Science, December 1969.

Eriksson, A. W. (1971). Polymorphism in Lapps in Finland. *Excerpta Med. International Congress Series*, **233**, 64.

Eriksson, A. W. (1973). Genetic polymorphisms in Finno-Ugrian populations. *Israel J. med. Sci.*, **9**, 1156–70.

Greene, J. C. & Vermillion, J. R. (1960). The oral hygiene index: a method for classifying oral hygiene status. *J. Amer. dent. Assoc.*, **61**, 172–9.

Hanihara, K. (1968). Mongoloid dental complex in the permanent dentition. *Proc. 8th int. Congr. Anthrop. Ethnol. Sci.*,Tokyo & Kyoto, **1**, 298–300.

Hanihara, K. (1973). Dentition of the Ainu and the Australian Aborigines. *Paper for the 9th int. Congr. Anthrop. Ethnol. Sci., Chicago.*

Hanihara, K., Masuda, T., Tanaka, T. & Tamada, M. (1975). Comparative studies of dentition. In *Anthropological and genetic studies on the Japanese*, ed. S. Watanabe, S. Kondo & E. Matsunaga, pp. 256–64. JIBP Synthesis Volume 2. Tokyo: University Press.

Hansson, H. (1971). A survey of roentgenpathologic findings in orthopantomograms of the Skolts. *Proc. 2nd int. Symp. circumpolar Health, Oulu, Finland, 21–4 June 1971.*

Hedegard, B. (1971). Frequency of oral diseases in the Skolt population. *Proc. 2nd int. Symp. circumpolar Health, Oulu, Finland, 21–4 June 1971.*

Helkimo, E., Helkimo, J. & Carlsson, G. E. (1971). Chewing efficiency in a Lapp population in Northern Finland. *Proc. 2nd int. Symp. circumpolar Health, Oulu, Finland*, 21–4 June 1971.

Helkimo, M., Carlsson, G. E., Hedegard, B., Helkimo, E. & Lewin, T. (1971). Functional disturbances in the masticatory system in a Lapp population. *Proc. 2nd int. Symp. circumpolar Health, Oulu, Finland, 21–4 June 1971.*

Helkimo, M., Carlsson, G. E., Hedegard, B., Helkimo, E. & Lewin, T. (1971). Funktion och funktionsrubbninger i tuggaparaten hos samer i norra Finland. Preliminar rapport.

Hrdlicka, A. (1928). Anthropological survey in Alaska. US Bureau of American Ethnology, 46th Annual Report, pp. 19–374.

Hrdlicka, A. (1940). Lower jaw. *Amer. J. phys. Anthrop.*, **27**, 281–308.

Hylander, W. L. (1972). The adaptive significance of Eskimo craniofacial morphology. PhD dissertation, Department of Anthropology, University of Chicago.

Jakobsen, J. (1970). Odontologiske undersogelser i Grønland. *Tandlaegebladet*, **74**, 626–39.

Jakobsen, J. & Hansen, E. R. (1974a). Cariessituationem i Grønland 1973–4. *Tandlaegebladet*, **78**, 839–47.

Jakobsen, J. & Hansen, E. R. (1974b). Tandplejen i Grønland 1974. *Tandlaegebladet*, **78**, 848–52.

Karlsson, U. (1971). Tooth extraction and alveolar bone changes. *Proc. 2nd int. Symp. circumpolar Health, Oulu, Finland, 21–4 June 1971.*

Kimura, K. (1961). The Ainus, viewed from their finger and palm prints. *Z. morph. Anthrop.*, **52**, 176–98.

Kirveskari, P. (1971). Some features of the dental morphology of the Skolt Lapps. *Proc. 2nd int. Symp. circumpolar Health, Oulu, Finland, 21–4 June 1971.*

Kirveskari, P. (1974). Morphological traits in the permanent dentition of living Skolt Lapps. Academic Dissertation, Institute of Dentistry, University of Turku, Turku, Finland.

Kirveskari, P., Hedegard B. & Dahlberg, A. A. (1972). Bulging of the lingual aspects of buccal cusps in posterior teeth of Skolt Lapps from Northern Finland. *J. dent. Res.*, **51**, 1513.

Kristoffersen, T. & Bang, G. (1973). Peridontal disease and oral hygiene in an Alaskan Eskimo population. *J. dent. Res.*, **72**, 791–6.

Lehmann, W., Eriksson, A. W., Jürgens, H. W. & Forsius, H. (1970). Hudlinjesystemet hos skoltsamer. *Nord. Med.*, **83**, 717–19.

Lewin, T. & Hedegard, B. (1970). An anthropometric study of head and face of mature adults in Sweden. *Acta Odont. Scand.*, **28**, 935–45.

Lewin, T., Hedegard, B. & Kirveskari, P. (1971). Odontological conditions among the Lapps in Northern Fenno-Scandia. *Suomen Hammaslaakariseuran Toimituksia*, **67**, 99–104.

McPhail, C. W. B., Curry, T. M., Williamson, R. G., Hazelton, R. & Paynter, K. J. (1971). Geographic pathology of dental disease among Central Arctic populations. *Proc. 2nd int. Symp. circumpolar Health, Oulu, Finland, 21–4 June 1971.*

Mayhall, J. T. (1970). The effect of culture change upon the Eskimo dentition. *Arctic Anthrop.*, **7**, 117–21.

Mayhall, J. T. (1971*a*). Odontological studies at Igloolik and Hall Beach, N. W. T., Canada. *Proc. 2nd int. Symp. circumpolar Health, Oulu, Finland, 21–4 June 1971.*

Mayhall, J. T. (1971*b*). Dental studies: a progress report. In *IBP–HA Project Report no. 3*, ed. D. R. Hughes, pp. 49–61. University of Toronto Anthropological Series 8.

Mayhall, J. T. (1972*a*). Dental studies: a progress report, 1971. In *IBP–HA Project Report no. 4*, pp. 21–32. University of Toronto Anthropological Series 11.

Mayhall, J. T. (1972*b*). Dental morphology of Indians and Eskimos; its relationship to the prevention and treatment of caries. *J. Canad. dent. Assoc.*, **28**, 152–4.

Mayhall, J. T. (1973). Progress report 1972–3. Department of Anthropology, University of Toronto.

Mayhall, J. T. (1976). Inuit culture change and oral health: a four-year study, pp. 414–20. In *Proc. 3rd int. Symp. circumpolar Health, Yellowknife, Canada, 8–11 July 1974.*

Mayhall, J. T., Dahlberg, A. A. & Owen, D. G. (1970*a*). Dental caries in the Eskimos of Wainwright, Alaska. *J. Dent. Res.*, **49**, 886.

Mayhall, J. T., Dahlberg, A. A. & Owen, D. G. (1970*b*). Torus mandibularis in an Alaskan Eskimo population. *Amer. J. phys. Anthrop.*, **33**, 57–60.

Misawa, S. & Hayashida, Y. (1972). On the blood groups of the Ainu in Hidaka, Hokkaido. *56th Conference of the Medico-Legal Society of Japan*, abstract.

Möller, I. J., Poulsen, S. & Orholm Nielsen, V. (1972). The prevalence of dental caries in Godhavn and Scoresbysund districts, Greenland. *Scand. J. dent. Res.*, **80**, 169–80.

Omoto, K. (1972). Polymorphisms and genetic affinities of the Ainu of Hokkaido. *Hum. Biol. Oceania*, **1**, 278–88.

Oschinsky, L. (1974). *The most ancient Eskimos*. Ottawa: The Canadian Research Center for Anthropology, University of Ottawa.

Owen, D. G. n.d. Saliva analysis for secretion of blood group substances in the Wainwright, Alaskan Eskimos. Ms., Dental Anthropology files, University of Chicago.

Rosenzweig, K. A. & Silberman, Y. (1967). Dental morphology of Jews from Yemen and Cochin. *Amer. J. phys. Anthrop.*, **26**, 15–22.

Russell, A. L., Consolazio, C. F. & White, G. L. (1961). Dental caries and nutrition in Eskimo Scouts of the Alaska National Guard. *J. dent. Res.*, **40**, 594–603.

Selmer-Olsen, R. (1949). An odontometrical study on the Norwegian Lapps. Skrifter utgitt av Det Norske Videnskaps-Akademi i Oslo. I. Mat. Naturv. Klasse No. 3, Oslo.

Smith, P. (1973). Variations in dental traits within populations. *Paper for 9th int. Congr. Anthrop. Ethnol. Sci., Chicago*.

Steegman, A. T. (1967). Frostbite of the human face as a selective force. *Hum. Biol.*, **39**, 131–44.

Steegman, A. T. (1970). Cold adaptation and the human face. *Amer. J. phys. Anthrop.*, **32**, 243–50.

Turner II, C. G. (1967). Dental genetics and microevolution in prehistoric and living Koniag Eskimo. *J. dent. Res.*, **46**, 911–17.

Turner II, C. G. (1969). Microevolutionary interpretations from the dentition. *Amer. J. phys. Anthrop.*, **30**, 421–6.

Turner II, C. G. & Scott, G. R. (1973). Peopling of the Pacific: The dentition of living Easter Islanders, eastern Polynesia. Paper for *9th int. Congr. Anthrop. Ethnol. Sci., Chicago*.

Yamaguchi, B. (1967). A comparative study of the Ainu and the Australian Aborigines. Australian Institute of Aboriginal Studies, Occasional Papers No. 10.

Zoubov, A. A. (1971). Some morphologic dental traits in several northern populations of the USSR and Finland. *Proc. 2nd int. Symp. circumpolar Health, Oulu, Finland, 21–4 June 1971*.

Zoubov, A. A. (1972). Einige Angabe der dentalen Anthropologie über die Bevölkerung Finnlands. *Ann. Acad. Sci. Fenn. A5*, No. 150.

6. Ophthalmology

HENRIK FORSIUS

Ophthalmology has developed remarkably during the last decade as a branch of Arctic medical research. In addition to intriguing experts, some of the ophthalmological observations are of general interest, for example, the Eskimo susceptibility to acute glaucoma, the striking recent increase in the frequency of myopia among Eskimo children and adolescents and the findings related to climatic ocular lesions.

Ophthalmological investigations which included studies of dyschromatopsia (color blindness), the color of the iris and the shape of the palpebral fissure (the space between the two margins of the eyelids) formed part of the anthropological project of the Human Adaptability component of the International Biological Program (IBP–HA). The Scandinavian HA team investigating Lapps in north Finland included a group of ophthalmologists from Oulu University under the direction of the author. Invited to participate in the Danish, American and Canadian IBP–HA expeditions, I have also investigated Eskimos at Aupilagtoq, northwest Greenland, under the guidance of Dr J. Balslev Jørgensen, at Wainwright, Alaska, under Dr F. A. Milan and at Igloolik, Canadian Northwest Territories under Dr David Hughes. I considered it appropriate also to include the results of investigations undertaken by others, for example, the extensive Canadian ophthalmological survey of Eskimos coordinated by Professor S. T. Adams from the Department of Ophthalmology at McGill University in Montreal. Valuable studies, referred to here, were also performed without any connection to the IBP–HA projects. The Japanese IBP–HA Program on the Ainu of Hokkaido also included investigations of the eye.

The eye region

Theoretically one would expect changes of the face, the eye region in particular, as an adaptation to an Arctic climate. All other parts of the body may be effectively protected against the climate by clothing, but free sight is of vital importance. Changes ought to be the most conspicuous in those ethnic groups who have lived longest in the Arctic, i.e., the Eskimos, Lapps and native Siberian populations. Even Kant, the German philosopher, believed that the flat face of the Mongols resulted from adaptation to the climate. Theoretically, a flat or semi-spherical face with a minimum of projections should best tolerate the cold and the wind. And these features are typical of

The human biology of circumpolar populations

Mongoloid peoples. Laughlin (1963), however, stated that there is no experimental evidence to suggest that the climate has shaped the Mongoloid face.

Bunak (1972) compiled the results of anthropological measurements performed in Siberia by Soviet anthropologists. In the populations investigated, no correlation was observed between temperature and nasal index, facial flatness or height and width of face. According to Bunak there is no ground for the assumption that the Siberian type of face has developed in response to a cold climate. Experimental investigations by Steegman (1972), however, have shown that the temperature of the face during cold exposure decreases most rapidly at prominent sites, and my observations in the field indicate that Eskimos with a prominent nose and prominent cheek bones are more liable to frost bite than others. But these circumstances apparently have not influenced natural selection. In any event, the Eskimo face, with a small nose and flat eye region would seem more advantageous in an Arctic climate as compared to the Lapp face, which exhibits many Caucasoid features, including salient cheek bones.

The size or height of the face varies in Arctic ethnic groups. It is large in the Canadian and Alaskan Eskimos, smaller in the purest Greenland Eskimos at Ammassalik (Skeller, 1954) and in the Lapps (Lewin, Skrobak-Kaczinsky & Karlberg, 1972).

Lapps, many Siberian peoples and Eskimos have thick eyelids, particularly as children.

In spite of the missing experimental evidence, Laughlin (1963) still assumed that a thick eyelid offered good protection against the weather. The eyelids

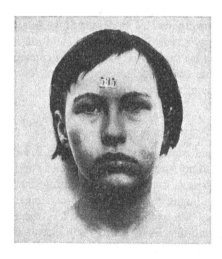

Fig. 6.1. 12 yr-old Lappish girl from Sevettijärvi, Finland, with European appearance but thick eyelids.

194

Fig. 6.2. Flattened face and thick eyelids in 9 yr-old girl from Aupilagtoq.

are not easily injured by the cold, and if frost bite occurs on them and on the eyeball, severe damage to the face is most certainly present also.

Eyebrows and beards are scanty in the Lapps, Eskimos and Siberian peoples, while abundant in the Ainus and in Caucasoid peoples. Eyelashes, which might be expected to protect the eyes from flying ice crystals and snow particles, are not particularly dense in native Arctic populations.

A narrow palpebral fissure (between the upper and the lower lid) was mentioned by Martin (1928) as typical of Mongoloid ethnic groups (Chukchis, Tunguses, Ostjaks, Yakuts, Samoyeds, Lapps, Eskimos) as compared to high values, 10–14 mm, for Europeans. A value of 8.78 mm was given for the Japanese. Fox (1966) in a series representing the general population in New York City, found that the majority had a palpebral fissure measuring less than 9 mm. In a large series of Eskimos from Ammassalik the height of the palpebral fissure ranged from 7.28 to 7.53 mm (Skeller, 1954). Systematic measurements of the fissure on standard photographs of Finns, Skolt Lapps, Eskimos in Greenland and Icelanders all taken by the same person (Heikki Nieminen), revealed no statistically significant differences between the populations with one exception. Finnish boys and girls aged 0–9 yr have a significantly narrower palpebral fissure than Eskimo and Skolt Lapp children of the same ages. (Differences Finn–Eskimos: males, $P = 0.01$, females, $P = 0.05$; Finns–Skolts: males, $P = 0.02$, females, $P = 0.05$ (Forsius, 1972.)

The subjective impression that the palpebral fissure is wider in Europeans than in Mongoloids is probably attributable to the presence of the Mongol fold.

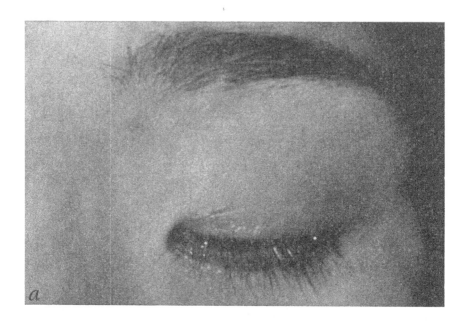

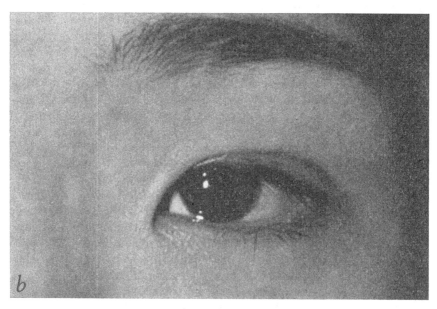

Fig. 6.3. *a.* Eskimo eye with eyefold in 20 yr-old female, Igloolik, Canada, closed; *b.* the same eye as in *a*, open.

196

The width of the palpebral fissure is known to decrease with age, and this we confirmed in all the populations we investigated. The frequency of the Mongoloid Fold, as well as the epicanthus, which is not the same thing, is reduced with age. This observation pertains to all ethnic groups examined including Eskimos. The skin becomes more wrinkled and subcutaneous fatty tissue decreases. The palpebral fissure is directed upward and outwards at an angle of 0–10° in 91% of the Eskimos. This according to Skeller (1954) corresponds to European values.

According to Mark (1970) the prevalence of the epicanthus in Lapps of the Kola Peninsula is also low.

Prominence of the eyeball

Eyes of Eskimos are no more prominent than those of Caucasians when measured in the sagittal direction from the *arcus zygomaticus* to the middle of the cornea (Skeller, 1954). Measurements of a Chinese series performed by Lee (1930, cited by Skeller, 1954) constitute similar evidence.

The cornea

Diameter

The corneal diameter is 5–10% smaller in Eskimos than Europeans (Skeller, 1954 and Alsbirk, personal communication). This may be an anatomical advantage, the risk of freezing of the eye being thus reduced.

According to our investigations, the Lapps have the same corneal diameters as Finns and other European ethnic groups.

Radius

The corneal radius was measured in the HA investigations in Finland, Greenland, Alaska and Igloolik, Canadian Northwest Territories (Forsius & Eriksson, 1973). All results were of the same magnitude. A vertical decrease of the corneal radius with age was observed in all populations investigated.

Thickness

Corneal thickness was measured in Skolt Lapps, Inari Lapps and Eskimos at Wainwright, Alaska, and Igloolik. The corneal thickness for 47 Alaskan Eskimos was 0.524 mm and for 91 Igloolik Eskimos was 0.502. The Skolt Lapps were found to have a significantly thinner cornea than other Lapps, who have a cornea of Caucasoid thickness. The thickness of the cornea in 117 Skolts was 0.495 mm, S.D. 0.032 (Forsius *et al.*, 1967, 1971).

Depth of the anterior chamber

It seems a reasonable assumption that the depth of the anterior chamber would be a significant factor in an Arctic climate, considering that the temperature of the cornea depends on heat coming from within or from other parts of the eye, mainly the circulation from the posterior chamber. Cold tests (Rysä & Sarvaranta, 1973) have shown that the corneal temperature decreases more rapidly if the anterior chamber is flat than if it is deep, since the amount of fluid is larger in a deep anterior chamber and thus cooled more slowly. The minimum corneal temperature, however, was not influenced by the depth of the anterior chamber. Anthropological cold adaptation ought to have resulted in a deep anterior cold chamber.

Measured with a Haag–Streit pachymeter, an optical device for measuring thickness, the chamber depth was within the Caucasoid range in Lapps, the Cheremisses in the USSR and Icelanders (own investigations) while extremely low values were obtained in Greenlandic and Canadian Eskimos (Alsbirk & Forsius, 1973). The low chamber depth in the Canadian Eskimos was confirmed by Drance *et al.* (1973). At Aupilagtoq, the readings obtained for anterior chamber depth and corneal thickness were too low throughout owing to a mistake in the pachymeter adjustment (Forsius, 1970).

It would appear that the low anterior chamber of the eye in Eskimos cannot be considered an adaptation.

The depth of the anterior chamber decreases with age. This leads to an increased risk of developing acute closure-angle glaucoma in old age. Glaucoma is a disease of the eye characterized by increased intraocular pressure due to restricted outflow of the aqueous through the aqueous veins and Schlemm's canal, excavation and degeneration of the optic disk, and nerve bundle damage producing arcuate defects in the field of vision. This disease is extremely frequent among the Eskimos in Greenland and the main cause of blindness among them (Clemmesen & Alsbirk, 1971). The depth of the anterior chamber is smaller in women than men, and acute glaucoma is consequently less frequent in the latter. Persons developing acute glaucoma also have a smaller axial length and a flatter cornea than the general population (Lowe, 1972).

An extensive investigation in Greenlandic Eskimos and the effect of chamber depth on ocular pressure has been published by Alsbirk (1974). Statistically, the anterior chamber is deeper in myopic subjects. The marked increase of myopia observed in young Eskimos in Alaska and northern Canada may be expected to result in a future decrease of the incidence of closure-angle glaucoma (Wyatt, 1973).

Axial length of the eyeball

The axial length of the eyeball was measured by Young *et al.* (1969) in

Eskimos in Alaska. He found that the subjects with myopia also have a greater bulblength than those with hyperopia. The majority of the myopic subjects were young and the mean age for the hyperopic was in the adult age groups, by Alsbirk (personal communication) in Eskimos on the west coast of Greenland, by Luukka (personal communication) in Finnish Lapps and by Young & Leary (1969) in Eskimos in Alaska. Young and his group found that children with myopia had a greater bulb-length than their parents.

Refraction

According to most ophthalmological texts, simple myopia is clearly an hereditary trait with usually irregular dominant inheritance. Other causes such as excessive accommodation of the lens, poor school illumination, malnutrition, etc., have also been mentioned. It is generally believed that native peoples who live by hunting and who are dependent on good vision for their survival have good eyesight and a low prevalence of myopia. Myopia is frequent among populations with an old civilization (in central and southern Europe, India, China, Japan). Previously, the prevalence of myopia was very low among Lapps, Eskimos and Indians.

More recent investigations, first undertaken by Cass (1966) have shown that the prevalence of myopia in the age groups under 30 has increased markedly among both Eskimos and Indians in northern Canada (Cass, 1969) and in Alaska (Young et al., 1969), while this is not the case among Eskimos in Greenland (Alsbirk & Forsius, 1973). Among the Lapps, myopia has increased only slightly (Luukka, 1974), and this condition seems to be rare in many ethnic groups in northeast Siberia as judged by their good eyesight (Prof. Rychkov, personal communication).

In a Myopia Conference Report (1972) Michaleva from the Soviet Union stated, 'The highest percentage of myopia was observed in school children of the Yakut region of Siberia'.

Myopia is a great problem among the Eskimo populations living under primitive conditions, for in the winter it is difficult for Arctic hunters to wear spectacles because they get misted. The increased prevalence of myopia may be attributed to an increased demand for accommodation resulting from altered habits of living ('school myopia', Young et al., 1969; Morgan, 1974), to the transition to a high carbohydrate diet (Luukka, 1974) or to the dietary change in combination with poor teeth (Cass, 1966). It may also be suggested that a genetic factor predisposing to myopia is involved, which has appeared in Eskimos and Indians owing to their changed habits of living.

Iris color

In general, the Eskimos have a dark-brown iris. Stefansson's report of fair-haired and blue-eyed Eskimos in Central Canada was disproved by Noice

Table 6.1. *Iris color types according to Martin Saller's scale in Finnish Lapp populations and in Finns from northern Finland and in Eskimos (1, light blue; 8, dark brown)*

Population	Sex	1	2	3	4	5	6	7	8	N
Skolt Lapps, Svettijärvi and Nellim	M	1.9	22.7	15.0	10.6	32.9	11.6	2.9	2.4	207
	F	1.3	16.1	19.7	6.7	33.2	13.0	8.5	1.3	223
	M + F	1.6	19.3	17.4	8.6	33.0	12.3	5.8	1.9	430
Half-breed Skolts (Skolt Finns)	M	3.7	37.0	24.1	5.6	20.4	3.7	3.7	1.9	54
	F	3.8	47.5	17.0	5.7	15.1	3.8	9.4	3.8	53
	M + F	3.7	39.3	20.6	5.6	17.8	3.7	6.5	2.8	107
Fisher Lapps (Nellim)	M	–	16.7	19.0	7.1	11.9	14.3	26.2	4.8	42
	F	–	10.4	4.2	12.5	27.1	16.7	22.9	6.3	48
	M + F	–	13.3	11.1	10.0	20.0	15.6	24.4	5.6	90
Finns	M	1.0	31.6	42.1	3.3	12.2	3.4	4.7	1.8	675
	F	1.9	39.9	37.3	1.9	11.4	0.3	4.7	2.5	316
	M + F	1.3	34.2	40.6	2.8	11.9	2.4	4.7	2.0	991
Alaska–Eskimos	M	–	–	–	–	1.0	0.0	6.1	92.9	98
	F	–	–	–	–	0.0	2.3	1.1	96.6	88
	M + F	–	–	–	–	0.5	1.0	3.8	94.6	186
Greenlandic–Eskimos	M	–	–	–	1.1	4.2	4.2	13.6	76.8	95
	F	–	–	–	1.2	5.9	1.2	13.1	78.6	84
	M + F	–	–	–	1.1	5.0	2.8	13.4	77.6	179
Canadian–Eskimos	M	–	–	–	–	–	–	6.3	93.6	94
	F	–	–	–	–	–	–	4.3	95.6	114
	M + F	–	–	–	–	–	–	5.3	94.7	208

(1922). At Aupilagtoq we saw individuals with eyes of mixed color and eyes partially blue in color, but in these instances a Caucasian ancestor is the most probable explanation. In the Lapps, mixed eye colors predominate, extremely dark as well as bright blue eyes being infrequent.

The eyes of the Lapps differ from the eyes of the Finns which are among the palest in the world. The Skolt Lapps have a somewhat more faintly colored iris, on the average, than the Inari Lapps. In Siberia there is a strong preponderance of dark iris colors, but it has been reported that a large proportion (up to 52.4%) of certain populations (the Selkups) have irises of mixed color (Levin, 1963). Over 90% of the world's population have dark eyes. Considering that the Eskimos as well as the Siberian Arctic peoples are of Mongoloid origin it is not surprising that they have a dark iris. The observations are tabulated in Table 6.1.

As a rule, the pigmentation of the eyeground and the iris are of the same intensity. However, in a striking fashion, many Lapps have a dark background despite a pale iris (Forsius *et al.*, 1970). A significant correlation between the pigmentation of the eyeground and the iris was found in female Skolt Lapps but not in males according to my unpublished data.

As a pale iris allows more light to pass through than a dark one, a dark-colored iris and eyeground are advantageous in the adaptation to extremely bright light, as for instance on ice and on forestless, snow-covered terrain in the spring. Although the origin of the Lapps is obscure, they exhibit many Caucasoid but only a few Mongoloid traits. Their darker iris, compared to the Finns, may have been a result of selection in the past. Experience has shown, however, that Caucasoid individuals with a scantier pigmentation are also capable of enduring life in the Arctic. Today the majority of people living north of the Arctic Circle are white: Norwegians, Finns, Swedes, immigrants in Alaska and north Russia. Murmansk alone has over 300 000 inhabitants.

Climatic ocular lesions

Dark and light adaptation

Populations living in Arctic and sub-Arctic regions are exposed to extreme weather conditions in winter and to extreme variations in light. Those living north of the Arctic Circle enjoy very little daylight during the winter months, while in summer they are continually exposed to solar radiation. We found a normal dark adaptation in 50 Lapps at Svettijärvi whom we investigated with the Goldmann adaptation equipment in 1966.

Members of expeditions passing the winter in the Arctic have reported that their companions had a yellowish complexion when they first saw them in sunlight early in the spring. In connection with a Swedish polar expedition in

1883 (Gyllencreutz & Holmgren) it was clarified that the yellow color was due to poor hygienic conditions indoors and not to a disturbance of vision!

Ultraviolet radiation

The ultraviolet component of the sun's radiation is considered to be the cause of several pathological conditions of the eye: acute snow blindness, climatic keratopathy, pterygium and pinguecula.

Keratopathy is a disorder of corneal tissue. Pterygium, from the diminutive of *pteryx*, Greek 'wing', is a disorder of the eye in which a triangular patch of hypertrophied bulbar conjunctival tissue extends from the inner canthus to the border of the cornea or beyond with the apex pointing toward the pupil. Pinguecula, from the Latin *pinguiculus*, 'fattish', is a yellowish spot sometimes observed on either side of the cornea. It is connective tissue, not fat.

A comparative investigation of the degree of pinguecula, and the frequency of climatic keratopathy and pterygium in some Arctic and sub-Arctic ethnic groups has been published by Forsius (1972, 1974). More detailed data as well as references are to be found in these papers, and some results are given in Table 6.2.

Both pterygium corneae, climatic keratopathy and pinguecula are common in equatorial regions, where the amount of ultraviolet radiation reaching the earth is abundant. These conditions are not always correlated, however. Impaired vision and blindness due to climatic keratopathy are common on certain islands in the Red Sea with white sand strongly reflecting the sunlight,

Table 6.2. *Populations investigated for pterygium, climatic keratopathy and degree of pinguecula*

	Males	Females	Males and females	No. over 20 yr old
Skolt Lapps	233	234	467	243
Mountain Lapps, N. Finland	40	43	83	64
Finns, Kökar	238	286	524	385
Finns, Oulu	98	122	220	220
Icelanders, Husavik 1972	318	316	634	419
Reykjavik 1973	142	156	298	298
Eskimos				
Upernavik, Greenland	104	87	191	92
Wainwright, Alaska	103	91	194	85
Igloolik, Canada	95	114	209	158
Cheremisses, USSR	326	–	326	323
Total no. investigated	1697	1449	3146	2297

while the prevalence of pterygium is low. By contrast, a very high prevalence of pterygium has been observed in older age groups on some tropical islands and in regions with a hot, arid climate. The ultraviolet content of sunlight decreases more abruptly when the sun is low in the sky. Consequently, the northernmost Arctic areas are less exposed to ultraviolet radiation than sub-Arctic regions. In contrast to the tropics, however, the ultraviolet radiation has a biological effect in Arctic regions (Sydow, Riemerschmid & Tiedemann, 1939) at an angle of incidence of 5–10° owing to the pure and dry air. The eyes are therefore less protected from direct ultraviolet radiation by the eyelids than in the tropics, where the sunlight is almost free from ultraviolet energy when the sun is low. Theoretically, worse and more frequent damage from ultraviolet radiation is to be expected in southern, cold Arctic regions when the sun is high in the sky but the terrain is still snow covered. This assumption is confirmed by experience (Wyatt, 1973).

Acute snow blindness

Snow blindness results from excessive exposure of the eyes to ultraviolet radiation. The epithelium in the middle of the cornea may be destroyed and completely exfoliated. Less severe damage occurs in the conjunctiva. There is negligible risk of developing this condition in the northernmost inhabited regions (Spitzbergen; Thule, Greenland), where the sun is low in the sky early in the spring, while the risk is great in the central regions of North America and Asia, where the terrain is still snow covered in the spring although the sun is high in the sky. The Arctic peoples – Eskimos, Indians, Lapps and various Siberian ethnic groups – have learned to protect themselves against acute snow blindness by wearing slit-goggles, visors, etc.; today they use modern sun-glasses.

Pterygium

Pterygium was observed in 23 out of 359 male Eskimos investigated by Skeller (1949); among the investigated females the condition was only half as frequent. A high prevalence of pterygium was observed by Norman-Hansen (1911) in Greenland. The prevalence seems to be higher in south than in north Greenland (Hertz, 1929). Wyatt (1973) discovered pterygium in only 0.9% of 4450 Canadian Eskimos and Indians. Reed & Hildes (1959) noted eight examples of pterygium in 503 Eskimos of whom only a small minority were over 40 years of age.

The prevalence of pterygium is markedly higher in Lapps and Eskimos than in urbanized populations in Central Europe and the USA, but it is low compared to the prevalences reported in tropical as well as in hot, arid regions (Cameron, 1965).

Our own results were published earlier (1974). The highest frequencies of pterygium were noted in Lapps in North Finland and Eskimos in Greenland.

Pinguecula

A particularly marked pinguecula is seen in subjects with pterygium and climatic keratopathy. On the whole, it is more marked in outdoor than indoor workers, and more conspicuous in men than in women. In the oldest age groups the pinguecula atrophies, particularly in subjects who no longer lead an active outdoor life. It was as marked in Lapps as in Eskimos (Forsius, 1974).

Climatic keratopathy

The first to observe climatic keratopathy in Arctic regions was Freedman (1965), who called this disease 'Labrador blindness'. He also found that it was caused by ultraviolet radiation and established its resemblance with the band-shaped keratitis occurring in tropical regions (1973*a*, *b*).

Climatic keratopathy is so typical in appearance that the present author's attention was drawn to this lesion, without any previous knowledge of its existence, at the first day of investigation of Skolt Lapps at Svettijärvi in 1966.

This disease is less common in north Canada than in Labrador (Freedman, 1973*b*; Wyatt, 1973). The prevalence of climatic keratopathy is the same among whites, Eskimos and Indians at the same latitude in Newfoundland. In some regions the frequency is 100% in males and 64% in females over 50 years of age (Johnson & Ghosh, 1975).

The results of the IBP investigations on Lapps and Eskimos and, for the sake of comparison, the figures obtained in some other populations were published earlier (Forsius, 1974). In all cases, a clear zone was observed laterally or medially between the limbus and the corneal opacity in at least one eye.

In those regions where our IBP investigations were carried out, climatic keratopathy had seldom led to impaired vision. This was the case in only two Eskimos and three Lapps. As a rule, those affected showed a pre-stage of the disease, a slight opacity peripherally in the cornea, at the level of the palpebral fissure (Forsius, 1974). In all populations investigated by us the frequency of pterygium and climatic keratopathy was higher in males than in females, and the pinguecula was also more marked in the males. This seems to be explained by the preponderance of outdoor activities among the males.

In our material there was no significant difference in the frequency of climatic keratopathy and pterygium between Eskimos and Lapps, nor between Icelanders and Lapps or Eskimos. The frequency of pterygium, on the other hand, was much lower in the Icelanders. Consequently, these diseases cannot be due to the same cause. Although both pterygium and climatic keratopathy are caused by ultraviolet radiation, the two diseases mostly do not affect the same individual. This problem has been analyzed by the present author in another context (Forsius, 1974).

Acute climatic corneal lesions

Epithelial lesions in the form of small erosions may develop in skiers during competitions in severe cold (Kolstad & Opsahl, 1969). We have noticed similar changes in extreme degree in Eskimos who have been exposed to the wind and flying ice crystals while driving a motor-sledge for several hours (Forsius, 1972).

Disease

Fritz (1970) has surveyed ocular disease in Alaska; Cass (1973) and Adams (1973) in Canada. The diseases directly caused by the Arctic climate have already been discussed in the foregoing section, i.e. pterygium, climatic keratopathy, snow blindness and pinguecula.

Color blindness

The prevalence of color blindness is low among the Eskimos in Alaska. According to Adam (1973) it is only 1.1% in pure native boys, but somewhat higher in Greenland, where 2.5% was obtained in 279 men on the east coast. On the west coast, where the Eskimos are mixed with whites, 6.8% were color blind (Skeller, 1954). Our own investigations show a low prevalence of color blindness. At Wainwright, Alaska, we found two strong deutans, or green-blind subjects, among 67 males. In a group of 92 males investigated at Igloolik there was one strong deutan and two mild protans, or red-blind subjects, and at Aupilagtoq we found eight cases of dyschromatopsia, of which there were seven deutans and one protan. In Greenland an investigation using Pickford's anomaloscope was unsuccessful owing to linguistic difficulties. The prevalence of color blindness in Finns is 7.9% (Forsius, Eriksson & Fellman, 1968). In connection with an IBP investigation involving 185 men at Fort Chimo in Labrador, only one case of color blindness was observed (Auger, personal communication).

The frequency of dyschromatopsia in Fisher and Mountain Lapps was found to be relatively low, or 6.42% of 981 males (Lahti, personal communication). This result has been confirmed by us. A series of 250 male Skolt Lapps included ten who were color blind, and five of these belonged to the same family.

In 167 Ainu males, Ochi, Fujiyama & Sueyoshi (1940) found only two cases

205

of red-green color deficiency. The Japan IBP investigation in 1966 in Shizumai showed the color blindness in Ainus to be much less than in Japanese males. The investigations in 1968 and 1969 confirmed these studies (Itoh, personal communication).

Strabismus

Squinting is rare among the Eskimos in Greenland and Canada (Skeller, 1949; Wyatt & Boyd, 1973; Cass, 1973). Fritz (1970) found 214 cases of esotropia (cross-eyes) among 1880 hospitalized Alaska Eskimos.

A reliable assessment of distance is necessary for a population living on hunting, and binocularity may therefore have influenced selection.

Our investigations on Lapps revealed manifest strabismus in 2.9% of 521 pure Lapps examined. Divergent squinting was observed in 11 of these 15 individuals. In addition, latent strabismus was noticed in 3.1%, among whom the convergent form predominated. In Caucasoid populations the frequency of manifest squinting has been estimated at 4%. An investigation involving 12 000 children in north Finland suggests a prevalence of the same magnitude (Krause, personal communication). In Wyatt & Boyd's (1973) survey in northern Canada the frequency of esotropia was comparable to Danish values but there was much less esotropia in Canadian natives, than in Denmark.

Glaucoma

The prevalence of closure-angle glaucoma is extremely high in Eskimos (Clemmesen & Alsbirk, 1971; *see* section on the depth of the anterior chamber).

An extremely low intra-ocular pressure seems to be common in Eskimos (Adams, personal communication) and in Lapps, according to our own observations (Table 6.3). Values $\leqslant 10$ mm Hg are seldom noted in normal European populations.

Among 396 Greenlanders over 40 yr old, investigated by Clemmesen & Alsbirk (1971), 15 had open-angle glaucoma.

In Iceland, the prevalence of closure-angle glaucoma is remarkably low, while the prevalence of open-angle glaucoma is one of the highest in the world. Blindness due to glaucoma is four times more common in Iceland than in the other Nordic countries (for reference, *see* Forsius *et al.*, 1974).

Pseudoexfoliation of the lens

Pseudoexfoliation of the lens is reported in highly varying frequencies in different countries. Its coincidence with open-angle glaucoma may worsen

206

Table 6.3. *Intraocular pressure by applanation tonometer in Skolts*

	Age of subjects (yr)											
	40–49		50–59		60–69		70–79		80–89		Total	
	Ta	No. eyes examined	Ta	No. eyes examined	Ta	No. eyes examined	Ta	No. eyes examined	Ta	No. eyes examined	Ta	No. eyes examined
Men	12.6	37	12.1	44	13.3	30	12.4	8	18.8	4	12.8	123
Women	13.4	40	13.3	47	13.5	32	10.7	14	10.0	4	13.0	137
Average	13.0		12.7		13.4		11.3		14.4		12.9	
Total		77		91		62		22		8		260
Tension ≤ 10 mm Hg												
Men		6		12		3		3		–		
Women		5		7		3		5		2		
Total		11		19		6		8		2		

the prognosis for the eye. We have not seen this condition in Eskimos, but we have encountered it in high frequency in the Scandinavian countries and Iceland. The prevalence is very high all over Finland as well as among the Finnish Lapps, or up to 30% in the oldest age groups (Forsius *et al.*, 1974).

Corneal scarring associated with tuberculosis and trauma

The association between tuberculosis and phlyctenular keratoconjunctivitis, frequently abbreviated as PKC, in Arctic regions has been the object of many investigations. Philip & Comstock (1965) and Philip, Comstock & Shelton (1965) found corneal opacities in 1587 Eskimos in the Bethel area in Alaska, where the prevalence of tuberculosis was also high. This problem has also been extensively studied in Canada. References may be found in a paper by Wyatt and other papers in the Arctic Symposium (*Canad. J. Ophthal.*, 1973, April number).

We have investigated the prevalence of corneal scars in Lapps, Eskimos and Finns using the same methods (Forsius & Eriksson, 1973). While the frequency of traumatic corneal lesions was not very variable in the males in all populations studied, corneal opacities due to diseases were more common in middle-aged and elderly Eskimos of both sexes than in Finns and Lapps. The prevalence of tuberculosis has been extremely high among both Lapps and Eskimos, and phlyctenular keratoconjunctivitis is no doubt the main cause of opacities due to disease and leading to impaired vision.

The low resistance to tuberculosis may in part have a genetic origin, but social factors seem to be more significant.

An intensive and successful anti-tuberculosis campaign among Eskimos and Lapps has led to a marked decrease of the incidence of this disease and to a parallel decrease in the frequency of phlyctenular keratoconjunctivitis. Corneal opacities are therefore rare in adolescents.

Trauma to the eye accounts for a large percentage of hospital admission in Alaska Eskimos (Fritz, 1970). Traumatic corneal lesions were also found by Wyatt (1973) to be common among the Eskimos in Canada.

Trachoma

Trachoma does not occur among the Eskimos in North America or among the Lapps.

Arcus senilis

The occurrence of arcus senilis was investigated in connection with the IBP–HA program (Forsius & Eriksson, 1973). The Lapps have an arcus senilis more often than the Eskimos and somewhat less often than the Finns.

Both the Finns and the Lapps have a high level of serum cholesterol, while a lower level has been noted in Eskimos. Sinclair (1969), too, reported that an arcus was less often seen in Eskimos than in whites, and Skeller (1954) found that an arcus was less frequent in pure than in less pure Eskimos in Greenland.

On the other hand, in the Canadian Ophthalmological Survey by Wyatt (1973) whites in Canada were reported to have a much lower prevalence of arcus senilis than Indians and Eskimos. The diverging results of this point are difficult to explain.

A valuable presentation of Arctic ophthalmology is to be found in the transactions of the Arctic Ophthalmology Symposium held in Canada in 1972 (*Canad. J. Ophthal.* (1973), **8**, 183–310). Many of these papers have been cited in this chapter. The transactions are supplied with an additional bibliography of Arctic ophthalmology (pp. 309–10).

References

Adam, A. (1973). Color blindness and gene flow in Alaskans. *Amer. J. hum. Genet.*, **25**, 564–6.

Adams, S. T. (1973). Editor's miscellany. *Canad. J. Ophthal.*, **8**, 306–8.

Alsbirk, P. H. (1974). Anterior chamber depth in Greenland Eskimos. I. A population study of variation with age and sex. II. Geographical and ethnic variation. *Acta Ophthal.*, **52**, 551–80.

Alsbirk, P. H. (1976). Primary Angle-Closure Glaucoma. Oculometry, epidemiology, and genetics in a high risk population. *Acta Ophthal.* Suppl. 127.

Alsbirk, P. H. & Forsius, H. (1973). Anterior chamber depth in Eskimos from Greenland, Canada (Igloolik) and Alaska (Wainwright). A preliminary report. *Canad. J. Ophthal.*, **8**, 265–9.

Bunak, V. V. (1972). *Climatic–regional and ethnical differences in face and head composition in the aborigines of North Asia.* Soviet IBP–Human Adaptability, pp. 25–37. Leningrad: Nauka Publishing House.

Cameron, M. E. (1965). *Pterygium throughout the World.* Springfield, Illinois: Charles C. Thomas.

Cass, E. E. (1966). Ocular conditions amongst the Canadian Western Arctic Eskimo. XX Concilium Ophthalmologicum Germania 1966. Pars. II, pp. 1041–53, International Congress Series No. 146. Amsterdam: Excerpta Medica Foundation.

Cass, E. E. (1969). The effects of civilization on the visual acuity of the Eskimo. In *The Eskimo people today and tomorrow*, ed. J. Malaurie. Paris: Mouton.

Cass, E. E. (1973). A decade of northern ophthalmology. *Canad. J. Ophthal.*, **8**, 210–17.

Clemmesen, V. & Alsbirk, P. H. (1971). Primary angle-closure glaucoma (a.c.g.) in Greenland. *Acta Ophthal.*, **49**, 47–58.

Drance, S. M., Morgan, R. W., Bryett, J. & Fairclough, M. (1973). Anterior chamber depth and gonioscopic findings among the Eskimos and Indians in the Canadian Arctic. *Canad. J. Ophthal.*, **8**, 255–9.

Forsius, H. (1970). Ophthalmological characteristics of Eskimos in Aupilagtoq. *Arctic Anthrop.*, **7**, 9–16.

Forsius, H. (1974). Pterygium, climatic keratopathy and pinguecula of the eyes in

The human biology of circumpolar populations

arctic and subarctic populations. In *Circumpolar Health*, ed. R. J. Shephard &
S. Itoh. University of Toronto Press.

Forsius, H. (1972). Climatic changes in the eyes of Eskimos, Lapps and Cheremisses. *Acta Ophthal.*, **50**, 532–8.

Forsius, H. & Eriksson, A. W. (1973). The cornea at northern latitudes – corneal
power, arcus senilis and corneal scars in Lapps, Eskimos and Finns. *Canad. J.
Ophthal.*, **8**, 280–5.

Forsius, H., Eriksson, A. W. & Fellman, J. (1968). Colour blindness in Finland.
Acta Ophthal., **46**, 542–52.

Forsius, H., Luukka, H. & Eriksson, A. W. (1971). Ophthalmo-genetic studies on
the Skolt Lapps. *Acta Ophthal.*, **49**, 498–502.

Forsius, H., Luukka, H., Fellman, J. & Eriksson, A. W. (1967). Corneal thickness
and its heredity in the population in north Finland. *Bull. Europ. Soc. hum.
Genet.*, **1**, 81–3.

Forsius, H., Luukka, H., Lehmann, W., Fellman, J. & Eriksson, A. W. (1970).
Irisfärg, korneabrytningsförmåga och korneatjocklek bland skoltsamer och
finnar. *Nordisk Medicin*, **84**, 1559–61.

Forsius, H., Sveinsson, K., Als, E. & Luukka, H. (1974). Pseudoexfoliation of the
lens capsule and depth of anterior chamber in northern Iceland. *Acta Ophthal.*,
52, 421–8.

Fox, S. A. (1966). The palpebral fissure. *Amer. J. Ophthal.*, **62**, 73–8.

Freedman, A. (1965). Labrador Keratopathy. *Arch. Ophthal.*, **74**, 198–202.

Freedman, A. (1973a). Climatic Droplet Keratopathy. I. Clinical Aspects. *Arch.
Ophthal.*, **89**, 193–7.

Freedman, A. (1973b). Labrador keratopathy and related diseases. *Canad. J.
Ophthal.*, **8**, 286–90.

Fritz, M. H. (1970). The eye and adnexal disease of the Alaska native – 1970.
Alaska Medicine, **12**, 70–5.

Gyllencreutz, R. & Holmgren, F. (1883). Undersökningar till förklaring af hud-
färgens anmärkta förändring efter öfvervintring i polartrakterna. Upsala
Läkarefören. Förhandl. 19, 190–230.

Hertz, V. (1929). In Meddelelser om Øjensygdomme i Grønland. *Ugesk. f. Laeger*,
91, 805.

Johnson, G. J. & Ghosh, M. (1975). Labrador keratopathy: clinical and pathological findings. *Canad. J. Ophthal.*, **10**, 119–35.

Kolstad, A. & Opsahl, R. Jr. (1969). Cold injury to corneal epithelium. A cause of
blurred vision in cross country skiers. *Acta Ophthal.*, **47**, 656–9.

Laughlin, W. S. (1963). Eskimos and Aleuts: Their Origins and Evolution. *Science,
Washington*, **142**, 633–45.

Levin, M. G. (1963). *Ethnic origins of the peoples of Northeastern Asia*, ed. H. N.
Michael, p. 134. Toronto: University of Toronto Press.

Lewin, T., Skrobak-Kaczinsky, J. & Karlberg, J. (1972). *Anthropometry of adult
Lapps. Table compilation (Hektogramm)*. Goethenburg: Anatomiska Insti-
tutionen. Sweden.

Lowe, R. F. (1972). Primary angle-closure glaucoma. Inheritance and environment.
Brit. J. Ophthal., **56**, 13–20.

Luukka, H. (1974). The incidence of myopia in Finnish Lapland. In *Circumpolar
Health*, commentaries, p. 382. Third International Symposium on Circumpolar Health, Yellowknife, Northwest Territories.

Martin, R. (1928). *Lehrbuch der Anthropologie*. 2. Aufl., Bd 1–2, Jena.

Mark, K. (1970). Zur Herkunft der Finnisch-Ugrischen Völker vom Standpunkt
der Anthropologie. Verlag 'Eesti Raamat', Tallinn.

Michaleva, M. (1972). In *The conference on prevention, pathogenesis and treatment of eye disease in children*, Moscow, 1–3 December 1971. *J. Paediat. Ophthal.*, **9**, 122.

Morgan, R. W. (1974). A pedigree study of myopia in Eskimos. *Proc. 3rd int. Symp. circumpolar Health, Yellowknife, Canada, 8–11 July 1973.*

Noice, H. H. (1922). In *The 'Blond' Eskimo – a question of method. Discussion and Correspondence*, Sullivan, L. R. *American Anthropologists*, **24**, 225–32.

Norman-Hansen, C. M. (1911). Oftalmologiska iakttagelser hos ett arktiskt folk. *Finska Läkaresällskapets Handlingar*, **53**, no. 7, 2–3.

Ochi, S., Fujiyama, H. & Sueyoshi, T. (1940). Augenkrankheiten bei den Aino. *Acta Soc. Ophthal. Japon*, **44**, 1–21.

Philip, R. N. & Comstock, G. W. (1965). Phlyctenular keratoconjunctivitis among Eskimos in southwestern Alaska. II. Isoniazid prophylaxis. *Amer. Rev. Resp. Dis.*, **91**, 188–96.

Philip, R. N., Comstock, G. W. & Shelton, J. H. (1965). Phlyctenular keratoconjunctivitis among Eskimos in Southwestern Alaska. I. Epidemiologic Characteristics. II. Isoniazid Prophylaxis. *Amer. Rev. Resp. Dis.*, **91**, 171–87.

Reed, H. & Hildes, J. A. (1959). Corneal scarring in Canadian Eskimos. *Canad. Med. Assoc. J.*, **81**, 364–6.

Rysä, P. & Sarvaranta, J. (1973). Thermography of the eye during cold stress. *XXI Meeting of Nordic Ophthalmologists, Turku, Finland, 13–16 June, pp. 234–9. Acta Ophthal.*

Sinclair, H. M. (1969). In *Metabolic and nutritional eye diseases*, ed. F. C. Rodger & H. M. Sinclair, p. 135. Springfield, Illinois: C. C. Thomas.

Skeller, E. (1949). Øjensygdomme i Grønland (Thule, Upernavik og Kutdligssat). *Ugeskr. f. Laeger*, **111**, 529–32.

Skeller, E. (1954). Anthropological and ophthalmological studies on the Angmagssalik Eskimos. Reprinted from 'Meddelelser om Grønland', Bind 107 Nr. 4. Bianco Lunos Bogtrykkeri A/s, København.

Steegman, A. T. Jr. (1972). Cold Response, Body Form and Craniofacial Shape in Two Racial Groups of Hawaii. *Am. J. phys. Anthrop.*, **37**, 193–222.

Sydow, E., Riemerschmid, G. & Tiedemann, M. (1939). Messungen der Ultraviolettstrahlung in Lappland und Spitzbergen. *Bioklimatische Beiblätter*, **6**, 29–33.

Wyatt, H. T. (1973). Corneal disease in the Canadian North. *Canad. J. Ophthal.*, **8**, 298–305.

Wyatt, H. T., Balisky, L. A. & Singh, O. (1974). Acute glaucoma in the Arctic. Studies of the difference between the Eskimo and Indian peoples and of the effects of myopia in each group. Third International Symposium on Circumpolar Health, Yellowknife, Northwest Territories, Canada.

Wyatt, H. T. & Boyd, T. A. S. (1973). Strabismus and strabismic amblyopia in Northern Canada. *Canad. J. Ophthal.*, **8**, 244–51.

Young, F. A. & Leary, G. A. (1969). Comparative oculometry of Caucasians, Eskimos and chimpanzees. *Ultrasonographia medica.* II, 595–612. *Proc. 1st World Congress on Ultrasonic Diagnostics in Medicine and Sidou III*, Vienna, 2–7 June 1969.

Young, F. A., Leary, G. A., West, D. C., Box, R. A., Harris, E. & Johnson, C. (1969). The transmission of refractive errors within Eskimo families. *Arch. Amer. Acad. Optom.*, **43**, 676.

7. Anthropometry of circumpolar populations

FRANKLIN AUGER, PAUL L. JAMISON,
J. BALSLEV-JØRGENSEN, THORD LEWIN,
JOAN F. DE PEÑA &
JERZY SKROBAK-KACZYNSKI

The Human Adaptability (HA) section of the International Biological Program focussed on the biological and cultural processes which characterized the adaptation of circumpolar peoples to their climatic and ecological systems and to their remote and scattered settlements. Anthropometric examinations of growth and adult morphology were important aspects of the research programs. Based on common anthropometric techniques, comparative statements can now be made about the different ethnic and genetic groups studied. This chapter on the anthropometry of circumpolar peoples will present the main body-build characteristics of the study populations as well as an introduction to further and more penetrating analyses of their biological relationships.

Although IBP–HA studies were carried out on Aleut, Ainu, Eskimo, Lapp and Siberian Mongoloid populations, comparative data are available for only the Eskimos and Lapps. The groups included and their investigators are: northwestern Alaskan Eskimos (Paul L. Jamison); northern Foxe Basin Eskimos of Canada (Joan F. de Peña); northern Quebec Eskimos of Canada (Franklin Auger); west Greenlandic Eskimos (Jørgen Balslev-Jørgensen); Norwegian and Finnish Lapps (Thord Lewin and Jerzy Skrobak-Kaczynski).

Genetically admixed (hybrid) individuals are not separated from non-admixed (non-hybrids) in this presentation. The intent is to describe population variation under the impact of both gene flow and socio-cultural influences. In addition the hybrids and non-hybrids live together sharing the same culture and environment of their respective populations.

Relatively few variables are included in this chapter in addition to stature and body weight. The intent is to reach readers with a broad, general interest in circumpolar population studies.

Subjects

Alaskan Eskimos

Five villages in northwestern Alaska contributed one Eskimo sample whose data appear in this presentation. The villages are: Wainwright, Point Hope,

Table 7.1. *Sample size: Alaskan Eskimo anthropometric data*

	Wain-wright	Point Hope	Anaktu-vuk Pass	Kaktovik	Barrow	Totals
			Children 1–20 yr			
Males	87	79	17	20	50	253
Hybrids	27	45	0	12	27	111
Non-Hybrids	60	34	17	8	23	142
Females	83	68	15	12	57	235
Hybrids	33	36	0	10	20	99
Non-Hybrids	50	32	15	2	37	136
Totals	170	147	32	32	107	488
			Adults 20–60 yr			
Males	22	16	12	0	41	91
Hybrids	3	5	0	0	13	21
Non-Hybrids	19	11	12	0	28	70
Females	23	38	21	0	53	135
Hybrids	5	12	1	0	16	34
Non-Hybrids	18	26	20	0	37	101
Totals	45	54	33	0	94	226

Anaktuvuk Pass, Kaktovik (Barter Island) and Barrow. Table 7.1 lists
sample sizes by sex, village, hybrid/non-hybrid status and two categories of
age: 1–20 yr old and 20–60 yr old. The hybrids are those Eskimos with
European or Old American White, African Black, or rarely, South Pacific
Islander genetic admixture. These were identified through genealogies collected by F. A. Milan and P. L. Jamison and through census records. Most of
the admixture resulted from matings two to three generations ago between
Eskimos and members of whaling crews who sailed to the western Arctic in
the latter part of the nineteenth century (Jamison, 1976; Jamison, Zegura &
Milan, 1978).

Foxe Basin Eskimos

A second group of Eskimos are the Igluligmiut residents of the modern communities of Igloolik and Hall Beach, located in the northern Foxe Basin of
the eastern Canadian Arctic (de Peña, 1972). These communities contain
recent migrants from some 42 identifiable coastal hunting sites, inhabited
between 1930 and 1960, scattered over a 700-mile stretch of Melville Peninsula
and the western slope of Baffin Island. During the research period the average
Igluligmiut population of these two villages was 780, of which 671 were
measured. 17% of the subjects were hybrids (all having White admixture)
based on memory genealogies, birth records and preliminary results of sero-

Table 7.2. *Sample size: Foxe Basin anthropometric data*

	Males	Females	Totals
	Children 1–19 yr		
Hybrids	48	51	99
Non-Hybrids	159	164	323
Totals	207	215	422
	Adults 20–60 yr		
Hybrids	6	12	18
Non-Hybrids	129	102	231
Totals	135	114	249

logical testing carried out for genetic marker studies. Table 7.2 presents sample sizes for these Canadian Eskimos.

Quebec Eskimos

The northern Quebec Eskimos included here are residents of Fort Chimo on the west shore of the Kosoak river south of Ungava Bay (Auger, 1976). The sample in this study represents about 90% of the males in the population (*see* Table 7.3). Fort Chimo's Eskimo population is rather heterogeneous since marriages have involved individuals from several villages around Ungava Bay. Nevertheless, a major proportion of the population was born at Fort Chimo or in nearby camps. Hybrids comprise about 50% of the children and 40% of the adults and they include Eskimos with White, or to a lesser extent Indian, genetic admixture.

Table 7.3. *Sample size: Quebec Eskimo anthropometric data*

	Males	Totals
	Children 5–23 yr	
Hybrids (with White admixture)	36	36
Hybrids (with White and Indian admixture)	24	24
Non-Hybrids	57	57
Totals	117	117
	Adults 20–60 yr	
Hybrids (with White admixture)	26	26
Hybrids (with White and Indian admixture)	10	10
Non-Hybrids	56	56
Totals	92	92

215

Table 7.4. *Sample size: Greenland Eskimo anthropometric data*

	Males	Females	Totals
	Adults 21–60 yr		
Aupilagtoq	23	22	45
Kraulshavn	22	17	39
Totals	45	39	84

Greenland Eskimos

Two villages in West Greenland, Aupilagtoq and Kraulshavn, contributed an adult sample for whom anthropometric data are reported here (Jørgensen & Skrobak-Kaczynski, 1972). These Eskimos still live in the traditional way, depending upon hunting of sea mammals for their subsistence. Genealogies extending back to the middle of the nineteenth century provided information on hybridization. During the last 100 years very little mating has occurred between Greenlanders and Europeans. Therefore, genetic admixture (which undoubtedly exists and may be of the nature of 20% admixture from Europeans) originated further back in time. Only adult data are currently available for Greenland Eskimos (Table 7.4).

Norwegian Lapps

The Norwegian Lapps in this study live in the Kautokeino area of Finnmark

Table 7.5. *Sample size: Lapp anthropometric data*

	Males	Females	Totals
	Children 7–16 yr		
Norwegian Lapps	402	404	806
	Children 1–25 yr		
Finnish Lapps			
Lapps with Finnish admixture	225	187	412
Skolt Lapps	121	121	242
Inari Lapps	52	49	101
Totals	398	357	755
	Adults 20–60 yr		
Lapps with Finnish admixture	44	57	101
Skolt Lapps	112	100	212
Inari Lapps	86	92	178
Totals	242	249	491

in northern Norway. About half of the sample are non-admixed individuals belonging to the genetic isolate of Kautokeino Lapps. Hybrids within the sample derive from marriages between Lapps and Scandinavians (i.e. Norwegians). The majority of the individuals belong to families whose economic base is reindeer breeding. Table 7.5 lists sample sizes by sex for the Norwegian Lapp children. Adult anthropometric data were not available (Lewin & Hedegård, 1971).

Finnish Lapps

The Finnish Lapps of concern here live in the northernmost part of Finland in the commune of Inari (Table 7.5). The group examined comprises 1152 individuals or about 40% of the total Lappish population in Finland. Genealogically about 450 individuals of the total sample were Skolt Lapps genetically isolated not only from the Finns but also from other Lapps at Inari (Lewin *et al.*, 1970). Some 275 subjects belonged to Lapp groups having about 10% Finnish admixture, but less than 1% Scandinavian admixture. The other 500 subjects of the sample showed about 50% Finnish admixture according to their genealogies.

Some additional information will also be presented on the growth of Icelandic children living under sub-Arctic conditions in Iceland (T. Lewin, personal communication). This information will be used for comparative purposes representing another population living in a rather harsh climate.

Methodology

The Eskimos in these studies were measured in light, indoor clothing with upper-arm circumference and triceps skinfold usually being measured directly on nude surfaces of the upper arm. In contrast to the Eskimos, Lapps generally wore only shorts during the measurement sessions. Sometimes Eskimos as well as Lapps would not remove items of clothing and in order not to disturb relationships with the groups under study these personal values were respected. Hence, in some instances measurements were impossible to take. In all cases the influence of clothes was taken into account with respect to body weight.

All measurements except stature were taken according to the methods outlined in IBP–HA Handbook No. 9 (Weiner & Lourie, 1969). For stature the subjects stood in an erect posture but not in a rigid military 'attention' stance. An anthropometer with straight bars was used to measure stature to the nearest millimeter. Investigators recorded body weights using scales calibrated to tenths of a kilogram or quarters of a pound. Pound weights were transformed to kilograms for the present comparisons. Upper-arm circumference and triceps skinfold were measured on the relaxed arm. In taking the

217

circumference of the mid-upper arm the measurer read a steel tape to the nearest millimeter. Triceps skinfold (i.e. skinfold thickness on the posterior of the mid-upper arm) was taken with either a Harpenden or Lange skinfold caliper, calibrated to 10 g/cm^2 pressure. Both biacromial and bicristal breadths and sitting height were measured to the nearest millimeter with an anthropometer.

Chronological age will be used throughout this chapter. Investigators subtracted the subject's birth date from the examination date (using the decimal of the year method of Weiner & Lourie) to determine age. Hence, all age groups include subjects born between mid-year intervals (e.g. five-year-olds are between 4.500 and 5.499 yr of age and adults aged 30–40 yr are between 30.500 and 40.499 yr of age).

Anthropometry of growth

Stature and weight

Tables 7.6–7.9 display descriptive statistics for stature and body weight by age for male and female children of the Eskimo and Lapp groups mentioned above. Figs. 7.1–7.4 show the age group means graphically for the same groups with the exception of the Norwegian Lapps.

Up to the age of 10 yr, in boys a range of about 2 cm includes the mean differences in stature between all the Eskimo and Lapp samples (Fig. 7.1). Thereafter this range increases to 5–6 cm, but at ages 18–20 yr the stature means again converge closer together. With the exception of ages 14–17 yr, Finnish Lapp boys have mean statures between those of the Alaskan and Foxe Basin Eskimo boys. Quebec Eskimo and Norwegian Lapp boys coincide with or are close to Finnish Lapp boys at ages 6–12 yr (Table 7.6) but after this age period both have nearly the same mean stature as Alaskan Eskimo boys. Within populations the Norwegian and Finnish Lapp boys are more similar in their growth in stature than is any intra-Eskimo group comparison. For the latter the Foxe Basin and Quebec Eskimos are the most similar and the Foxe Basin and Alaskan Eskimos are the most different in statural growth.

Finnish Lapp girls have mean statures 1–2 cm below those of Alaskan Eskimo girls but some 4–5 cm above the Foxe Basin Eskimos prior to age 10 yr (Fig. 7.2). After this age the mean stature lines of Alaskan Eskimo girls and Finnish Lapp girls intersect several times. Table 7.7 indicates that Norwegian Lapp girls' stature is similar to that of the Finnish Lapps.

Average stature for 18–20 yr-old males in these Eskimo and Lapp groups is between 166 and 168 cm. Eskimo and Lapp females of the same age average below 160 cm (155–6 cm). For purposes of comparison, Icelandic children living in the sub-Arctic region of Iceland have mean statures 5 cm or more above those of the Eskimo and Lapp children seen here and this difference

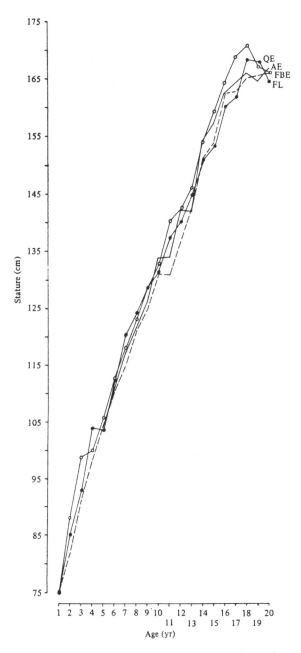

Fig. 7.1. Mean stature versus age for all males aged 1–20 yr. AE, Alaskan Eskimos; FBE, Foxe Basin Eskimos; QE, Quebec Eskimos; GE, Greenland Eskimos; FL, Finnish Lapps; NL, Norwegian Lapps.

Table 7.6. *Means and standard deviations of stature in male children (cm)*

Age (yr)	Alaskan Eskimos			Foxe Basin Eskimos			Quebec Eskimos			Norwegian Lapps			Finnish Lapps		
	N	Mean	S.D.	N	Mean	S.D.	N	Mean	S.D.	N	Mean	S.D.	N	Mean	S.D.
1	4	75.0	5.7	14	72.9	5.6							4	74.9	3.8
2	7	88.2	6.6	13	81.8	2.8							6	85.0	4.9
3	6	98.6	4.6	15	91.2	5.0							12	92.7	5.1
4	10	100.0	7.2	13	98.1	5.4							11	103.9	4.9
5	13	105.8	4.7	14	104.3	3.5	3	104.0	–				8	103.6	2.7
6	20	112.7	2.8	23	110.4	3.5	7	110.4	2.7				12	112.5	5.5
7	14	118.1	4.1	16	115.0	4.0	9	117.0	5.6	17	116.6	4.8	20	120.7	6.1
8	13	123.3	6.8	9	121.3	4.2	9	121.6	4.1	39	123.7	5.9	23	124.4	8.3
9	23	128.7	4.2	11	125.3	4.5	12	126.1	4.6	54	127.4	5.0	22	128.8	6.7
10	21	133.0	5.8	9	131.5	4.0	6	133.8	6.0	56	132.0	4.3	28	131.7	5.9
11	17	140.5	6.3	7	131.0	6.9	7	134.0	4.1	37	137.3	4.2	24	137.6	5.8
12	22	142.8	5.4	5	137.1	9.5	8	142.3	6.1	52	140.7	5.9	21	140.4	8.9
13	14	146.3	5.4	7	145.2	6.6	6	142.0	9.2	50	145.7	7.5	32	145.1	6.3
14	13	154.2	7.2	10	151.4	9.4	6	154.4	4.9	57	153.2	8.7	40	151.1	9.9
15	14	159.7	8.4	12	154.1	6.9	7	157.5	5.8	37	159.3	8.4	32	153.4	10.3
16	12	164.9	7.9	6	162.6	5.2	6	162.6	3.2	3	163.9	–	13	160.2	10.3
17	7	169.3	7.1	11	163.1	4.5	–	–	–				14	162.1	5.5
18	8	171.1	7.6	5	165.5	4.0	3	166.2	–				18	168.7	6.9
19	3	167.4	–	6	167.2	6.1	7	164.9	8.9				10	168.2	6.6
20	7	166.4	4.0				11	167.0	5.1				8	164.9	9.3
21							6	163.9	5.2				6	170.8	7.8
22							1	170.9	–				6	166.7	6.3
23							3	168.5	–				10	166.2	4.9
24													7	170.6	4.2
25													11	163.6	6.5

220

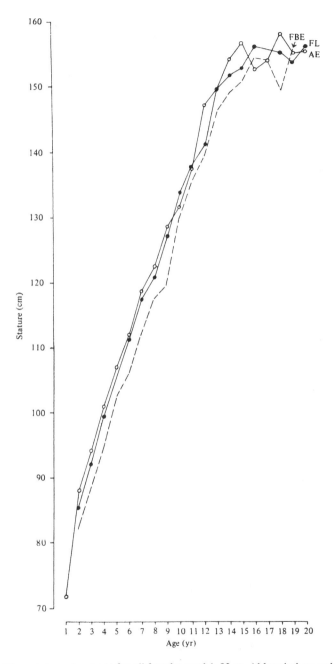

Fig. 7.2. Mean stature versus age for all females aged 1–20 yr. Abbreviations as in Fig. 7.1.

221

Table 7.7. *Means and standard deviations of stature in female children (cm)*

Age (yr)	Alaskan Eskimos			Foxe Basin Eskimos			Norwegian Lapps			Finnish Lapps		
	N	Mean	s.d.	N	Mean	s.d.	N	Mean	s.d.	N	Mean	s.d.
1	4	71.6	2.6	18	71.3	5.5				2	–	–
2	6	88.1	3.5	23	82.3	4.7				7	85.4	1.7
3	11	94.2	3.9	20	92.0	4.1				10	92.1	3.4
4	5	100.9	5.2	11	94.5	5.5				9	99.5	2.9
5	11	107.0	4.1	19	102.6	3.4				3	–	–
6	16	112.2	2.8	11	106.0	4.8				5	111.4	6.7
7	16	118.7	4.4	13	112.3	2.6	13	119.0	5.9	23	117.3	5.1
8	22	122.4	6.1	11	117.6	4.0	40	123.6	5.8	24	120.9	5.9
9	18	128.5	6.3	8	119.7	5.4	49	126.2	6.7	29	127.3	7.0
10	24	131.6	5.7	8	129.0	7.8	45	130.0	7.3	25	133.9	6.9
11	11	137.3	4.4	8	135.4	7.0	36	134.8	6.5	27	137.6	7.2
12	21	147.2	6.1	13	139.3	5.7	47	141.0	7.7	29	141.2	9.8
13	19	149.6	6.3	8	146.1	2.1	57	147.3	7.9	21	149.6	5.5
14	7	154.2	4.0	4	149.1	2.2	63	152.3	7.4	35	151.8	7.2
15	9	156.8	4.6	15	150.8	6.4	39	155.1	8.2	26	152.9	6.2
16	7	152.6	2.8	6	154.3	7.5	13	154.7	4.1	20	156.1	5.8
17	10	153.9	5.6	6	153.9	6.6				13	153.8	7.0
18	9	158.0	4.9	5	149.4	5.1				16	155.2	5.6
19	5	155.2	4.4	8	155.5	5.2				17	153.6	5.8
20	3	155.4	–							7	156.0	6.2
21										8	152.1	7.4
22										7	152.3	5.1
23										8	154.0	5.2
24										9	156.0	5.4
25										7	152.8	4.0

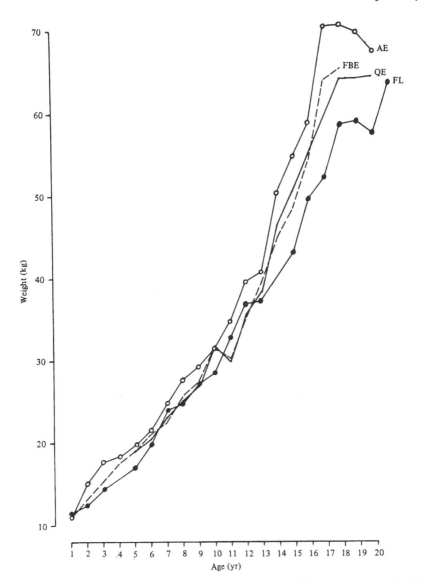

Fig. 7.3. Mean weight versus age for all males aged 1–20 yr. Abbreviations as in Fig. 7.1.

holds up to the time of the adolescent growth spurt. After adolescence the Icelanders are 10 cm or more taller on average.

Table 7.8 and Fig. 7.3 show mean body weights for male children. Both Foxe Basin and Quebec Eskimo boys are intermediate in weight between Alaskan Eskimo and Finnish Lapp boys. Up to age 12 yr, Finnish Lapp boys

Table 7.8. *Means and standard deviations of body weight in male children (kg)*

Age (yr)	Alaskan Eskimos			Foxe Basin Eskimos			Norwegian Lapps			Finnish Lapps		
	N	Mean	S.D.	N	Mean	S.D.	N	Mean	S.D.	N	Mean	S.D.
1	7	10.5	2.0	14	10.9	1.7				4	11.3	1.5
2	8	15.1	1.5	13	13.1	1.4				5	12.4	1.9
3	6	17.7	2.2	15	15.3	1.9				9	14.3	1.0
4	10	18.2	3.1	12	17.5	1.9				11	17.2	1.6
5	12	19.6	1.9	14	18.9	1.6				8	16.9	1.4
6	20	21.4	1.7	23	21.0	2.3				12	19.6	2.1
7	12	24.7	1.7	16	22.4	1.9	16	20.9	2.0	19	23.8	5.0
8	12	27.6	4.4	9	25.7	1.6	38	24.3	3.1	23	24.3	4.1
9	19	29.1	2.7	10	27.3	3.9	52	26.9	3.9	22	27.0	5.8
10	18	31.4	4.2	9	31.5	5.0	50	29.7	3.4	27	28.3	3.5
11	17	34.8	4.5	7	30.0	4.1	34	31.9	4.0	24	32.7	4.6
12	19	39.4	6.3	5	35.0	5.6	44	36.0	8.7	21	36.9	8.7
13	14	40.6	4.5	7	39.5	3.1	45	39.0	7.0	32	37.3	5.6
14	12	50.3	7.8	10	44.5	8.0	56	44.8	9.2	38	40.2	7.9
15	13	54.7	10.2	12	48.1	6.3	36	50.6	8.1	31	43.0	9.4
16	12	58.7	9.5	6	54.2	3.2	3	55.1	–	13	49.4	8.5
17	7	70.4	9.2	11	60.8	6.9				14	52.0	6.1
18	7	70.5	4.9	5	63.9	6.8				18	58.5	6.9
19	2	69.7	–	6	65.4	9.1				10	58.8	8.0
20	7	67.4	6.0							8	57.4	6.1
21										5	63.5	7.9
22										6	63.4	7.6
23										10	60.3	5.1
24										7	66.1	5.4
25										11	59.4	6.7

224

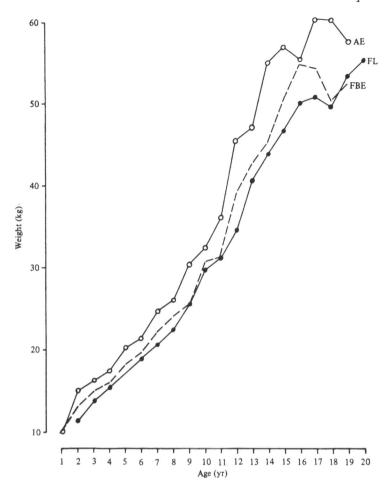

Fig. 7.4. Mean weight versus age for all females aged 1–20 yr. Abbreviations as in Fig. 7.1.

have mean body weights 2–4 kg below those of Alaskan Eskimos and this difference increases to 10–12 kg in the 16–20 yr age range. From age 7 to 16 yr old the body weight of Norwegian Lapps is of about the same magnitude as that of the Foxe Basin boys, i.e. between the mean body weights of Alaskan Eskimo boys and Finnish Lapps.

Body weight information for females may be seen in Table 7.9 and Fig. 7.4. As for males, the Foxe Basin Eskimos fall between the Alaskan Eskimo body weights (above) and Finnish Lapps (below). The difference between the Lapps and the Alaskans increases from about 2 kg in early childhood to about 10 kg at 15–18 yr of age, and then the difference diminishes again. Norwegian Lapp girls between the ages of 7 and 13 yr have about the same body

225

Table 7.9. *Means and standard deviations of body weight in female children (kg)*

Age (yr)	Alaskan Eskimos			Foxe Basin Eskimos			Norwegian Lapps			Finnish Lapps		
	N	Mean	S.D.	N	Mean	S.D.	N	Mean	S.D.	N	Mean	S.D.
1	5	9.9	1.8	18	9.8	1.8						
2	6	14.9	1.1	23	13.1	1.3				7	11.4	1.0
3	11	16.1	1.8	20	16.0	2.1				10	13.8	1.2
4	5	17.2	2.3	11	15.8	1.9				9	15.4	1.1
5	11	20.2	2.6	19	18.1	2.3						
6	16	21.3	2.3	11	19.6	2.0				5	18.9	2.6
7	14	24.6	2.3	12	22.3	1.7	13	22.3	2.7	21	20.6	2.1
8	19	25.9	3.5	11	24.0	2.3	38	24.9	3.3	24	22.4	3.5
9	16	30.3	5.0	8	25.5	3.0	44	26.3	3.6	29	25.4	4.2
10	23	32.3	6.8	8	30.5	4.4	38	28.8	4.9	25	29.6	6.7
11	10	35.9	4.5	8	31.0	4.2	36	31.7	5.9	27	31.1	6.3
12	17	45.3	10.5	13	39.0	6.4	39	35.9	7.6	29	34.5	8.2
13	17	47.0	8.8	8	42.7	4.7	49	42.5	7.3	21	40.6	4.9
14	7	54.8	7.4	4	45.4	3.7	61	46.4	7.3	35	43.8	7.1
15	9	56.8	7.4	14	50.7	6.6	38	51.8	8.1	26	46.7	9.8
16	7	55.1	6.2	6	54.7	8.4	13	50.2	5.2	20	50.0	6.0
17	8	60.1	8.4	6	54.1	5.2				13	50.8	8.5
18	9	60.1	4.8	5	50.5	8.5				16	49.5	7.6
19	5	57.2	6.9	8	52.3	8.0				17	53.2	5.7
20	3	56.3	–							6	55.3	7.3
21										8	50.6	4.6
22										7	55.3	7.1
23										8	52.4	6.6
24										7	54.6	5.3
25										5	54.5	4.8

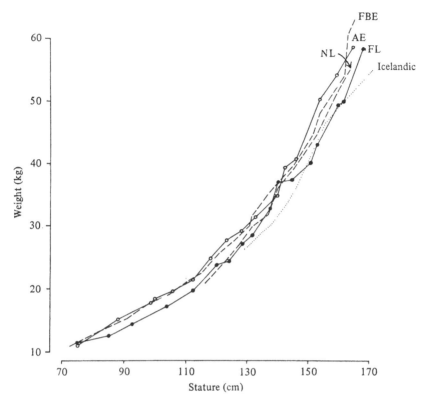

Fig. 7.5. Mean weights plotted as functions of mean statures for all males aged 1–20 yr. Abbreviations as in Fig. 7.1.

weight as Finnish Lapps. Again using Icelandic children as a comparative framework, these sub-Arctic inhabitants have body weights for both boys and girls that fall within the above ranges up to age 16 yr. After this age the Icelanders forge 10 kg or more ahead of all the other groups in Figs. 7.3 and 7.4.

In order to provide an overall impression of body size variation Figs. 7.5 and 7.6 show weight against stature using the age group means. This comparison generally demonstrates higher body weight per unit stature among Eskimos than among the Lapps. This holds true for both males and females. For statures above 130 cm in both boys and girls there is somewhat greater spread among the groups with Foxe Basin Eskimos and Norwegian Lapps having body weights that fall between those of Alaskan Eskimos and Finnish Lapps. Sub-Arctic Icelandic children generally have weight-to-stature values below the Lapp data given in Figs. 7.5 and 7.6.

To summarize, the differences between these Eskimo and Lapp groups with regard to stature and weight are not particularly marked. Greatest differences

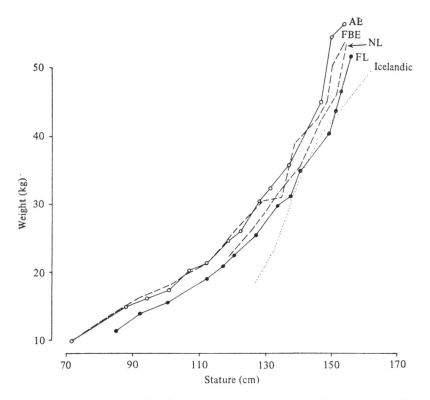

Fig. 7.6. Mean weights plotted as functions of mean statures for all females aged 1–20 yr Abbreviations as in Fig. 7.1.

tend to be found between Alaskan Eskimos and Finnish Lapps. Lower body weights relative to stature are rather consistently found for Lapps when compared to the Eskimos. Icelandic children by comparison tend to have higher statures and body weights but lower weight-to-stature ratios than either Eskimos or Lapps.

Trunk measurements and body proportions

Sitting height, biacromial and biiliac breadths are included in these comparisons to examine variation in body proportions. Arithmetic means and standard deviations by sex and age for each group are compiled in Tables 7.10–7.15.

As a general statement the Eskimo and Lapp groups are quite similar in their sitting heights with around a 2 cm variation in most age categories. There is a tendency among both boys and girls for one of the Eskimo groups

228

Table 7.10. *Means and standard deviations of sitting height in male children (cm)*

Age (yr)	Alaskan Eskimos			Foxe Basin Eskimos			Quebec Eskimos			Norwegian Lapps			Finnish Lapps		
	N	Mean	s.d.	N	Mean	s.d.	N	Mean	s.d.	N	Mean	s.d.	N	Mean	s.d.
3				9	54.0	2.9							9	54.3	3.0
4				12	56.1	2.8							8	60.3	2.6
5	2	62.8	—	13	59.2	1.9	3	60.6	—				8	58.0	5.3
6	16	63.5	2.8	22	62.4	3.3	7	62.5	2.1				10	62.1	3.7
7	13	64.1	2.0	16	64.1	2.5	9	66.1	2.8	9	64.6	2.8	17	64.9	5.9
8	13	67.4	3.2	9	67.3	2.5	9	67.7	2.5	30	67.1	2.9	21	67.0	3.8
9	23	69.5	2.5	11	68.4	2.7	12	69.1	2.1	43	69.0	2.6	19	68.9	3.4
10	21	71.4	3.4	9	70.6	2.8	6	72.8	3.1	52	70.7	2.1	19	70.4	3.2
11	17	74.6	2.9	7	71.0	3.7	7	72.2	1.9	36	72.7	2.1	21	71.3	4.2
12	22	74.9	2.5	5	72.8	5.1	8	76.7	3.6	52	74.1	3.2	17	72.6	5.7
13	14	76.5	3.2	7	72.5	9.2	6	76.4	4.5	50	76.0	3.6	26	76.0	3.2
14	13	79.9	4.2	10	80.3	4.6	6	80.3	3.3	53	79.0	4.9	38	78.5	4.9
15	14	83.0	4.5	12	80.5	3.5	7	84.5	4.1	37	82.9	4.7	30	79.8	5.4
16	12	85.4	5.1	6	85.6	2.6	6	87.2	2.1	3	87.1	—	12	84.0	5.4
17	7	90.1	5.0	11	86.1	2.0	—	—	—				11	85.5	3.5
18	7	89.4	2.5	5	88.2	2.2	3	88.4	—				13	89.1	3.2
19	3	88.6	—	6	89.5	4.9	7	90.6	4.5				8	89.6	3.9
20	7	88.5	1.9				11	91.3	3.1				6	89.7	3.4
21							6	90.6	3.1				4	91.2	1.7
22							1	94.5	—				5	89.6	2.0
23							3	92.9	—				8	89.3	3.3
24													5	89.4	3.7
25													5	88.7	1.8

Table 7.11. *Means and standard deviations for sitting height in female children (cm)*

Age (yr)	Alaskan Eskimos			Foxe Basin Eskimos			Norwegian Lapps			Finnish Lapps		
	N	Mean	s.D.	N	Mean	s.D.	N	Mean	s.D.	N	Mean	s.D.
2				6	50.5	3.3				6	54.1	1.4
3				16	53.6	2.4				6	56.4	1.6
4				9	55.0	3.0					–	–
5	1	60.6	–	19	58.1	2.7				–	–	–
6	12	62.2	2.0	11	60.3	2.4				4	61.7	0.8
7	14	64.0	3.3	13	63.1	2.6	5	67.6	1.7	18	63.7	2.9
8	22	66.3	3.2	11	65.5	1.6	35	67.3	2.9	24	65.3	3.9
9	18	68.2	2.9	8	65.5	3.0	41	68.0	2.8	27	68.1	4.3
10	24	70.0	2.8	8	69.1	4.9	37	70.4	3.7	23	70.0	4.9
11	11	73.0	2.5	8	72.6	4.3	35	72.3	2.9	24	73.2	4.9
12	21	76.7	3.3	13	75.4	4.0	44	75.0	3.4	25	74.4	4.7
13	19	78.8	4.1	8	78.0	2.5	55	78.0	3.9	20	78.6	3.2
14	7	82.4	2.9	4	80.3	2.2	62	80.5	3.8	29	79.2	2.6
15	9	82.8	3.8	15	82.6	3.3	40	82.6	3.9	23	81.2	3.9
16	7	81.2	2.4	6	84.2	3.9	13	83.2	2.3	16	82.6	3.5
17	10	82.2	4.0	6	83.9	4.2				10	81.6	2.4
18	9	83.7	2.6	5	83.2	2.5				11	83.4	3.4
19	5	82.6	2.4	8	85.3	4.4				14	83.1	3.0
20	3	84.2	–							4	83.0	1.8
21										6	84.3	2.3
22										7	81.1	3.6
23										8	83.9	2.7
24										6	84.3	2.6
25										4	84.3	1.9

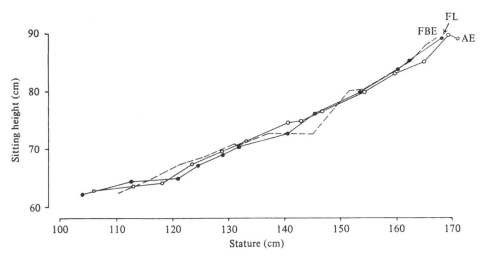

Fig. 7.7. Mean sitting heights plotted as functions of mean stature for all males aged 1–20 yr. Abbreviations as in Fig. 7.1.

to have the greatest sitting heights but either an Eskimo or a Lapp group may have the shortest. Fig. 7.7 shows the relationship between sitting height and stature for Alaskan and Foxe Basin Eskimo boys and Finnish Lapp boys. The relationship of these variables is essentially similar across all three groups. Quebec Eskimo boys (not illustrated) average 1–2 cm higher sitting height at the same stature than the other groups. For females, the Alaskan Eskimos and Finnish Lapps share the same sitting height-to-stature ratios while the Foxe Basin Eskimos have higher sitting heights for statures above 140 cm.

Biacromial breadth comparisons present a more clear cut picture of population differences than sitting height. Tables 7.12 and 7.13 provide descriptive statistics for biacromial breadth by age and sex and they indicate that the Eskimos have broader shoulders (2–3 cm) on average than the Lapps. This difference is also apparent in Fig. 7.8 showing the relationship between biacromial breadth and stature for two Eskimo groups and two Lapp groups. Fig. 7.8 displays the relationship for males but females also clearly demonstrate broader shoulders per unit stature among Eskimos than among Lapps. Icelandic children are quite similar to Lapps in their biacromial breadth-to-stature ratios and thereby also different from the Eskimos.

Much the same comparative picture holds for biiliac breadth as is discussed above for biacromial breadth. Tables 7.14 and 7.15 indicate that the Eskimos generally have broader hips than the Lapps but the difference (about 1.5 cm) is less than that seen for shoulder breadths. Fig. 7.9 shows larger biiliac dimensions per unit stature among Eskimo males than those seen for

231

Table 7.12. Means and standard deviations for biacromial breadth in male children (cm)

Age (yr)	Alaskan Eskimos N	Mean	s.D.	Foxe Basin Eskimos N	Mean	s.D.	Quebec Eskimos N	Mean	s.D.	Norwegian Lapps N	Mean	s.D.	Finnish Lapps N	Mean	s.D.
3				4	22.9	0.9							8	20.7	1.5
4													8	23.4	1.5
5	2	24.8	–	18	26.2	1.3	3	24.3	–	17	24.8	1.3	7	23.5	1.2
6	16	26.6	0.9	15	26.6	1.3	7	26.0	0.6	39	26.7	1.4	10	24.7	0.8
7	13	27.5	1.3	9	28.1	0.8	9	27.2	1.5				16	26.2	1.6
8	13	28.9	1.5				9	28.7	1.2				22	27.2	1.6
9	23	30.0	1.1	10	29.2	1.5	12	28.9	0.9	54	27.8	1.4	20	28.2	2.0
10	21	30.6	1.4	9	29.1	2.2	6	30.8	1.7	55	28.6	1.4	26	28.5	1.6
11	17	31.9	1.7	7	29.7	1.6	7	30.3	1.5	37	29.2	1.4	23	29.8	1.4
12	22	32.8	1.5	5	30.4	2.1	8	32.9	1.8	52	30.3	1.7	20	30.2	2.3
13	14	33.0	1.5	6	33.3	2.8	6	32.6	2.0	50	31.2	2.2	29	31.1	1.3
14	13	35.8	2.0	10	34.2	2.9	6	35.5	1.9	53	33.3	2.3	39	32.2	2.1
15	14	36.1	2.6	11	34.9	2.7	7	36.7	1.5	37	34.7	2.1	29	33.4	2.5
16	12	38.0	2.6	6	37.1	1.6	6	38.2	1.0	3	35.0	–	13	34.6	2.5
17	7	40.4	2.5	11	38.3	1.8	–	–	–				13	35.6	1.8
18	7	40.7	1.7	5	39.2	1.2	3	40.3	–				17	37.3	2.1
19	3	40.2	–	6	38.3	1.5	7	39.1	1.4				10	36.6	1.9
20	7	40.7	1.3				11	40.3	1.7				8	37.6	2.3
21							6	39.9	1.2				5	37.6	1.5
22							1	40.1	–				6	39.0	2.1
23							3	38.6	–				10	38.3	1.4
24													5	39.4	1.4
25													9	37.5	1.6

Table 7.13. Means and standard deviations for biacromial breadth in female children (cm)

Age (yr)	Alaskan Eskimos			Foxe Basin Eskimos			Norwegian Lapps			Finnish Lapps		
	N	Mean	S.D.	N	Mean	S.D.	N	Mean	S.D.	N	Mean	S.D.
2				4	22.8	1.9				7	20.2	0.6
3				4	22.2	1.4				7	21.2	1.0
4										7	21.8	1.1
5	1	27.1	–	4	24.1	1.2				–	–	–
6	12	25.4	1.0	9	25.0	1.4				3	–	–
7	14	23.6	1.4	13	26.4	1.2	12	25.3	2.0	21	25.2	1.1
8	22	23.6	1.4	11	27.1	1.1	40	27.1	1.2	23	26.4	1.8
9	18	29.9	1.6	8	26.9	2.3	49	26.9	1.7	29	26.9	2.2
10	24	30.3	1.3	8	29.4	2.1	44	27.9	1.7	25	28.7	1.8
11	11	31.7	1.0	8	31.2	1.3	37	28.8	2.3	25	29.8	2.0
12	21	33.8	1.5	13	32.6	1.5	47	30.4	2.0	27	30.4	1.9
13	19	34.0	1.6	8	33.2	0.9	57	31.2	2.5	19	32.2	1.6
14	7	35.6	1.0	4	33.8	2.0	62	32.8	2.1	34	32.4	2.1
15	9	36.7	1.9	15	34.5	1.7	39	33.5	2.4	25	33.6	1.4
16	7	34.9	1.3	6	35.4	1.5	13	33.5	1.5	19	33.6	1.1
17	10	36.3	1.9	6	34.7	1.3				11	34.1	2.1
18	9	36.4	1.0	5	35.4	1.1				14	34.0	1.6
19	5	36.8	1.2	8	34.2	2.2				17	34.1	2.6
20	3	37.5	–							4	34.8	3.0
21										7	34.1	1.6
22										7	34.1	2.9
23										8	34.6	1.5
24										7	35.0	1.8
25										7	35.1	2.1

233

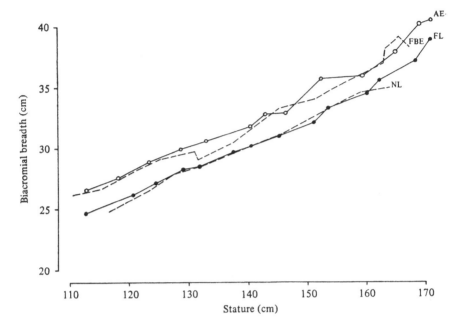

Fig. 7.8. Mean biacromial breadths plotted as functions of mean stature for all males aged 1–20 yr. Abbreviations as in Fig. 7.1.

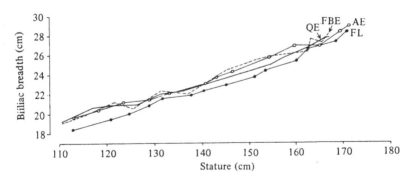

Fig. 7.9. Mean biiliac breadths plotted as functions of mean statures for all males aged 1–20 yr. Abbreviations as in Fig. 7.1.

234

Table 7.14. *Means and standard deviations for biiliac breadth in male children (cm)*

Age (yr)	Alaskan Eskimos N	Mean	S.D.	Foxe Basin Eskimos N	Mean	S.D.	Quebec Eskimos N	Mean	S.D.	Norwegian Lapps N	Mean	S.D.	Finnish Lapps N	Mean	S.D.
3				4	18.0	1.1							9	15.5	1.1
4													8	17.6	1.2
5	2	19.8	–	–	–	–	3	18.7	–				7	17.3	1.0
6	16	19.7	0.9	18	19.1	1.1	7	19.2	1.0	17	18.6	0.9	10	18.4	0.9
7	13	20.4	0.9	15	19.9	1.2	9	20.6	1.0	39	20.1	1.0	16	19.5	1.6
8	13	21.2	1.0	9	21.2	1.0	9	20.8	0.5				22	20.0	1.4
9	23	21.4	1.0	9	20.7	0.6	12	20.9	0.9	54	20.8	1.1	20	20.9	2.5
10	21	22.1	1.3	9	22.4	1.2	6	22.3	0.8	55	21.4	1.0	26	21.2	1.1
11	17	23.1	1.5	7	22.3	2.0	7	22.5	1.0	37	22.0	1.1	23	21.9	1.4
12	22	23.7	1.4	5	22.0	1.3	8	23.3	1.1	52	22.6	1.8	20	22.4	1.6
13	14	24.3	1.6	6	24.4	1.2	6	23.5	1.7	50	23.6	1.7	29	23.0	1.1
14	13	25.8	1.8	10	25.5	2.1	6	25.0	1.1	52	24.7	1.9	39	23.8	1.9
15	14	27.0	2.2	11	25.5	1.5	7	25.8	1.4	37	25.8	1.7	29	24.4	1.9
16	12	27.0	2.0	6	26.4	1.3	6	26.7	1.1	3	25.6	–	13	25.4	2.4
17	7	28.5	1.6	11	27.7	2.0	–	–	–				13	26.4	1.3
18	7	29.0	1.7	5	27.3	1.9	3	27.7	–				17	27.5	1.6
19	3	28.3	–	6	28.3	1.2	7	27.9	1.7				10	28.0	1.4
20	7	28.3	1.2				11	27.8	1.3				8	27.5	1.7
21							6	27.8	0.8				5	28.4	0.7
22							1	27.2	–				6	27.4	2.5
23							3	27.6	–				10	27.3	1.3
24													5	27.7	1.2
25													9	27.4	1.4

Table 7.15. *Means and standard deviations for biiliac breadth in female children (cm)*

Age (yr)	Alaskan Eskimos			Foxe Basin Eskimos			Norwegian Lapps			Finnish Lapps		
	N	Mean	s.d.	N	Mean	s.d.	N	Mean	s.d.	N	Mean	s.d.
2				4	16.9	1.4				5	14.5	0.7
3				4	16.9	1.0				6	17.1	2.0
4										7	16.4	1.2
5	1	19.7	–	4	17.8	1.4				–	–	–
6	11	19.7	1.1	9	18.8	1.1				–	–	–
7	14	20.5	0.9	13	20.0	1.2	12	19.9	1.0	21	18.7	1.0
8	22	21.0	1.4	11	20.4	1.2	40	20.4	1.1	23	19.6	1.2
9	18	21.3	2.0	8	20.6	1.9	49	20.4	1.2	29	20.4	1.3
10	24	22.4	1.0	8	22.3	1.0	44	21.3	1.2	25	21.4	1.4
11	11	23.2	1.0	8	22.4	1.4	37	22.1	1.5	25	22.1	2.0
12	21	25.3	1.8	13	23.6	1.5	47	23.4	1.6	27	23.0	2.0
13	19	25.5	1.4	8	24.8	1.6	57	24.4	2.2	19	24.2	1.2
14	7	26.3	1.0	4	24.3	1.6	62	25.9	1.7	34	25.4	1.5
15	9	27.3	2.1	15	26.2	2.1	39	26.8	1.7	25	26.0	1.7
16	7	27.1	1.7	6	27.4	1.5	13	26.3	1.8	19	26.3	1.3
17	10	27.4	2.1	6	27.3	2.1				11	26.4	1.5
18	9	28.2	1.0	5	26.3	3.1				14	26.5	1.3
19	5	27.5	1.5	8	27.1	2.1				17	26.6	1.9
20	3	28.0	–							4	28.4	3.4
21										7	26.0	1.8
22										7	26.6	2.2
23										8	27.9	1.4
24										7	27.5	0.9
25										7	27.9	1.3

Lapp males. Sub-Arctic Icelandic children, by comparison, have biiliac breadth-to-stature ratios lower still than Lapps (1–2 cm lower).

In summary, sitting heights and sitting height-to-stature ratios do not markedly discriminate Eskimos and Lapps or either group from Icelanders. On the other hand, shoulder breadths and to a lesser extent hip breadths clearly demonstrate greater dimensions for the Eskimo children than either Lapp or Icelandic children. The previous finding of greater body weight per unit stature among Eskimos than Lapps can now be determined to be partially due, at least, to broader trunks among the former groups.

Upper-arm circumference and triceps skinfold

Age group means and standard deviations for the measurements of upper arm circumference and triceps skinfold are presented in Tables 7.16–7.19. These data will allow general comparisons to be made of muscularity and subcutaneous fat.

Based on the data in Table 7.16, the growth curve for upper-arm circumference shows that Alaskan Eskimo boys have larger circumferences at all ages when compared to the other groups. A very similar finding is seen in Table 7.17 for Alaskan Eskimo girls. In the methodology section above this measurement was described as being taken on the relaxed arm. For the Alaskan Eskimo sample this measurement was taken on arms that were in a semiflexed position. Therefore this finding could be a result of slight contraction of the biceps muscle rather than a biological difference among the populations. The other populations are quite uniform in upper-arm circumferences for both boys and girls. There is a tendency for the Lapps to have slightly greater and particularly the Foxe Basin Eskimos to have slightly smaller arm circumferences across the age groups listed. For comparison, sub-Arctic Icelandic children have arm circumferences as large or larger than the data listed for Alaskan Eskimos in these two tables.

Tables 7.18 and 7.19 provide age group data for triceps skinfolds on boys and girls, respectively. In these tables the most striking population difference is the low values for the Foxe Basin Eskimo skinfold measurements when compared to the other groups. For males the Alaskan and Quebec Eskimos have triceps values similar to those of the Lapps until ages 12–16 yr when the Norwegian Lapps forge somewhat ahead of the others. After age 16 yr the Alaskan Eskimos tend to have the largest skinfold measurements among the boys. The female data are more variable with Alaskan Eskimo girls and Norwegian Lapp girls tending to have the largest values. Again after age 16 yr the Alaskans begin a consistent pattern of greater triceps measurements. Among Icelanders, male triceps skinfold measurements are markedly larger than any of the groups compared here and for females the Icelandic values would be at the upper end of the distribution.

Table 7.16. *Means and standard deviations for upper-arm circumference in male children (cm)*

Age (yr)	Alaskan Eskimos			Foxe Basin Eskimos			Quebec Eskimos			Norwegian Lapps			Finnish Lapps		
	N	Mean	S.D.	N	Mean	S.D.	N	Mean	S.D.	N	Mean	S.D.	N	Mean	S.D.
3				9	15.7	1.3							7	17.0	0.7
4				12	16.7	1.3							8	16.5	1.3
5	1	17.1	–	12	16.6	1.2	3	16.5	–				7	16.6	1.3
6	14	17.5	0.8	20	17.2	1.8	7	15.8	1.1				11	16.8	1.3
7	12	18.2	0.9	16	16.7	1.3	9	16.8	1.2	17	16.6	1.1	16	17.6	2.4
8	12	19.1	1.1	19	17.8	1.3	9	17.4	1.0	39	17.4	1.2	20	17.2	1.5
9	23	19.3	1.2	10	17.8	1.9	12	17.9	1.1	54	17.9	1.8	19	18.4	2.8
10	19	19.9	1.4	9	18.9	1.6	6	18.9	0.8	56	18.6	1.5	18	18.2	1.3
11	17	20.7	1.5	7	18.1	1.0	7	17.3	0.6	36	19.2	1.7	19	19.8	1.8
12	22	22.0	2.2	5	18.2	1.3	8	19.9	1.8	52	20.3	1.9	19	19.8	2.2
13	14	22.1	1.5	7	20.7	1.0	6	20.0	2.6	50	21.1	2.5	22	20.9	1.9
14	12	23.9	1.8	10	21.3	2.2	6	21.5	3.7	57	22.1	2.9	39	21.1	2.2
15	14	24.4	2.8	11	21.5	2.0	7	21.8	2.6	37	23.3	2.8	28	22.0	2.8
16	11	25.8	2.2	6	22.6	1.1	6	23.9	0.9	3	25.3	–	11	23.1	2.9
17	7	29.7	1.6	11	25.2	2.1	–	–	–				12	24.4	1.7
18	7	29.4	3.0	5	26.2	2.0	3	26.3	–				16	25.6	2.3
19	3	29.6	–	6	27.3	2.8	7	26.4	1.9				10	26.3	1.6
20	7	29.5	1.2				11	25.9	1.6				8	25.7	2.6
21							6	26.2	1.2				5	26.9	2.9
22							1	25.5	–				6	27.1	1.8
23							3	26.4	–				9	26.7	1.9
24													5	27.8	1.9
25													10	25.9	2.3

Table 7.17. *Means and standard deviations for upper-arm circumference in female children (cm)*

Age (yr)	Alaskan Eskimos			Foxe Basin Eskimos			Norwegian Lapps			Finnish Lapps		
	N	Mean	S.D.	N	Mean	S.D.	N	Mean	S.D.	N	Mean	S.D.
2				5	15.9	1.1				4	15.5	1.2
3				10	16.9	1.3				6	16.3	0.7
4				7	16.8	1.9				7	15.7	0.8
5	1	17.5	–	12	16.9	1.0				–	–	–
6	10	17.8	1.2	10	16.6	0.8				–	–	–
7	11	18.4	1.2	13	16.6	1.1	12	17.5	1.2	20	17.0	1.1
8	21	18.8	1.4	11	17.2	1.3	40	18.0	1.2	20	17.3	1.5
9	16	20.0	2.1	7	17.0	0.9	49	18.3	1.3	22	18.1	2.0
10	23	20.0	2.1	8	18.4	2.1	45	18.9	1.9	24	19.9	2.5
11	9	21.3	1.8	8	18.9	0.9	36	19.4	2.1	23	19.5	1.9
12	21	23.4	3.1	13	20.1	2.6	47	20.6	3.1	24	20.3	2.6
13	19	23.3	2.5	8	21.3	1.9	57	21.9	2.5	19	21.1	1.6
14	6	24.4	3.1	4	21.2	2.4	62	23.0	2.3	31	22.3	2.3
15	9	26.5	1.6	15	23.5	2.1	38	24.1	2.3	23	22.9	2.4
16	7	26.3	2.0	6	23.9	2.0	13	24.5	2.3	18	24.0	2.1
17	10	27.4	2.2	6	23.2	1.4				10	24.4	1.9
18	9	27.2	1.7	5	24.1	1.2				12	24.5	3.7
19	5	26.3	1.6	8	22.6	1.7				15	25.7	2.2
20	3	25.0	–							4	26.7	2.6
21										7	25.1	2.3
22										–	–	–
23										9	25.8	2.0
24										8	26.1	1.4
25										6	25.8	1.4

Table 7.18. *Means and standard deviations for triceps skinfolds in male children (mm)*

Age (yr)	Alaskan Eskimos			Foxe Basin Eskimos			Quebec Eskimos			Norwegian Lapps			Finnish Lapps		
	N	Mean	s.d.	N	Mean	s.d.	N	Mean	s.d.	N	Mean	s.d.	N	Mean	s.d.
3				9	6.4	2.4							7	9.2	2.4
4													7	9.1	1.7
5	2	6.0	2.0	11	5.5	2.1	3	8.6	–				7	9.2	1.2
6	14	7.0	2.0	20	5.6	2.3	7	7.6	1.3				9	7.6	2.1
7	12	7.0	2.0	16	5.2	1.5	9	7.3	1.6	17	7.3	1.1	12	7.8	1.9
8	13	7.0	1.0	9	6.8	2.3	9	7.5	0.8	39	7.8	2.1	15	7.7	2.1
9	23	7.0	2.0	10	4.9	1.2	12	8.4	2.3	54	8.0	2.8	18	6.7	1.7
10	21	7.0	2.0	8	5.9	2.9	6	8.2	1.5	55	8.5	2.6	14	6.9	1.9
11	17	7.0	1.0	7	5.9	1.3	7	6.1	1.3	37	8.3	2.2	18	8.6	2.6
12	22	8.0	4.0	5	4.4	1.7	8	7.5	2.0	52	9.7	3.1	13	8.3	3.1
13	14	8.0	2.0	6	5.6	2.9	6	8.0	1.4	50	9.5	3.7	21	7.7	1.9
14	12	8.0	3.0	9	5.5	2.9	6	7.1	2.5	53	9.3	4.5	38	7.5	2.6
15	14	6.0	2.0	11	4.0	1.3	7	6.0	2.9	37	8.5	4.3	27	7.5	2.1
16	11	7.0	2.0	6	3.3	1.0	6	6.9	1.9	3	8.1	–	11	5.9	2.1
17	7	11.0	3.0	11	5.0	1.9	–	–	–				10	5.4	1.3
18	7	8.0	2.0	5	5.2	1.3	3	7.8	–				12	6.7	2.1
19	3	11.0	–	6	3.5	1.4	7	6.5	1.4				8	6.4	2.5
20	7	8.0	2.0				11	5.5	0.9				6	6.2	1.3
21							6	5.9	1.1				4	7.0	2.8
22							1	6.0	–				5	6.0	2.4
23							3	6.4	–				7	5.6	1.2
24													5	6.1	1.3
25													4	5.8	2.3

Table 7.19. Means and standard deviations for triceps skinfolds in female children (mm)

Age (yr)	Alaskan Eskimos			Foxe Basin Eskimos			Norwegian Lapps			Finnish Lapps		
	N	Mean	s.d.	N	Mean	s.d.	N	Mean	s.d.	N	Mean	s.d.
3				9	6.9	1.9				4	10.2	1.9
4				5	7.4	1.6				6	9.2	0.5
5	1	9.0	–	11	7.6	2.1				–	–	–
6	11	8.0	2.0	10	5.0	2.9				–	–	–
7	13	8.0	3.0	12	5.9	3.3	12	8.7	2.0	16	7.4	2.1
8	21	9.0	2.0	11	5.8	2.7	40	10.2	2.1	16	8.6	3.5
9	18	10.0	3.0	7	4.1	1.3	49	10.1	2.5	19	9.3	2.5
10	23	10.0	4.0	8	5.5	1.9	44	10.6	3.3	17	9.8	4.3
11	9	10.0	4.0	8	5.8	2.1	37	10.7	3.5	21	8.9	3.3
12	21	13.0	8.0	13	6.5	3.8	47	10.5	3.3	21	9.7	3.6
13	18	14.0	6.0	8	6.8	3.5	57	13.2	8.7	19	9.5	3.4
14	7	12.0	5.0	4	5.4	4.0	62	13.9	4.7	29	11.7	3.5
15	9	18.0	5.0	15	8.0	4.2	39	15.4	4.7	22	12.0	4.1
16	7	18.0	3.0	6	8.8	6.2	13	14.8	6.7	15	15.3	5.0
17	8	19.0	6.0	6	8.7	2.7				10	12.9	4.4
18	9	20.0	6.0	5	9.2	2.4				10	13.2	5.1
19	5	17.0	7.0	8	8.8	4.1				13	15.3	5.2
20	3	11.0	–							–	–	–
21										6	13.8	1.6
22										7	16.2	4.2
23										8	13.2	2.6
24										6	12.5	3.2

If the triceps skinfold is taken as a measure of nutritional status, the Foxe Basin Eskimos are in the poorest state. The other Eskimo groups and all of the Lapp groups seem better off in this respect as they attain similar means at all ages, although for boys at least, the means are still lower than those found among Icelandic children. An alternative (and perhaps more likely) explanation in light of nutritional findings on the Foxe Basin Eskimos is that their diet may be implicated in this result. Lower carbohydrate intakes have been specifically mentioned in this regard (Schaefer, 1973, 1977).

To the extent that information on growth rates can be gleaned from cross-sectional data such as these, the Eskimo and Lapp children appear to have similar growth rates for most dimensions and both have slower rates when compared to Icelandic children. The latter live in a relatively harsh climate but under better socio-economic conditions than either Lapps or Eskimos. For the Eskimos and Lapps the slower rate of growth appears established by 7–8 yr of age. The Icelandic children's growth would suggest that a simple climatic explanation for the slower growth of Eskimos and Lapps cannot be implicated. Environmental effects and a possible adaptational response via natural selection are both likely to be involved in these growth patterns.

The only consistent Eskimo versus Lapp differences appear to be higher weight-for-stature ratios and sturdier trunks among Eskimos at nearly all ages. These may be genetically controlled body build differences as they occur in Eskimo children who are both at and below Lapp children nutritionally (as indicated by triceps skinfold averages). Body weight relative to stature and upper-arm circumference also point to a generally higher muscularity of Eskimo children when compared to Lapp children. Here, however, it must be noted that Eskimo children generally have higher protein intakes than Lapps, and the latter often receive less protein than their calculated needs.

Adult morphology and secular trends

Tables 7.1–7.5 provide sample sizes by sex for the adult anthropometric data. Only male stature was available for Quebec Eskimos and no data were included for Norwegian Lapps. On the other hand, this aspect of the analysis does include data on adult West Greenlandic Eskimos in two 20-yr age categories.

Stature and body weight

Descriptive statistics for stature and body weight by age category and sex can be seen in Tables 7.20 and 7.21. The lowest mean statures recorded for the 21–40 yr age group of males occur among the West Greenland Eskimos at about 161 cm. By comparison, young adult male statures of Alaskan, Foxe Basin and Quebec Eskimos and Finnish Lapps average 165 cm or more (*see*

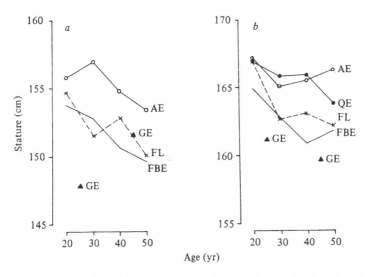

Fig. 7.10. Mean statures for adults aged 20–50 yr. Abbreviations as in Fig. 7.1. *a*, females; *b*, males.

Fig. 7.10). Much the same holds true for the females, among whom West Greenland Eskimo women of 21–40 yr of age have mean stature of 148 cm compared to the young adult statures of 154–156 cm for the females of all other groups. Except for these Greenlandic Eskimos, all groups exhibit secular stature increases influenced by age differences. West Greenland Eskimo males aged 21–40 yr show a 1.5 cm higher average stature than those at ages 41–60 yr; the expected stature decrease through normal ageing processes levels out this minor difference, leaving no indicated secular trend for these males. For Greenland females the 21–40 yr olds demonstrate a lower average stature (about 3 cm) than those aged 41–60 yr.

Comparisons of male age categories 20–30 yr and 31–40 yr show stature increases of about 1 cm per decade for Alaskan Eskimos, increases of about 2 cm per decade for both Foxe Basin and Quebec Eskimos, and as much as 4 cm per decade increases for Finnish Lapps. Among females of the same age groups the data show a 1 cm increase for Foxe Basin Eskimos, 3 cm for Finnish Lapps, but a 1 cm decrease at ages 20–30 yr for Alaskan Eskimo women.

When statures for the succeeding decades are compared (31–40 yr relative to 41–50 yr) little or no stature changes occurred in Alaskan and Quebec Eskimos or Finnish Lapp males, but the Foxe Basin Eskimo males increased their stature by about 2 cm during this period. In the same decade Alaskan and Foxe Basin Eskimo females increased in mean stature with Finnish Lapp females showing a stature decrease of about 1.5 cm.

If the stature means for adults aged 51–60 yr are increased by about 1.5 cm

Table 7.20. *Means and standard deviations for measurements on adult males*

Age (yr)	Alaskan Eskimos			Foxe Basin Eskimos			West Greenland Eskimos			Finnish Lapps		
	N	Mean	s.d.	N	Mean	s.d.	N	Mean	s.d.	N	Mean	s.d.
Stature (cm)												
20-30	28	167.2	6.1	69	164.9	5.1	32	161.2	4.8[a]	80	167.0	6.2
31-40	27	165.1	5.9	34	162.8	6.4				62	162.7	7.0
41-50	20	165.5	6.0	21	160.9	5.1	13	159.7	7.9[a]	47	163.1	6.0
51-60	16	166.3	4.2	10	161.8	5.6				51	162.2	5.9
Weight (kg)												
20-30	24	71.0	6.5	68	66.4	5.9				79	62.3	6.8
31-40	21	69.8	12.9	33	66.4	10.7				61	63.3	8.7
41-50	17	75.3	13.5	21	66.4	7.1				46	62.3	10.5
51-60	14	75.2	10.2	10	61.8	3.4				50	63.6	10.6
Sitting height (cm)												
20-30	28	89.4	2.7	65	89.5	2.9	31	87.2	4.2[a]	71	89.2	3.2
31-40	27	88.3	2.9	32	87.9	4.2				53	86.2	4.8
41-50	20	88.3	3.0	21	86.0	4.0	13	85.2	3.9[a]	38	87.6	3.5
51-60	16	87.6	3.1	10	86.7	2.5				48	86.6	3.6
Biacromial breadth (cm)												
20-30	28	41.5	2.4	62	38.1	2.1	30	37.6	1.7[a]	75	37.9	1.5
31-40	27	39.6	1.6	29	38.3	3.0				54	37.2	2.3

Age (yr)	n			n			n			n		
41–50	20	38.7	1.8	20	37.5	2.4	13	36.4	1.8[a]	39	36.6	1.7
51–60	16	39.3	1.2	10	37.2	1.2				47	36.7	1.7
Biiliac breadth (cm)												
20–30	28	28.4	1.2	62	28.9	2.2	30	27.9	1.4[a]	75	27.7	1.5
31–40	27	29.7	1.8	28	29.2	2.9				54	27.2	1.6
41–50	20	33.0	2.2	20	30.0	1.4	12	28.8	2.0[a]	39	27.5	1.6
51–60	16	30.2	1.5	10	29.6	1.3				47	28.1	1.8
Upper-arm circumference (cm)												
20–30	28	30.3	2.0	64	27.8	2.0	31	28.3	2.4[a]	72	27.1	2.2
31–40	27	30.7	3.1	30	28.2	1.9				49	27.9	2.7
41–50	20	31.6	2.7	20	28.4	2.7	12	29.1	2.4[a]	37	27.2	2.8
51–60	16	30.5	2.2	10	26.9	2.1				47	27.5	2.9
Triceps skinfold (mm)												
20–30	28	8.0	4.0	66	4.8	1.3				62	6.9	3.5
31–40	27	8.0	6.0	30	5.1	2.2				42	8.5	6.3
41–50	20	9.0	7.0	20	4.5	2.1				32	8.2	4.0
51–60	16	8.0	4.0	9	5.6	3.0				42	9.5	4.9

Stature data for Quebec Eskimos is as follows: 20–30 yr (33), 166.9 ± 5.1; 31–40 yr (30), 165.8 ± 6.5; 41–50 yr (14), 165.9 ± 6.1; and 51–60 yr (15), 163.8 ± 5.0

[a] Data reported for 20-yr intervals: 21–40 yr and 41–60 yr.

Table 7.21. *Means and standard deviations for measurements on adult females*

Age (yr)	Alaskan Eskimos			Foxe Basin Eskimos			West Greenland Eskimos			Finnish Lapps		
	N	Mean	S.D.	N	Mean	S.D.	N	Mean	S.D.	N	Mean	S.D.
					Stature (cm)							
20–30	41	155.8	4.4	53	153.8	4.9	26	147.9	4.4[a]	75	154.7	6.1
31–40	34	157.0	6.5	31	152.8	5.6				65	151.5	6.1
41–50	30	154.8	4.3	16	150.6	5.2	15	151.6	4.9[a]	52	152.8	5.3
51–60	29	153.4	5.4	14	149.6	3.9				56	150.0	5.7
					Weight (kg)							
20–30	35	60.4	11.8	52	55.8	6.9				72	55.2	8.6
31–40	28	65.1	14.0	31	57.5	12.0				64	57.2	10.6
41–50	23	70.0	17.4	16	55.6	6.9				52	61.6	12.9
51–60	25	67.3	15.3	14	57.2	12.3				55	60.7	11.1
					Sitting height (cm)							
20–30	41	84.2	2.6	53	84.4	3.4	24	80.8	3.3[a]	67	83.7	3.0
31–40	34	84.4	3.4	30	80.8	4.0				62	82.0	3.3
41–50	30	82.9	2.4	16	80.3	4.1	15	80.8	3.2[a]	47	82.4	3.2
51–60	29	82.2	3.4	14	80.7	3.3				51	80.6	2.6

	n	Mean	SD	n	Mean	SD	n	Mean	SD	n	Mean	SD
Biacromial breadth (cm)												
20–30	41	36.6	1.4	51	34.6	1.8	24	33.4	1.5[a]	69	34.9	2.0
31–40	34	36.0	1.5	30	34.7	1.8				62	34.1	1.8
41–50	30	36.2	1.6	16	35.3	1.6	15	34.1	2.0[a]	46	34.5	1.9
51–60	29	35.7	1.8	12	34.4	1.5				50	34.0	2.3
Biiliac breadth (cm)												
20–30	41	23.8	1.8	51	27.6	1.6	22	27.4	1.8[a]	69	27.3	1.9
31–40	34	29.0	1.6	30	28.4	2.9				59	27.4	3.2
41–50	30	29.9	2.1	16	28.6	3.0	14	29.2	1.2[a]	46	28.0	2.1
51–60	29	29.5	3.0	14	29.2	2.3				50	28.5	1.9
Upper-arm circumference (cm)												
20–30	41	27.4	3.1	51	24.4	2.4	24	28.6	2.6[a]	68	26.4	2.5
31–40	34	29.5	3.7	30	25.5	3.7				60	27.5	1.6
41–50	29	30.6	4.2	16	25.5	2.3	15	26.0	3.9[a]	46	29.0	3.9
51–60	29	30.4	4.9	14	27.0	4.1				50	28.6	3.6
Triceps skinfold (mm)												
20–30	41	16.0	8.0	51	7.7	4.9				62	15.0	4.5
31–40	33	20.0	10.0	30	10.5	8.8				58	17.3	7.1
41–50	31	22.0	11.0	15	9.8	4.4				44	19.0	8.5
51–60	28	20.0	9.0	13	14.5	9.1				48	19.6	7.2

[a] Data reported for 20-yr intervals: 21–40 yr, 41–60 yr.

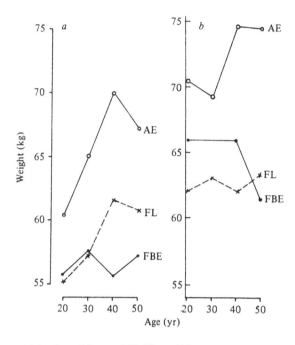

Fig. 7.11. Mean weights for adults aged 20–50 yr. Abbreviations as in Fig. 7.1. *a*, females; *b*, males.

to compensate for ageing, then no changes in stature seem to have occurred in the 1920s among Quebec Eskimo and Finnish Lapp males and Alaskan and Foxe Basin Eskimo females. Finnish Lapp females do appear to have increased their average stature by about 1 cm, and both Alaskan and Foxe Basin Eskimo males underwent a decrease in average stature of about 2–3 cm during this decade.

Body weights of adult age groups may be seen in Tables 7.20 and 7.21 and in Fig. 7.11. Alaskan Eskimos of both sexes have body weights 5 kg or more in excess of those of both Foxe Basin Eskimos and Finnish Lapps. The Alaskans also have higher mean body weight than the Quebec Eskimos. Young adult male Alaskan Eskimos average about 71 kg, Foxe Basin young males average about 65 kg, and Finnish Lapps about 62 kg. Among females, Alaskan Eskimos have mean body weights of 60 kg in young adults compared to weights of about 55 kg for both Foxe Basin Eskimos and Finnish Lapps. The adult male and female Alaskan Eskimos also increase average body weight with age as do Finnish Lapp females. The other groups either remain relatively stable in weight as they age or have decreasing average body weights.

248

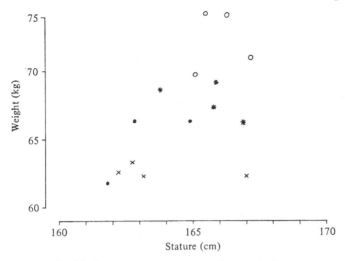

Fig. 7.12. The relationship between mean statures and mean body weights for all adult males. Alaskan Eskimos ○, Foxe Basin Eskimos ●, Quebec Eskimos ✳, and Finnish Lapps ×.

Relationships between adult body weight and stature in males (Fig. 7.12) indicate that the higher mean body weights of Alaskan Eskimo males are partially related to overall higher mean statures. Finnish Lapp males of the same average stature as Alaskan Eskimo males have lower average body weight by 5–8 kg. Foxe Basin and Quebec Eskimos are intermediate in this respect, at similar statures they average about 3 kg less than the Alaskan Eskimo males. For females, the relationship between stature and weight again demonstrates Alaskan Eskimos with the highest ratios but the Foxe Basin Eskimos and Finnish Lapp women have similar ratios.

To summarize, little stature change seems to have taken place among West Greenland adult males born since the 1920s while West Greenland adult females may have even decreased somewhat in stature during this period. For individuals born between 1920 and 1930, there were minimal stature increases among Quebec Eskimo and Finnish Lapp males and both the Alaskan and Foxe Basin Eskimo males showed some stature decline. Stature increases occurred for Finnish Lapp females in this period and there was a minimal decrease in stature among the Alaskan and Foxe Basin Eskimo females. The decade from 1930 to 1940 reflects relative stability in stature among the Alaskan Eskimo, Quebec Eskimo and Finnish Lapp men and a noticeable increase in average stature among Foxe Basin Eskimo men. All females born during this decade underwent an increase in stature as adults except for Finnish Lapp women who decreased in average stature. During the last adult decade (1940–50) all groups with the exception of Alaskan Eskimo females and West Greenland Eskimo females showed stature increases, on average, of 1–2 cm

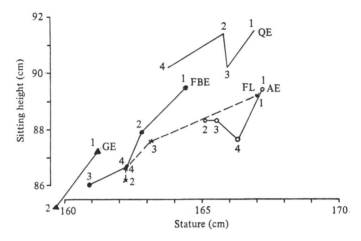

Fig. 7.13. Relationships between mean sitting heights and mean statures for all adult males. 1, 20–30 yr; 2, 30–40 yr; 3, 40–50 yr; 4, 50–60 yr.

for Eskimos and 3–4 cm for Lapps. Only the Alaskan Eskimo and Finnish Lapp females showed an increased weight relative to age. Alaskan Eskimos have high body weights associated with their high statures when compared to the other groups, but generally speaking for the same statures Eskimos have higher weights than do the Lapps.

Trunk measurements and body proportions

Sitting height, biacromial and biiliac breadths are included in this analysis of adults in order to examine variations in trunk dimensions and body proportions. Tables 7.20 and 7.21 include means and standard deviations for these variables by age groups. Variation in sitting height by age group reflects the changes found in stature relative to age. All groups showing increased stature over and above the effects of ageing also showed increased sitting height. Finnish Lapp age groups with low mean statures had sitting height-to-stature proportions similar to those of Eskimos of low mean statures (Fig. 7.13). At higher mean statures the Finnish Lapps and Alaskan Eskimos have comparable ratios. However, at average statures around 165 cm, both Foxe Basin and Quebec Eskimos have relatively higher sitting heights than Alaskan Eskimos or Finnish Lapps.

Among male samples, all of the Eskimo groups have greater biacromial breadths than do the Finnish Lapps (Table 7.20). The same holds true for females with the exception of West Greenlanders (Table 7.21). The differences range from 1–2 cm for the comparison between Foxe Basin Eskimos of

250

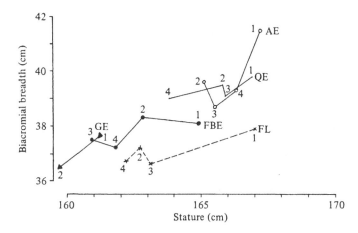

Fig. 7.14. Relationships between biacromial breadths and mean statures for all adult males. Abbreviations as in Figs. 7.1 and 7.13.

either sex and Finnish Lapps to 3–5 cm between Alaskan Eskimos and the latter groups. When biacromial breadth is related to stature (Fig. 7.14) it is clear that within Eskimos higher average biacromial breadths are associated with greater stature in both sexes. However, at comparable statures among males the Finnish Lapps have consistently smaller biacromial breadths. Among females, Foxe Basin and Alaskan Eskimos have broader shoulders at comparable statures than either Finnish Lapps or West Greenland Eskimos. In terms of secular change, most groups show decreasing biacromial breadths relative to age.

Tables 7.20 and 7.21 demonstrate rather small but consistent increases in biiliac breadth with age. Finnish Lapps of both sexes have slightly smaller hip breadths than do the Eskimos. Relative to stature (Fig. 7.15) the Eskimos have greater biiliac breadths than the Lapps although the difference is more marked for males than for females. There is also a tendency for hip breadths to decrease with increasing stature among all of the Eskimo samples except West Greenland females. Finnish Lapp males show a similar trend but the females do not.

The differences, then, between the groups with respect to sitting height and sitting height-to-stature relationships are equivocal. Alaskan Eskimos and Finnish Lapps show similar relationships while Foxe Basin and Quebec Eskimos suggest greater ratios than the other two groups. For the other trunk dimensions – shoulder and hip breadths – the population differences appear less equivocal. Eskimos tend to have broader biacromial and biiliac breadths than the Finnish Lapps. Shoulder breadths appear to increase and hip

251

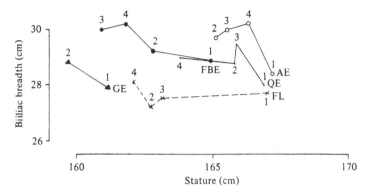

Fig. 7.15. Relationships between mean biiliac breadths and mean statures for all adult males. Abbreviations as in Figs. 7.1 and 7.13.

breadths appear to decrease with increasing stature for most of the groups and both sexes.

Upper-arm circumference and triceps skinfold

Descriptive statistics for upper-arm circumferences and triceps skinfolds of adult males and females can be seen in Tables 7.20 and 7.21, respectively. Except for Finnish Lapp males, Quebec Eskimo males and Greenlandic Eskimo females, all groups display a tendency for upper-arm circumference values to increase to 41–50 yr of age. The former two groups appear to maintain the same averages throughout all their adult years. Within the 51–60 yr age interval most of the groups have decreased upper-arm circumference means compared to their 41–50 yr levels. The exception to this is the Foxe Basin Eskimo females who increase upper-arm circumferences in later years. Although West Greenland Eskimo males have higher averages for this circumference at ages 41–60 yr than at the earlier 21–40 yr interval, West Greenlandic females reverse this relationship. Alaskan Eskimo males and females have the highest upper-arm circumferences of any of the groups. The difference is 2–3 cm for males while among females the Alaskans are about 1 cm above Finnish Lapp women at younger ages (2 cm in later years) and these Lapps are about 2–3 cm above the Foxe Basin Eskimo females in arm circumference. As was mentioned above regarding upper-arm circumference measurements on the children, the Alaskan Eskimo result may be partly attributed to a different measurement technique rather than being a true population difference.

Quebec Eskimo and Alaskan Eskimo males along with Finnish Lapp males have similar values for triceps skinfolds in all the age categories. The Foxe Basin Eskimo males fall about 3 mm less on average but all four groups show no tendency for the skinfold to increase with increasing age. The situation is somewhat different for females. Here Alaskan Eskimos and Finnish Lapps share similar values but the difference between them and Foxe Basin Eskimo females is very marked – the latter have approximately one-half the values of the former groups. Sexual dimorphism can also be seen to be rather pronounced for this measurement with no overlap in the observed values of the means except for Foxe Basin Eskimo females in the youngest age category. The females, of course, have higher values than males for the triceps skinfold.

In summary, with respect to average stature values of adults, the Lapps in this analysis cannot be separated from the Eskimos. Both sexes of all groups tend to show the common pattern of stature change with time, namely, stature increase (secular trend), stability, and stature decrease (ageing). The secular change among Lapps may be somewhat greater than that seen in other groups compared here. Body-weight differences found among the different Eskimo groups are to a large extent related to statural differences. The overall lower body weight of the Lapps, however, occurs in spite of similar stature averages. Thus, the Eskimos have a more sturdy body build than Lapps. Trunk dimensions indicate that this finding is also associated with a more sturdy skeleton among the Eskimos compared to Lapps. Higher muscularity among Eskimos is also suggested by the comparisons of both upper-arm circumference and triceps skinfold values. Sitting height-to-stature ratios are similar among both Eskimos and Lapps, making the broader shoulders and pelves of the Eskimos again reflect the factor of greater sturdiness. These body-build differences between the Eskimos and Lapps are more probably a reflection of genetic differences than simple environmental correlates.

Discussion

The various subgroups of Lapp and Eskimo children are similar with respect to stature and body weight. They both lag behind Icelandic children in the sub-Arctic region of Iceland in both overall stature and body weight attained, although the Icelandic children have the lowest body weight relative to stature. Icelandic children, genetically similar to Norwegian children, have much the same stature and body-weight means, ranges and relationships as Norwegian and Scandinavian children who are growing up under excellent socio-economic conditions (Pálsson *et al.*, 1972). A slower rate of growth among both Eskimo and Lapp children appears established by 7–8 yr of age and it becomes accentuated after that age. This is a typical pattern displayed when growth retarding factors are of environmental origin. Well-documented genetic differences between Eskimos and Lapps would suggest that the similarity of their growth patterns in this respect might not be under the same genetic control.

The human biology of circumpolar populations

The retrospective view of growth possible through an examination of adults placed in decade age-groups, makes possible a view of secular changes between roughly 1920 and 1950. More change can be seen among Lapps across this time period than can be seen for the Eskimo groups. In the last decade, Eskimos and Lapps have had a secular stature increase of 1–2 cm and 3–4 cm, respectively. Hybridization has occurred to differing extents in the groups compared here. It is interesting therefore that even those Lapp groups that have remained relatively genetically isolated have participated in the upward change (Mellbin, 1962; Lewin, Jürgens & Louekari, 1970; Skrobak-Kaczynski *et al.*, 1971; Skrobak-Kaczynski, Lewin & Karlberg, 1974).

Two decades earlier the data presented here demonstrate little or no secular change, while other studies have shown that secular increases in stature among Eskimos and Lapps may have begun in the 1920s (Lewin & Hedegård, 1971; Jørgensen & Skrobak-Kaczynski, 1972). Secular increases began much earlier (around 1850) for European and American populations (Bakwin & McLaughlin, 1964; Udjus, 1964; Tanner, 1968). The latter groups have had a rate of increase of approximately 1 cm/decade while the former groups have recently increased at a greater rate. Improvements in the level of health care and knowledge of hygienic conditions certainly must be involved in these secular trends. The contribution of hybridization is not measured but at least for some Lapps can be shown to have minimal impact. Reduction in dietary deficiencies or dietary changes must also be involved. The fact that obesity is becoming an increasing problem especially among women might suggest that the dietary changes that have occurred are not all improvements.

Eskimos, both as children and as adults, have broader shoulders and pelves, greater weight for stature, but similar sitting height-to-stature ratios to those found among Lapps. The sturdier trunk builds occur furthermore in Eskimo children who are both at and below Lapp children nutritionally (assuming triceps skinfold as a crude nutritional measure). These body-build differences appear more likely to be genetically than environmentally controlled. The comparison with Icelandic children might suggest that both of the former populations display a degree of adaptation to their cold, harsh climates through body build. Greater sturdiness implies greater mass per unit surface area which has been suggested as a means of reducing heat loss to the environment. Surely, the primary means of adaptation for both Eskimos and Lapps is through behavioral modes but particularly the Eskimos may also demonstrate evidence of the biological adaptation as well.

References

Auger, F. (1976). Growth patterns of Fort Chimo and Spotted Island Eskimos. In *Circumpolar Health*, ed. R. J. Shephard & S. Itoh, p. 266. Toronto: University of Toronto Press.

Bakwin, H. & McLaughlin, S. D. (1964). Secular increase in height. Is the end in sight? *Lancet*, **2**, 1195–6.

de Peña, J. (1972). Growth and development. In *IBP Annual Report No. 4, Human Adaptability Project, Univ. of Toronto Anthropological Series*, No. 11, ed. D. R. Hughes, pp. 47–69.

Jamison, P. L. (1976). Growth of Eskimo children in northwestern Alaska. In *Circumpolar Health*, ed. R. J. Shephard & S. Itoh, pp. 223–9. Toronto: University of Toronto Press.

Jamison, P. L., Zegura, S. L. & Milan, F. A. (1978). *Eskimos of Northwestern Alaska: A Biological Perspective*. Stroudsburg, Pennsylvania: Dowden, Hutchinson & Ross.

Jørgensen, J. B. & Skrobak-Kaczynski, J. (1972). Secular changes in the Eskimo community of Aupilagtoq. *Z. Morph. Anthrop.*, **64**, 12–19.

Lewin, T., Jürgens, H. W. & Louekari, L. (1970). Secular trend in the adult height of Skolt Lapps. *Arctic Anthrop.*, **7**, 53–62.

Lewin, T. & Hedegård, B. (1971). Anthropometry among Skolts, other Lapps and other ethnic groups in northern Fennoscandia. In *Introduction to the Biological Characteristics of the Skolt Lapps*, ed. T. Lewin, *Proc. Finnish Dent. Soc.*, **67**, Suppl. 1, 71–98.

Mellbin, T. (1962). The children of Swedish nomad Lapps. A study of their health, growth and development. *Acta paediat. Uppsala*, **51**, Suppl. 131.

Pálsson, J. O. P., Lewin, T., Sigholm, G., Karlberg, J. & zu Dohna, B. (1972). Somatometri hos skolbarn i nordöstra Island. I stencil, 49 pp. *Rapport till Nordiska humankologiska forskargruppen.*

Schaefer, O. (1973). The changing health picture in the Canadian North. *Can. J. Ophth.*, **8**, 196–204.

Schaefer, O. (1977). Are Eskimos more or less obese than other Canadians? A comparison of skinfold thickness and ponderal index in Canadian Eskimos. *Amer. J. clin. Nut.*, **30**, 1623–8.

Skrobak-Kaczynski, J., Torp, K., Vandbak, Ø. & Andersen, K. Lange. (1971). *Growth pattern of Lappish children in Kautokeino*. IBP–HA, Oslo, Report No. 11.

Skrobak-Kaczynski, J., Lewin, T. & Karlberg, J. (1974). Secular changes in body dimensions in a homogeneous population. *Nordic Council for Arctic Medical Research Report*, **8**, 17–46.

Tanner, J. M. (1968). Growth and physique in different populations of mankind, as a guide to health and economic progress, and a contribution to the science of human biology. *Materialy I Prace Anthropolog, Wroclav*, **76**, 29–46.

Udjus, L. G. (1964). *Anthropometrical changes in Norwegian men in the twentieth century*. Oslo: Univ. forl.

Weiner, J. S. & Louric, J. A. (1969). *Human Biology*. IBP handbook No. 9. Philadelphia: F. A. Davis Co.

8. Nutrition

H. H. DRAPER

This chapter contains a summary of nutritional studies on circumpolar peoples carried out under the auspices of the International Biological Program, from published and unpublished reports by IBP investigators. Emphasis is given to studies bearing on processes of adaptation to the aboriginal diet and to recent changes in the diet associated with modernization. Priority is also given to topics of research on which comparable data have been collected on several circumpolar populations. Of necessity, treatment of the data is brief and numerous significant findings have been omitted. For convenience, the studies reviewed have been divided into several categories according to whether they deal with dietetics, biochemistry or one of several specialized topics.

Dietary studies

Lapps

Recent diet surveys conducted on several circumpolar populations attest to the inroads of processed foods into the aboriginal diet. In 1969–70 Hasunen & Pekkarinen (1975) compared the diet of the Skolt Lapps of Svettijärvi in northeast Finland, among whom the traditional nomadic way of life is still practiced, with that of the Lapp inhabitants of Utsjoki and Karigasniemi. The proportions of protein, fat, and carbohydrate in the diets of the two groups were not significantly different (Table 8.1). However, the intake of

Table 8.1. *Sources of energy (% of calories) in the adult diet of two Finnish Lapp communities (Hasunen & Pekkarinen, 1975)*

	Protein	Fat	Carbohydrate
Svettijärvi (1970)			
Males, winter	17	36	47
Males, summer	15	29	56
Females, winter	16	31	53
Females, summer	15	29	54
Utsjoki–Karigasniemi (1969)			
Males	16	38	46
Females	15	35	50

257

The human biology of circumpolar populations

Table 8.2. *Mean daily intake of nutrients by Finnish Lapps* (*Hasunen & Pekkarinen, 1975*)

| | Svettijärvi | | | | Utsjoki–Karigasniemi | |
| | Male | | Female | | | |
	Winter	Summer	Winter	Summer	Male	Female
Protein, g	89	82	67	63	153	96
Fat, g	86	74	58	53	159	102
Carbohydrate, g	295	302	210	244	459	343
Energy, kcal	2164	2267	1714	1653	3800	2620
Calcium, mg	439	428	359	338	1564	1087
Vitamin A, IU	8419	1853	5901	1657	7736	7300
Thiamin, mg	1.2	1.1	0.9	0.9	2.6	1.7
Riboflavin, mg	1.6	1.3	1.1	0.8	3.9	2.6
Niacin, mg	16.8	16.7	11.5	10.4	28.9	17.0
Vitamin C, mg	47	31	34	29	88	85

protein, energy and several key nutrients was higher among the Lapps of Utsjoki–Karigasniemi (Table 8.2). Only the mean intakes of protein and vitamin A by the Skolt Lapps met the standards adopted for the study, whereas the nutrient intakes of the Utsjoki–Karigasniemi population were found to be generally adequate.

Some extraordinarily high intakes of vitamin A were recorded among Finnish Lapps. One-quarter of adult males consumed 12 000–53 000 IU per day in winter. Much of this intake was derived from reindeer liver. Even higher intakes of vitamin A (over 62 000 IU per day) were observed in some Norwegian Lapp farmers of the Kautokeino and Karasjok regions.

Diet surveys also were carried out in six Lapp communities in Northern Norway (Gassaway, 1969). The proportion of calories from fat (about 40%) was somewhat higher than that in Finnish Lapps and carbohydrate energy was slightly lower; in fact, calorie distribution in the diet resembled that of the mixed diet of industrialized countries. However, some migratory reindeer hunters derived up to 23% of their calories from protein. The intake of several nutrients was marked by wide variations which could be related to differences in occupation (dairy farming, reindeer herding, fishing, laboring).

Øgrim (1970) has investigated the pattern of food consumption by Lapps living in the municipalities of Polmak and Gamvik and on the Finnmark mountain plain of northern Norway. Adult males living on the plain consumed large quantities of reindeer meat, whereas those in Polmak and Gamvik relied heavily on fish. Protein intake was high in all three areas (34–52 g per 1000 kcal diet) and fat constituted 40–46% of dietary energy. Ascorbic-acid, calcium and vitamin-D intakes were below Norwegian standards, whereas other nutrients were generally in good supply. Lapps inhabiting the

Finnmark plain consumed more protein, iron, vitamin A and B-complex vitamins than did the general Norwegian population.

Eskimos

Diet surveys have recently been carried out in conjunction with biochemical assessment of nutritional status in Greenland, Canadian and Alaskan Eskimos.

In 1972, Bang, Dyerberg & Hjorne (1976) compiled seven-day diet records on seven subjects considered representative of the adult inhabitants of the coastal Eskimo settlement of Igdlorssuit in northwest Greenland. The diet, consisting of a mixture of indigenous foods (seal, whale and fish) and commercial items (bread, biscuits, sugar, margarine, potatoes and beer), contained over twice as much protein and substantially less carbohydrate than that of the general population of Denmark (Table 8.3). The authors cite data recorded in northwest Greenland in 1914 indicating that at that time protein provided 44% of diet calories, fat 47% and carbohydrate 8%. From hunting statistics it was estimated that in 1970 the average daily consumption of seal and/or whale meat by inhabitants of all ages in the Umanak region of northwest Greenland was about 400 g. In Igdlorssuit the *per capita* consumption of major commercial foods averaged 134 g of bread, biscuits and rye flour, 31 g of rice, 42 g of potatoes and 164 g of sugar. In an investigation of the fatty-acid composition of the diet, substantial differences from ordinary Danish food were found. The most pronounced differences were a larger amount of mono-unsaturated fatty acids and a lower amount of saturated acids though there was a similar amount of polyunsaturated acids. In contrast to Danish food there were relatively large concentrations of some long chain polyunsaturated fatty acids ($C_{20:5}$ and $C_{22:6}$) in the Eskimo diet.

In 1969 Auger (1974) carried out an anthropometric study on the Eskimo and Metis (French Canadian–Indian hybrids) of Fort Chimo, New Quebec, Canada. Body weight and skinfold measurements, expressed as a function of adult age, revealed a marked difference between the two groups (Figs. 8.1, 8.2). No increment in body weight could be detected in Eskimos between the ages of 25 and 55 yr, whereas a substantial increase was seen in Metis (Fig. 8.1). Major increases in subscapular, iliac and tricipital skinfold thickness

Table 8.3. *Sources of calories* (%) *in the diet of Igdlorssuit Eskimos and of the general population of Denmark* (*Bang* et al., *1976*)

	Eskimos	Danes
Protein	26.2	11.3
Fat	37.1	41.7
Carbohydrate	36.7	47.0

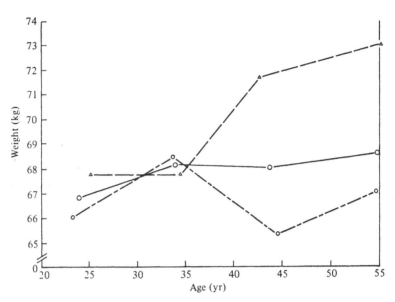

Fig. 8.1. Body-weight changes as a function of age in Metis and Eskimos of New Quebec (Auger, 1974). ○———○, total sample; △— —△, Metis; ○— – —○, Eskimos.

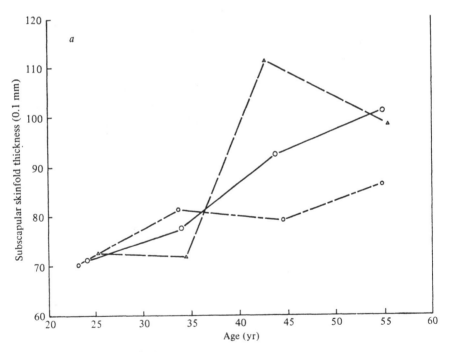

Fig. 8.2a. Subscapular skinfolds vs age in Metis and Eskimos of New Quebec (Auger, 1974). Total sample, ○———○; Metis, △— —△; Eskimos, ○— – —○.

260

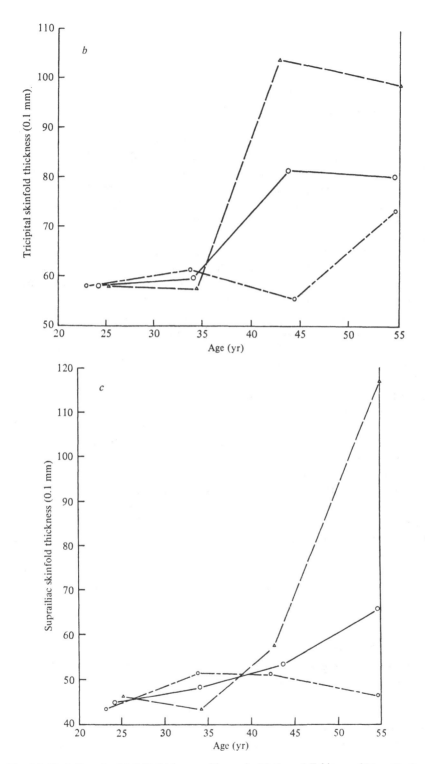

Fig. 8.2. Variations in skinfold thickness with age in Metis and Eskimos of New Quebec (Auger, 1974). *b*, tricipital skinfold; *c*, suprailiac skinfold. ○——○, total sample; △— —△, Metis; ○— – —○, Eskimos.

occurred in Metis as opposed to modest increases in Eskimos (Fig. 8.2). The tendency to adiposity in the Metis was associated with a more sedentary life style and with greater access to a regular supply of food.

During the summer of 1971 and the winter of 1972, three- to four-day dietary records were obtained on inhabitants of the northern Alaskan Eskimo villages of Wainwright and Point Hope (Bell & Heller, 1978). Nutrient intakes, calculated from published data on the composition of commercial and native foods, were compared with the Recommended Dietary Allowances (RDA) proposed by the National Research Council of the US National Academy of Sciences or, in the case of calcium, with FAO/ WHO standards.

The native diet at Wainwright consists largely of caribou, seal and beluga whale, and of lesser amounts of walrus, birds and fish. At Point Hope, baleen and beluga whale, seal and fish are the native staples. The estimated percentages of major nutrients derived from native foods at these two locations are shown in Table 8.4. The data indicate that Wainwright adults obtained somewhat less than half of their calories from native foods (women more than men) and children about one-quarter. At Point Hope indigenous foods supplied 22% of calories in the adult diet and only 8% in the diet of children. A comparison of the relative contributions of native and non-native foods to the intake of calories and other nutrients indicates that the native regimen was richer in protein, phosphorus, iron, vitamin A, thiamin, riboflavin and niacin, lower in calcium, and much lower in carbohydrate and vitamin C. There is also striking evidence of erosion of the traditional diet culture among children.

The contributions of protein, fat and carbohydrate to diet calories at Wainwright and Point Hope are shown in Table 8.5. The data provide additional evidence that the incursion of commercial foodstuffs has been greater at Point Hope than at Wainwright, and that this incursion has affected children more than adults. Considering the low carbohydrate content of native foods (Table 8.4), the magnitude of the carbohydrate contribution to total diet calories is impressive. The commercial items recently introduced into the Eskimo dietary appear to be higher in carbohydrate than the general US mixed diet.

Although nutrient intakes calculated from the diet records on Alaskan Eskimos indicated that significant portions of the population consumed less than the RDA for several nutrients, there was little biochemical evidence of undernutrition (Bergan & Bell, 1978). It must be remembered that (except for calories) the RDA exceeds the nutrient requirement for most individuals, and hence a value below the RDA does not necessarily reflect an inadequate intake. Although the diet records indicated that low calorie intakes were also prevalent, there was little evidence of substandard body weights or heights, suggesting that (as in many other diet surveys) food consumption was under-

Table 8.4. *Percent of nutrients obtained from native foods (Bell & Heller, 1978)*

Subjects	N	Calories	Protein	Fat	Carbo-hydrate	Calcium	Phos-phorus	Iron	Vitamin A	Vitamin C	Thiamin	Ribo-flavin	Niacin
Wainwright 1971													
Male adults	9	42.6	79.4	47.7	0.5	27.5	74.1	78.0	72.5	0	56.8	78.5	75.4
Female adults	7	52.2	83.3	62.7	0	29.7	76.8	83.4	53.9	0	59.4	77.7	74.4
Children	16	23.2	55.6	30.0	0.05	6.4	41.1	57.6	41.1	0	31.7	46.6	52.0
Wainwright 1972													
Male adults	17	33.0	68.1	44.7	0.5	24.4	49.0	78.1	76.3	0	43.5	63.4	57.9
Female adults	15	45.5	73.2	60.3	0.9	18.6	59.8	72.8	82.9	1.4	58.4	68.4	69.6
Children	21	25.3	55.6	42.0	0.7	18.1	37.8	78.6	69.7	1.1	22.6	45.9	51.3
Wainwright Average	–	34.1	66.0	45.8	0.5	19.2	51.6	73.8	67.0	0.5	41.5	59.6	60.4
Point Hope													
Male adults	16	23.9	62.0	25.9	0.5	10.3	39.5	70.3	40.9	19.1	36.7	30.3	63.2
Female adults	16	19.8	51.8	22.3	0	11.4	36.9	58.5	28.5	0	27.3	47.5	53.0
Children	12	8.1	22.0	12.5	0	2.5	9.5	38.6	11.0	0	7.8	17.7	26.2
Point Hope Average	–	18.1	47.4	20.9	0.2	8.6	30.4	57.4	28.2	6.9	25.4	33.1	49.4

263

Table 8.5. *Percentage of calories from protein, fat and carbohydrate in the diet of Alaskan Eskimos (Bell & Heller, 1978)*

	Protein	Fat	Carbohydrate
Wainwright			
Adults	25.0	43.1	31.9
Children	18.3	38.4	43.3
Point Hope			
Adults	21.6	35.5	42.8
Children	15.2	37.6	47.2

estimated. However, it was also apparent that calcium and vitamin C were frequently consumed by adults in less than recommended amounts. Children received additional supplies of these nutrients in milk and juice furnished at school.

Ainu

From 1967 to 1972 Koishi and coworkers (Koishi *et al.*, 1975) carried out a comparative study of the diets of the ordinary Japanese Wajin and Ainu inhabitants of the island of Hokkaido. The historic diet of the Ainu consisted of deer, bear, whale, seal and fish and the roots of wild plants.

Skinfold measurements taken at 10 sites revealed that Ainu male junior-high-school students and females were thinner than age-matched Wajin (Fig. 8.3). Ainu subjects under 20 yr of age were shorter and lighter than Wajin in general, but beyond this age there was no difference between the races. Food intake was estimated from four-day records and nutrient consumption

Table 8.6. *Daily nutrient intake of Wajin and Ainu students at the junior high*

Sex	Race	N	Height (cm)	Weight (kg)	Energy (kcal)	Protein (g)
Male	Wajin	41	153.8 ± 9.6	43.7 ± 8.3	2313 ± 385	73.7 ± 13.7
	Ainu	17	156.0 ± 6.7	44.7 ± 7.0	2433 ± 565	65.4 ± 17.2
Female	Wajin	20	154.4 ± 4.7	46.5 ± 6.5	1964 ± 404	60.0 ± 14.7
	Ainu	6	152.8 ± 4.1	45.7 ± 4.9	1798 ± 162	56.5 ± 10.6
Male	Japanese Standard	–	160.5	50.0	2700	90.0
Female		–	154.0	48.5	2450	75.0

[a] Mean and s.D. Average age of subjects 14 yr.

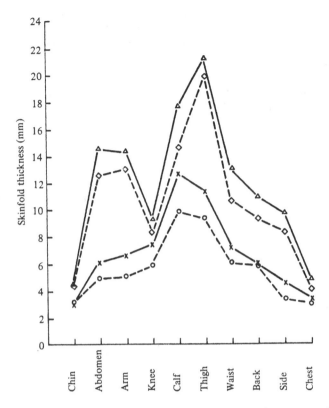

Fig. 8.3. Comparison of skinfold thickness in Ainu (x——x, male; o— —o, female) and Wajin (△——△, male; ◇— —◇, female) (H. Koishi *et al.*, 1975).

school in Nükappu and Usu in August 1969 and July 1971 (Koishi et al., *1975)*[a]

Fat		Calcium (mg)	Iron (mg)	Vitamins			
(g)	(% kcal)			A(IU)	B$_1$(mg)	B$_2$(mg)	C(mg)
50.2 ± 19.9	19.5	467 ± 352	12.5 ± 5.1	1169±1657	0.8 ± 0.2	0.9 ± 0.4	77 ± 60
44.4 ± 16.5	16.4	337 ± 186	11.6 ± 3.8	661 ± 521	1.1 ± 0.3	0.8 ± 0.4	50 ± 31
42.2 ± 13.3	18.2	386 ± 143	11.0 ± 2.3	762 ± 520	0.6 ± 0.2	0.7 ± 0.3	80 ± 49
21.8 ± 11.0	10.6	236 ± 58	8.9 ± 1.3	501 ± 96	0.5 ± 0.1	0.6 ± 0.1	57 ± 39
	20.0	900	12.0	2000	1.2	1.4	45
	20.0	900	15.0	2000	1.1	1.2	45

was calculated from standard tables of Japanese food composition (Table 8.6). No evidence of the traditional Ainu diet culture could be found; the modern Ainu food habits were qualitatively indistinguishable from those of low-income Japanese. The predominant foods of both groups were boiled white rice, potatoes, bread and vegetables. The Ainu diet was lower in protein (including animal protein) and fat, but was comparable to the Wajin diet in total calories, calcium, vitamin A, thiamin, riboflavin and vitamin C.

Biochemical studies

Assessment of nutritional status

Biochemical evaluations of nutritional status have been carried out in the course of IBP studies on Alaskan and Canadian Eskimos. A survey of Eski-

Table 8.7. *Thiamin and riboflavin status of Canadian and Alaskan Eskimos as assessed by urinalysis (Sayed et al., 1976; Bergan & Bell, 1978)*[a]

	Age (yr)	N	Status (%)		
			Acceptable	Low	Deficient
Hall Beach–Igloolik 1969–70					
Thiamin	4–15	16	100	0	0
	>15	27	100	0	0
Riboflavin	4–15	16	100	0	0
	>15	27	100	0	0
Wainwright 1971					
Thiamin	4–15	65	98	2	0
	>15	69	100	0	0
Riboflavin	4–15	69	95	4	1
	>15	65	100	0	0
Point Hope 1972					
Thiamin	4–15	77	97	3	0
	>15	81	99	0	1
Riboflavin	4–15	83	71	28	1
	>15	89	97	2	1
Kasigluk 1973					
Thiamin	4–15	79	94	5	1
	>15	41	85	10	5
Riboflavin	4–15	80	96	4	0
	>15	41	100	0	0
Nunapitchuk 1973					
Thiamin	4–15	85	100	0	0
	>15	53	100	0	0
Riboflavin	4–15	85	99	0	1
	>15	54	100	0	0

[a] Classified according to US Ten State Survey Standards.

mos living in Wainwright, Alaska, carried out in mid-winter 1969 indicated that, with the exception of iron-deficiency anemia in children, the residents were 'in a generally acceptable nutritional state' (Sauberlich *et al.*, 1970). Reflecting the high concentration of most B-complex vitamins in their native foods, the urinary excretion of these nutrients was also unusually high. One-third to one-half of the children up to 5 yr of age had low hemoglobin values.

Canadian Eskimos residing in Hall Beach and Igloolik, Northwest Territories, were all in adequate thiamin and riboflavin status according to current standards (Ellestád-Sayed *et al.*, 1975) (Table 8.7). This was also the general case with Alaskan Eskimos living in Wainwright and Nunapitchuk (Bergan & Bell, 1978). However, 28% of the Point Hope children sampled were found to be in low riboflavin status and there was evidence of inadequate thiamin nutriture in some children and adults at Kasigluk (Table 8.7). Urinary *N*-methylnicotinamide determinations on the Canadian cohort reflected an adequate niacin status (Ellestad-Sayed *et al.*, 1975).

Low hemoglobin values were observed by Bergan & Bell in about one-third of children and adults at Wainwright, Point Hope and Kasigluk, and in one-

Table 8.8. *Assessment of anemia in Canadian and Alaskan Eskimos* (*Sayed* et al., *1976; Bergan & Bell, 1978*)[a]

	Age (yr)	N	Status (%)		
			Acceptable	Low	Deficient
Hall Beach–Igloolik 1969–72					
Hemoglobin	2–5	5	100	0	0
	6–12	81	86	14	0
	13–16	35	88	9	3
	>16 (M)	92	63	37	0
	>16 (F)	46	94	6	0
Serum iron	2–5	7	71	29	–
	6–12	82	61	39	–
	>12	199	46	54	–
Wainwright 1971					
Hemoglobin	2–5	4	75	25	0
	6–12	41	59	39	2
	13–16	23	70	30	0
	>16	75	61	32	7
Point Hope 1972					
Hemoglobin	2–16	61	79	18	3
	>16	89	64	35	1
Kasigluk 1973					
Hemoglobin	>16	38	61	31	8
Nunapitchuk 1973					
Hemoglobin	2–16	95	42	46	12
	>16	53	40	53	7

[a] Classified according to US Ten-State Survey standards.

half at Nunapitchuk (Table 8.8). A smaller incidence of anemia was found in Hall Beach–Igloolik Eskimos, except among adult males, for whom the standard ($\geq$ 14 g/100 ml) appears to be relatively more stringent. Serum iron determinations revealed 'less than acceptable' levels in about half the adults and in about a third of the children. These findings indicate that moderate anemia, attributable largely to a low iron intake, is a prevalent nutritional problem among US and Canadian Eskimos, as it is among the non-Eskimo populations of these two countries. This problem undoubtedly has been aggravated by a declining consumption of iron-rich native foods.

Vitamin-C levels in blood plasma were found to be adequate in nearly all subjects at Wainwright and Point Hope, where ascorbic acid-fortified beverages were supplied at school and widely consumed at home. Plasma vitamin-A levels were also generally satisfactory in these villages, but at Nunapitchuk, where the processed diet clearly predominates, one-third of the children and adults were in 'less than acceptable' status.

An unexpected finding among Canadian Eskimos residing at Hall Beach and Igloolik was a high prevalence of substandard serum folacin concentra-

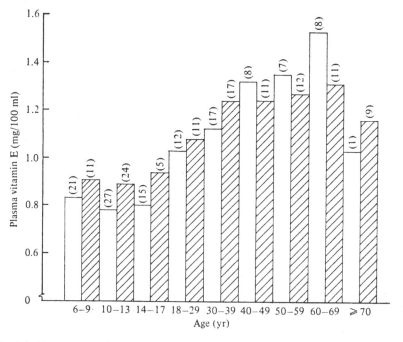

Fig. 8.4. Plasma α-tocopherol levels as a function of age in northern Alaskan Eskimos in Wainwright, □— and Point Hope⧄—. Number of subjects in parentheses. (Wo & Draper, 1975.)

tions. Of 82 subjects examined, only four had 'acceptable' serum folacin levels, 36 were classified as 'low' and 42 as 'deficient'. Similar results were obtained on Eskimos living in four northern Canadian settlements who were examined as part of a national nutrition survey (Forbes, 1974). However, there was little evidence of macrocytic anemia of the folacin deficiency type. The folacin status of Eskimos requires further investigation to reconcile the biochemical and hematological findings and to establish the prevalence of inadequacy.

In addition to folacin deficiency, the general nutritional status of the Canadian Eskimos examined during the national survey was inferior to that

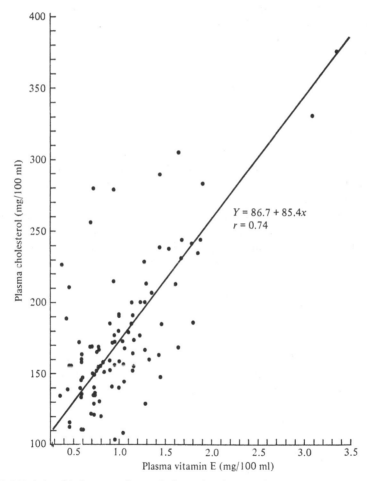

Fig. 8.5. Relationship between plasma cholesterol and α-tocopherol levels in north Alaskan Eskimos (Wo & Draper, 1975).

seen in the Canadian IBP cohort, as reflected in clinical signs of scurvy, widespread anemia and low vitamin A intakes. The nutrition of Eskimos in the national survey was observed to be inferior to that of Indians which, in turn, was inferior to that of the general Canadian population (Forbes, 1974).

The native diet of the Arctic Eskimo is virtually devoid of plant foods, which constitute the main source of vitamin E in the mixed diet of industrialized societies. In addition, marine mammals and fish contain a high proportion of polyunsaturated fatty acids which enhance the vitamin-E requirement. These considerations prompted an investigation of the vitamin-E status of Eskimos residing in the Alaskan villages of Wainwright, Point Hope, Kasigluk and Nunapitchuk (Wo & Draper, 1975). The data for Wainwright and Point Hope are summarized in Fig. 8.4. Plasma vitamin-E levels at all ages were at least equivalent to those reported for populations consuming a mixed diet. A linear relationship was observed between plasma vitamin-E and cholesterol concentrations (Fig. 8.5). Although the Eskimo native diet is lower in total vitamin E than the mixed diet, the vitamin is present almost exclusively as α-tocopherol, the most active naturally occurring form.

Lipid metabolism

Studies on various aspects of lipid metabolism recently have been carried out in several Arctic populations. In general, these studies have been designed to elucidate the role of dietary factors in the low incidence of hypercholesterolemia, cardiovascular disease and diabetes in these populations.

1. *Eskimos.* Bang & Dyerberg (1972) compared the plasma lipid and lipoprotein pattern of 130 adult Eskimos living on the west coast of Greenland with that of an age-matched sample of Eskimos and Danes living in Denmark. Significantly lower concentrations of total lipids, cholesterol, triglycerides, β-lipoproteins and pre-β-lipoproteins were found in Greenland Eskimos than in Danes (Table 8.9). Eskimos living in Denmark had values similar to those of Danes. The authors concluded that the low plasma lipid and lipoprotein concentrations in Greenland Eskimos are caused by environmental rather than genetic factors. They speculate that this lipid pattern, and the low incidence of coronary atherosclerosis associated with it, may be related to the higher amount of certain long-chain polyunsaturated fatty acids in the Eskimo native diet. They further speculate that the rare occurrence of diabetes in Greenland Eskimos may be associated with their low concentrations of plasma triglycerides and pre-β-lipoproteins.

In a continuation of these studies, Dyerberg, Bang & Hjorne (1975) compared the composition of plasma esterified fatty acids in Greenland Eskimos, Eskimos living in Denmark and Danes. Compared to Danes or Denmark Eskimos, Greenland Eskimos had higher proportions of palmitic, palmitoleic and timnodonic (eicosapentaenoic) acids but only one-third to one-half the

Table 8.9. *Plasma lipid and lipoprotein concentrations in Greenland Eskimo women and in Eskimo and Danish women living in Denmark* (*Bang & Dyerberg, 1972*)[a]

	Greenland Eskimos	Denmark Eskimos	Danes
Total lipids, g/l	5.93 ± 0.16	7.32 ± 0.27	6.55 ± 0.16
Cholesterol, mmol/l	5.58 ± 0.24	7.30 ± 0.25	6.77 ± 0.20
Triglycerides, mmol/l	0.43 ± 0.03	1.12 ± 0.10	0.98 ± 0.06
Phospholipids, mmol/l	2.92 ± 0.25	3.18 ± 0.09	2.91 ± 0.08
Chylomicrons, g/l	0.27 ± 0.03	0.24 ± 0.02	0.15 ± 0.02
β-lipoproteins, g/l	4.27 ± 0.16	5.00 ± 0.25	4.58 ± 0.19
Pre-β-lipoproteins, g/l	0.44 ± 0.05	1.10 ± 0.13	1.05 ± 0.08
α-lipoproteins, g/l	3.64 ± 0.23	4.24 ± 0.27	3.64 ± 0.16

[a] Mean and s.d. $N = 25–41$. Mean age = 38.0–41.1 yr.

proportion of linoleic acid. Although the concentration of total polyunsaturated fatty acids was lower in the Greenland subjects, certain highly unsaturated acids were present in much larger proportions. Timnodonic acid, which is found in negligible amounts in the plasma lipids of Danes and other Western societies, constituted up to 16% of the plasma esterified fatty acids of Greenland Eskimos. The Greenland residents had a higher proportion of saturated acids than Danes and a similar proportion of monoenes (Table 8.9). These findings suggest that the most important aspect of the plasma-lipid pattern of Eskimos from the standpoint of the prevention of atherosclerotic heart disease may not be the concentration of total lipids or the ratio of polyunsatutated to saturated fatty acids, but the composition of the polyunsaturated acids themselves.

A study of lipid and carbohydrate metabolism was conducted in Point Hope, Alaska by Feldman and associates (1972, 1975). A total of 168 subjects representing two-thirds of the population over the age of six yr was sampled. Table 8.10, which contains the data from that study, indicates that northern Alaskan Eskimos have, on the average, significantly lower fasting serum triglycerides (69 mg/100 ml) and very low density lipoproteins (35 mg/100 ml) than does the US white population. These findings were consistent with their relatively high fat and low carbohydrate diet. The hyper-free fatty acidemia, which rose to a mean of 49 mg/100 ml without ketosis in persons fasted more than 18 h, may represent an adaptation by the Eskimo to his habit of consuming only one meal per day.

Table 8.11 indicates the effect of a much higher carbohydrate and lower protein intake among a sample of Eskimo youth. 41 age-matched northern Alaskan Eskimos residing at a boarding school at Mt Edgecumbe, Alaska, were compared with their cohorts in their native village. There was marked elevation of serum triglycerides, a slight, albeit non-significant, rise in very

Table 8.10. *Mean fasting serum-lipid levels in Eskimos residing in Point Hope, Alaska (Feldman et al., 1972)*

Age (yr)	Total cholesterol (mg/100 ml)	Triglycerides (mg/100 ml)	Free fatty acids (mg/100 ml)	Very low density lipoproteins (mg/100 ml)	Low density lipoproteins (mg/100 ml)	High density lipoproteins (mg/100 ml)
6–12	200	61	52			
13–17	218	65	39			
18–24	219	86	34			
25–35	224	71	33			
36–60	231	81	34			
61–82	250	88	34			
Males, all ages	226	71	31	28 ± 9[a]	299 ± 18[a]	288 ± 34[a]
Females, all ages	217	69	41	36 ± 11[a]	345 ± 21[a]	310 ± 37[a]

[a] S.D.

Table 8.11. *Fasting lipid, glucose and insulin levels in the plasma of two groups of young adult Alaskan Eskimos (Feldman* et al.*, 1975)*

	Point Hope[a]	Boarding School[a]
Cholesterol, mg/100 ml	219.0 ± 49.1	144.3 ± 32.3[b]
Triglycerides, mg/100 ml	70.5 ± 9.7	112.5 ± 32.3[b]
Very low density lipoproteins, mg/100 ml	19.0 ± 8.5	26.0 ± 10.0
Low density lipoproteins, mg/100 ml	269.2 ± 30.4	279.4 ± 34.3
High density lipoproteins, mg/100 ml	214.0 ± 42.0	214.0 ± 51.0
Glucose, mg/100 ml	86.2 ± 18.1	71.8 ± 18.4
Insulin, μU/ml	18.2 ± 6.1	10.2 ± 4.8
Ratio of integrated insulin divided by glucose areas over two-hour period (glucose tolerance test)	0.38 ± 0.05	0.56 ± 0.04[b]

[a] All values are means ± S.D.
[b] Significantly different from comparable value ($P < 0.001$).

low density lipoproteins, and a fall in serum cholesterol among the boarding-school sample, all consonant with the dietary change. Levels of fasting serum glucose and insulin were not different between the two subject pools and each group exhibited similar and normal two-hour oral glucose tolerance test results. Tolbutamide-tolerance tests given the north slope sample revealed a prompt hypoglycemic effect suggesting adequate insulin reserves and normal sensitivity to the hormone.

The northern Eskimos exhibited serum cholesterol levels generally within the normal range (Table 8.10). The relationship between age (X) and serum cholesterol (Y) was $Y = 201.7 + 0.7 X$ (S.E.E. $= 4.3$). There was no significant sex difference. A cholesterol-balance study conducted among eight adult northern Eskimos revealed a high efficiency of absorption of dietary cholesterol (about 48%) which remained linear over a range of 420 to 1650 mg per day. Total cholesterol input (exogenous absorption plus endogenous synthesis) remained fairly constant at about 1315 mg/day although the turn-over rate of cholesterol did increase with the amount absorbed. The relation between serum cholesterol level (Y) and cholesterol absorbed (X) was cholesterol $Y = 211.3 + 0.048 X$. There was no direct evidence for suppression of endogenous synthesis at the levels of dietary intake observed during the study. Maximal suppression of cholesterol biosynthesis (an estimated 36%) was effected by the lowest amount of cholesterol consumed (420 mg/day).

A latitudinal gradient was observed in plasma-cholesterol concentration in Alaskan Eskimos which was associated with increasing degrees of dietary acculturation (Table 8.12) (Draper, 1976). In the twin villages of Kasigluk and Nunapitchuk in southwest Alaska, where processed foods are predominant, about 40% of the adult population exhibited hypercholesterolemia. A

Table 8.12. *Plasma cholesterol levels in Eskimos living in three Alaskan villages* (*Draper, 1976*)

Site	N	Total cholesterol (mean ± s.d.)			% Hypercholesterolemia[a]		
		Males	Females	Both	Males	Females	Both
		Children (7–16 yr)					
Wainwright	63	145 ± 21	151 ± 29	149 ± 46	0	2.4	1.6
Point Hope	56	164 ± 22	168 ± 27	166 ± 24	0	0	0
Nunapitchuk	83	201 ± 25	205 ± 30	202 ± 27	14.9	25.0	19.3
		Adults (18–74 yr)					
Wainwright	66	187 ± 41	205 ± 57	197 ± 51	6.7	13.8	10.6
Point Hope	82	212 ± 39	210 ± 38	211 ± 38	10.3	14.0	12.2
Kasigluk–Nunapitchuk	79	247 ± 36	260 ± 48	256 ± 45	38.9	42.6	41.8

[a] Defined as < 230 mg/100 ml in children and < 260 mg/100 ml in adults.

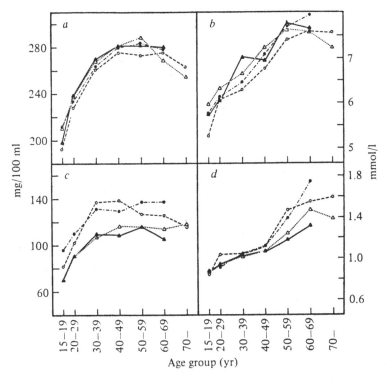

Fig. 8.6. Serum cholesterol (*a*, males; *b*, females) and triglyceride (*c*, males; *d*, females) levels in various Finnish populations ▲——▲, Lapps; △· · ·△, rural Finns; O– – –O, semi-urban Finns; ●– · – · –●, Finns in industry. (Bjorkstene *et al.*, 1975).

Table 8.13. *Fasting levels of plasma free fatty acids and ketone bodies in Ainu and Japanese* (*Itoh, 1974*)

	N	Free fatty acids μEq/1 (means $\pm$ S.D.)	Ketones μMol/1 (means $\pm$ S.D.)
Japanese			
Farmers	10	419 $\pm$ 36	612 $\pm$ 27
Students	17	421 $\pm$ 16	695 $\pm$ 43
Policemen	14	376 $\pm$ 32	680 $\pm$ 63
Fishermen	21	382 $\pm$ 37	528 $\pm$ 42
Ainu	11	285 $\pm$ 17	1063 $\pm$ 107

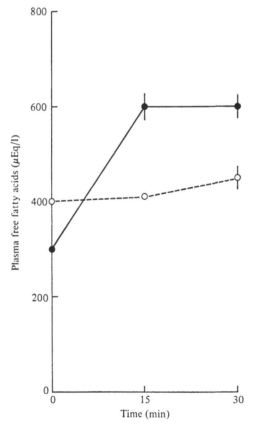

Fig. 8.7. Effect of norepinephrine (0.025 mg/10 kg) on plasma free fatty acid concentration in Asahikowa Ainu (●——●) and Sapporo students (○– – –○) (Itoh, 1974).

275

rural–urban gradient in serum cholesterol content among Alaskan Eskimos has also been reported by Maynard (1974). It is apparent that the historic pictogram of low blood-cholesterol levels and blood pressures in Alaskan Eskimos is no longer valid.

2. *Lapps*. Bjorksten and coworkers (Bjorksten *et al.*, 1975) have conducted a large scale survey of serum cholesterol and triglyceride levels in Lapps and Finns. There were no marked differences between the serum and lipid pattern of Lapps and that of Finns or most other Nordic populations, despite the fact that there are significant differences in mortality caused by coronary heart disease (Fig. 8.6). Finns were observed to have one of the highest mean cholesterol concentrations recorded for any adult population (about 260 mg/100 ml). The sources of calories in the diet of Lapps and rural, semi-urban or urban Finns were not appreciably different (37–39% from fats including 20–23% from saturated fatty acids and 46–50% from carbohydrates).

3. *Ainu*. As part of a comprehensive study of the physiology of cold-adapted man, Itoh and associates (Itoh, 1974) have compared the plasma lipid profile of the Ainu natives of Hokkaido with that of non-Ainu Japanese. Under basal conditions, no significant differences were found in the concentration of total lipids, triglycerides, phospholipids, total and free cholesterol,

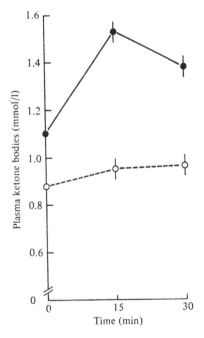

Fig. 8.8. Effect of norepinephrine (0.025 mg/10 kg) on the level of ketone bodies in the plasma in Asahikawa Ainu (●——●) and Sapporo students (○– – –○) (Itoh, 1974).

276

or esterified fatty acids. However, in the Ainu the fasting concentration of free fatty acids was markedly lower and the concentrations of ketones was higher (Table 8.13). The lower concentrations of plasma fatty acids rose markedly in the Ainu after norepinephrine administration, whereas no response was observed in the Japanese (Fig. 8.7). A concomitant response was seen in ketone bodies (Fig. 8.8) suggesting that these compounds may be more extensively utilized by the Ainu for metabolic fuel. Ketone bodies were also present at higher concentrations under basal conditions in the Ainu than in other Japanese.

In a study of environmental effects on the fatty acid composition of adipose tissue, Moriya & Itoh (1969) observed that subcutaneous fat of the extremities was lower in saturated acids and higher in monoenes than that of the chest and abdomen. In winter the fat of the forearm and leg became more unsaturated whereas that of areas better protected from cold remained unchanged.

Special Topics

Glucose tolerance

Several efforts have been made to determine whether introduction of processed foods into the Eskimo diet has affected their traditional resistance to diabetes mellitus. From a study of glucose tolerance in 320 Alaskan Eskimos living in the Bethel region, Mouratoff & Scott (1973) concluded that, although clinical diabetes is still rare, more Eskimos were intolerant of glucose in 1972 than was the case a decade earlier. A higher proportion of the population was also overweight (Table 8.14).

Table 8.14. *Comparison of glucose tolerance (means ± S.D.) in Alaskan Eskimos 40 yr and older in 1962 and 1972 (Mouratoff & Scott, 1973)*

	Men		Women	
	1962	1972	1962	1972
N	161	97	138	96
Mean age (yr)	51.4 ± 0.7	54.1 ± 0.8	50.6 ± 0.7	53.0 ± 0.9
% overweight[a]	3.0 ± 1.5	9.4 ± 3.0	6.1 ± 2.4	11.7 ± 3.3
Fasting glucose, mg/100 ml	95.7 ± 0.7	96.3 ± 1.1	96.1 ± 1.1	97.3 ± 1.3
% with blood glucose rise to > 150 mg/100 ml after 100 g glucose load	0.7 ± 0.7	5.2 ± 2.2	6.1 ± 2.4	10.4 ± 3.1

[a] > 13.6 kg above the average for white subjects of the same age, sex and height.

Feldman and co-workers (1975) compared fasting glucose and insulin levels in the plasma of 41 Eskimos aged 16–20 yr residing at Point Hope in northern Alaska with those of 32 age-matched students attending a boarding school at Mt Edgecumbe in southwest Alaska. The diet at the boarding school was of the high carbohydrate modern type whereas that at Point Hope was a mixture of native and processed foods. No significant differences in fasting glucose or insulin concentrations were found (Table 8.11). However, there were substantial differences in cholesterol and triglyceride concentrations which could be associated with differences in dietary cholesterol and carbo-hydrate, respectively. Tolbutamide-tolerance tests indicated that in both groups there was an adequate insulin reserve and a normal sensitivity to circulating insulin. Norman plasma-glucose and insulin responses were also observed by Bell, Draper & Bergan (1973) in a sample of 17 adult Eskimos living in Wainwright and Point Hope.

In contrast to these findings on Alaskan Eskimos, Schaefer, Crockford & Romanowski (1972) observed an impaired plasma glucose response to oral (but not intravenous) glucose in over half of a sample of 76 hospitalized adult Canadian Eskimos. This abnormal response was associated with a delay in insulin release and a prolonged plasma-insulin decay curve. The response was normal when glucose was preceded by a high protein meal. Whether these results reflected a true difference between Alaskan and Canadian Eskimos or some factor in the hospital environment is unknown.

An extraordinary incidence of diabetes has been recorded among Aleuts of the Pribiloff Islands (Dippe *et al.*, 1974). In a study involving 335 subjects, 10% of males and 27% of females over 39 yr old had plasma levels $\geq$ 200 mg/

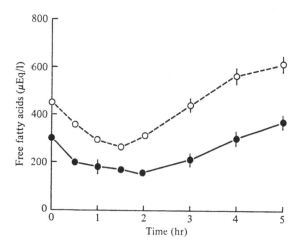

Fig. 8.9. Plasma free fatty acid levels after a glucose load in Japanese (○– – –○) and Ainu (●——●). (Itoh, 1974).

100 ml two hours after a glucose load. 46% of those exhibiting this glucose level were diabetic. 31% of the entire sample had a fasting serum-triglyceride level above 150 mg/100 ml. The prevalence of diabetes among Aleuts is remarkably greater than among Alaskan Eskimos and Indians, and approaches that of the Pima Indians of Arizona who have the highest reported prevalence in the world.

Kuroshima *et al.* (1972) compared the response to oral glucose in Japanese subjects with that in the Ainu, among whom the incidence of diabetes mellitus is very low. Following a glucose load, no significant difference in plasma glucose or insulin response was found. However, there was a difference in plasma free fatty acids. The fasting free fatty-acid level in the Ainu was lower and exhibited no significant rebound from the depression induced by glucose administration (Fig. 8.9). In the Japanese there was a greater depression following glucose administration and a marked rebound above initial levels. Evidence also was obtained for a lower fasting level of glucose in the Ainu (Itoh, 1974).

Lactose tolerance

Eskimos and Lapps resemble other cultures lacking a long history of dairying in having a limited capacity for lactose digestion. Lactose intolerance, which is due to a decline in the synthesis of lactase in the small intestine, has been extensively investigated in Lapps by Sahi and coworkers. Intolerance to a 50-g oral lactose load, appraised according to the subsequent rise in plasma glucose or galactose concentration, was estimated at 17% in the general Finnish-speaking population (Sahi, 1974) as opposed to 34% in Finnish Lapps (Sahi *et al.*, 1974). Intolerance was determined to be due to a single recessive autosomal gene (Sahi *et al.*, 1973).

Intolerance to lactose, which is manifested clinically in abdominal cramps, flatulence and diarrhea, was observed in over 50% of Greenland Eskimos given the standard 50-g oral load (McNair *et al.*, 1972). Duncan & Scott (1972) found that over 80% of a mixed sample of 36 Alaskan Eskimos and Indians were intolerant to a similar dose, and Bell *et al.* (1973) observed a 56–70% incidence of intolerance in Alaskan Eskimo children (Table 8.15). However, Eskimo children and adults were found to be capable of tolerating nutritionally significant amounts of lactose. Use of graduated lactose loads indicated that 19 of a sample of 20 Eskimo adults were asymptomatic after a 10-g load (equivalent to one cup of milk) and that 11 were tolerant to a 20-g load (Table 8.16).

It seems apparent that the standard 50-g lactose load (an amount necessary to produce a rise in plasma glucose that can be reliably monitored) yields a spuriously high estimate of intolerance to dairy foods in normal usage. Most adults, as well as children, appear to tolerate at least one cup of milk (or its

The human biology of circumpolar populations

Table 8.15. *Results of 50-g lactose load tests performed on 27 Alaskan Eskimo children (Bell et al., 1973)*

Age (yr)	Tested	Showing 20 mg/100 ml rise in plasma glucose	Showing symptoms of intolerance	Consuming milk
		Number of children		
7	2	0	0	2
8	4	2	3	4
9	5	3	2	5
10	4	4	2	3
11	3	2	3	3
12	3	2	2	3
13	4	4	2	4
14	2	2	1	2
Total	27	19	15	26
Percentage		70.4	55.6	96.7

Table 8.16. *Results of graduated lactose load tests performed on 20 adult Alaskan Eskimos (Bell et al., 1973)*

	Lactose load (g)		
	10	20	30
No. first experiencing symptoms of intolerance	1	8	4
No. intolerant (cumulative)	1	9	13
% intolerant (cumulative)	5	45	65

equivalent in other dairy products) at a time. If this quantity is ingested two or three times per day at intervals of several hours, the recommended consumption of dairy foods can be achieved by most children, adult males and non-pregnant females. Many women, however, will experience difficulty in consuming the additional amounts of dairy products commonly recommended during pregnancy. Dairy foods are an important source of calcium and protein as well as riboflavin (one of the nutrients most likely to be inadequate in the mixed diet of children of low economic status). Continued use of milk products in feeding programs for children has been recommended by the Protein Advisory Group of the United Nations.

Sucrose tolerance

A novel racial–ethnic form of primary sucrose deficiency has been described in Greenland and Alaskan Eskimos (McNair *et al.*, 1972; Bell *et al.*, 1973). The symptoms are similar to those of lactase deficiency. Typically they appear

280

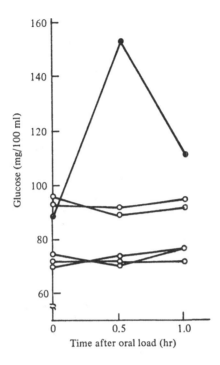

Fig. 8.10. Plasma glucose response to 50–g sucrose load in five sucrase deficient Eskimos (○——○) and one normal subject (●——●). (Bell, Draper & Bergan, 1973).

after consumption of a piece of cake, an ice-cream cone or a bottle of carbonated beverage. In intolerant subjects given a 50-g sucrose load, there is no plasma-glucose response (Fig. 8.10). There is a distinct familial pattern to the occurrence of sucrose intolerance.

The incidence of sucrase deficiency in Eskimos is not yet well documented and may vary with geography. McNair *et al.* (1972) found an incidence of 10.5% in a sample of Greenlandic Eskimos who included hospital patients and their relatives, some of whom had a history of chronic diarrhea. In the whole population of Wainwright, Alaska, the incidence was estimated at 2–3% (Bell *et al.*, 1973). No evidence of intolerance was found at Kasigluk or Nunapitchuk in southwest Alaska, where berries contribute significant amounts of sucrose to the aboriginal diet. It is possible that sucrase deficiency is confined to the Arctic region where sucrose has been virtually absent from the diet for many centuries. Under these conditions a tolerance for sucrose presumably was a negligible factor in natural selection.

Sucrose intolerance differs from lactose intolerance with respect to its prevalence, mode of inheritance and nutritional significance. Sucrase deficiency

The human biology of circumpolar populations

is present from birth, whereas lactase deficiency develops gradually during maturation. Sucrose intolerance is considerably less prevalent than lactose intolerance, but its nutritional implications for those affected are more serious. Whereas lactose intolerance can be managed by moderating the consumption of a limited number of foods, sucrose intolerance imposes many restrictions on the selection of foods from the modern diet.

Bone metabolism

Studies on the mineral content of the forearm bones of northern Alaskan Eskimos, estimated by direct photon absorptiometry, indicated that beyond age 40 yr both males and females have a mineral deficit of 10–15% relative to standards for Whites (Mazess & Mather, 1974). Similar results were obtained from studies on Canadian Eskimos of the northern Foxe Basin (Mazess & Mather, 1975). Bone loss, beginning at age 40–50 yr, was about 10% per decade in Canadian Eskimo males and 15% in females. Both values were about 5% greater than those seen in the general US population. These findings indicate that ageing bone loss is earlier in onset and greater in severity in Eskimos than in Whites.

The role of nutrition, if any, in the early onset and rapid progression of ageing bone loss in Eskimos is unclear. However, it is known that high-protein, high-phosphorus, low-calcium diets of the type traditionally consumed by Eskimos cause an acceleration of bone loss in adult animals. Further research will be required to establish which nutrient(s) in the Eskimo diet is of primary importance in bone homeostasis.

References

Auger, F. (1974). Poids et plis cutanes chez les Esquimaux de Fort Chimo (Nouveau-Québec). *Anthropologica*, XVI, 137–59.
Bang, H. O. & Dyerberg, J. (1972). Plasma lipids and lipoproteins in Greenlandic west coast Eskimos. *Acta med. Scand.*, **192**, 85–94.
Bang, H. O., Dyerberg, J. & Hjorne, N. (1976). The composition of food consumed by Greenland Eskimos. *Acta med. Scand.*, **200**, 69–73.
Bell, R. R., Draper, H. H. & Bergan, J. G. (1973). Sucrose, lactose and glucose tolerance in northern Alaskan Eskimos. *Amer. J. clin. Nutrition*, **26**, 1185–90.
Bell, R. Raines & Heller, C. A. (1978). Nutrition studies: an appraisal of the modern North Alaskan Eskimo diet. In *Eskimos of Northwestern Alaska: A Biological Perspective*, ed. P. L. Jamison, S. L. Zegura & F. A. Milan, pp. 145–56. Stroudsburg, Pa: Dowden, Hutchinson & Ross, Inc.
Bergan, J. G. & Bell, R. Raines (1978). Nutrition studies: biochemical assessment of nutritional status. In *Eskimos of Northwestern Alaska: A Biological Perspective*, ed. P. L. Jamison, S. L. Zegura & F. A. Milan, pp. 157–61. Stroudsburg, Pa: Dowden, Hutchinson & Ross, Inc.
Bjorksten, F., Aromaa, A., Eriksson, A. W., Maatela, J., Kirjarinta, M., Fellman, J. & Tamminen, M. (1975). Serum cholesterol and triglyceride concentrations

of Finns and Finnish Lapps: interpopulation comparisons and occurrence of hyperlipidemia. *Acta med. Scand.*, **198**, 23–33.

Dippe, S. E., Bennett, P. H., Dippe, D. W., Humphry, T., Burks, J. & Miller, M. (1974). Glucose tolerance among Aleuts on the Pribiloff Islands, p. 17. *Proc. 3rd Int. Symp. circumpolar Health, Yellowknife, Canada, 8–11 July 1974.*

Draper, H. H. (1976). A review of recent nutritional research in the Arctic. In *Circumpolar Health*, ed. R. J. Shephard & S. Itoh, pp. 120–9. Toronto: University of Toronto Press.

Duncan, I. W. & Scott, E. M. (1972). Lactose intolerance in Alaskan Indians and Eskimos. *Amer. J. clin. Nutrition*, **25**, 867–8.

Dyerberg, J., Bang, H. O. & Hjorne, N. (1975). Fatty acid composition of the plasma lipids in Greenlandic Eskimos. *Amer. J. clin. Nutrition*, **28**, 958–66.

Ellestad-Sayed, J., Hildes, J. A., Schaefer, O. & Lobban, M. C. (1975). Twenty-four hour urinary excretion of vitamins, minerals, and nitrogen by Eskimos. *Amer. J. clin. Nutrition*, **28**, 1402–7.

Feldman, S. A., Ho, K., Lewis, L. A., Mikkelson, B. & Taylor, B. C. (1972). Lipid and cholesterol metabolism in Alaskan Arctic Eskimos. *Arch. Pathology*, **94**, 42–58.

Feldman, S. A., Rubenstein, A., Ho, K., Taylor, C. B., Lewis, L. & Mikkelson, B. (1975). Carbohydrate and lipid metabolism in the Alaskan Arctic Eskimo. *Amer. J. clin. Nutrition*, **28**, 588–94.

Forbes, A. L. (1974). Nutritional status of Indians and Eskimos as revealed by Nutrition Canada, p. 17. *Proc. 3rd int. Symp. circumpolar Health, Yellowknife, Canada, 8–11 July 1974.*

Gassaway, A. R. (1969). Diet and environment in Finnmark. *Geog. Rev.*, **59**, 440–2.

Hasunen, K. & Pekkarinen, M. (1975). Nutrient intake of adult Finnish Lapps. *Nordic Council for Arctic Medical Research, Report 13, pp. 15–32.*

Itoh, S. (1974). *Physiology of cold-adapted man. Hokkaido University Medical Library Series Vol. 7.* Sapporo, Japan: Hokkaido University School of Medicine.

Koishi, H., Okuda, T., Matsudaira, T., Takaya, S. & Takemura, K. (1975). Nutrition and cold tolerance. In *Anthropological and Genetic Studies on the Japanese. Japanese IBP Synthesis Vol. 2 on Human Adaptability*, ed. S. Watanabe, S. Kondo & E. Matsunaga, pp. 309–19. Toronto: University of Toronto Press.

Kuroshima, A., Itoh, S., Azuma, T. & Agishi, Y. (1972). Glucose tolerance test in the Ainu. *Int. B. Biometeor.*, **16**, 193–7.

McNair, A., Gudmand-Hoyer, E., Jarnum, S. & Orrild, L. (1972). Sucrose malabsorption in Greenland. *Brit. med. J.*, **2**, 19–21.

Maynard, J. E. (1974). Coronary heart disease risk factors in relation to urbanization in Alaskan Eskimo men, p. 23. *Proc. 3rd int. Symp. circumpolar Health, Yellowknife, Canada, 8–11 July 1974.*

Mazess, R. B. & Mather, W. (1974). Bone mineral content of northern Alaskan Eskimos. *Amer. J. clin. Nutrition*, **27**, 916–25.

Mazess, R. B. & Mather, W. (1975). Bone mineral content in Canadian Eskimos. *Human Biology*, **47**, 45–63.

Moriya, K. & Itoh, S. (1969). Regional and seasonal differences in the fatty acid composition of human subcutaneous fat. *Int. J. Biometeor.*, **13**, 141–6.

Mouratoff, G. J. & Scott, E. M. (1973). Diabetes mellitus in Eskimos after a decade. *J. Amer. med. Assoc.*, **226**, 1345–6.

Øgrim, M. E. (1970). The nutrition of Lapps. *Arctic Anthropology*, **VII**, 49–52.

The human biology of circumpolar populations

Sahi, T. (1974). Lactose malabsorption in Finnish-speaking and Swedish-speaking populations in Finland. *Scand. J. Gastroent.*, **9**, 303–8.

Sahi, T., Eriksson, A. W., Isokoski, M. & Kirjarinta, M. (1974). Lactose malabsorption in Finnish Lapps, p. 18. *Proc. 3rd int. Symp. circumpolar Health, Yellowknife, Canada, 8–11 July 1974.*

Sahi, T., Isokoski, M., Jussila, J., Launiala, K. & Pyorala, K. (1973). Recessive inheritance of adult-type lactose malabsorption. *The Lancet*, **13**, 823–6.

Sauberlich, H. E., Goad, W., Herman, Y. F., Milan, F. & Jamison, P. (1970). Preliminary report on the nutrition survey conducted among the Eskimos of Wainwright, Alaska, 21–7 January 1969. *Arctic Anthropology*, **VII**, 122–4.

Sayed, J. E., Hildes, J. A. & Schaefer, O. (1976). Biochemical indices of nutrition of the Iglooligmiut. In *Circumpolar Health*, ed. R. J. Shephard & S. Itoh, pp. 130–4. Toronto: University of Toronto Press.

Schaefer, O., Crockford, P. M. & Romanowski, B. (1972). Normalization effect of preceding protein meals on 'diabetic' oral glucose tolerance in Eskimos. *Can. Med. Assoc. J.*, **21**, 733–8.

Wo, C. K. W. & Draper, H. H. (1975). Vitamin E status of Alaskan Eskimos. *Amer. J. clin. Nutrition*, **28**, 808–13.

9. Physiology of circumpolar people

SHINJI ITOH

In order to investigate the physiological characteristics of circumpolar peoples, a number of extensive studies were carried out by several groups of IBP investigators. However, due to the difficulties of experimenting on native populations, the results obtained were generally not enough to allow systematic comparisons to be made between ethnic groups. In this chapter, therefore, previous literature and non-IBP data have been included to give a fuller account of the traits of cold adaptability of circumpolar populations.

Basal metabolic rate

A high basal metabolic rate (BMR) in Eskimos, 13–33% above DuBois standard, has been reported by many investigators (Table 9.1). According to Rodahl (1952), the highest figures were found among the primitive Eskimos at Gambell and Anaktuvuk Pass and the lowest figures among the more civilized Eskimos at Kotzebue. He attributed the high BMR to apprehension and the high protein native diet. When these two factors were eliminated, the metabolism was almost exactly the same as in the white controls. It was therefore concluded that there are no racial differences between Eskimos and whites in basal heat production. Hart *et al.* (1962) confirmed the high BMR in Eskimo hunters from Cumberland Sound, Baffin Island, but in hospitalized Eskimos living in a white man's society in Edmonton the resting metabolism was found to be identical to DuBois standard. According to Milan, Hannon & Evonuk (1963) and Milan & Evonuk (1966), Wainwright Eskimos had essentially normal values for the BMR (38.9 kcal/m²/h), while the inland Eskimos living in the mountains of interior Alaska at Anaktuvuk Pass had mean values of 47.6 and 45.4 kcal/m²/h on successive mornings.

The BMR of the Ainu was on the normal level of the non-Ainu Japanese (Itoh, 1974).

Thyroid activity

The high BMR of the Eskimos might be associated with hyperactivity of the thyroid gland. Gottschalk & Riggs (1952) reported that the serum protein-bound iodine of Eskimos from Southampton Island and Chesterfield Inlet studied during summer months ranged from 4.2 to 9.0 μg/100 ml, with a mean of 6.5 μg/100 ml. The values were appreciably higher than those for normal

Table 9.1. *Basal metabolic rates of Eskimos and Ainu*

Investigator	Race	Locality	Sex	Number tested	BMR (%)	kcal/m²/h
Heinbecker (1928)	Eskimo	Cape Dorset, Baffin Island			+ 33	
Heinbecker (1931)		Pangnirtung, Baffin Island			normal	
Rabinowitch & Smith (1936)		Canadian eastern Arctic	M	10	+ 26	
Crill & Quiring (1939)			M	30	+ 14.5	
			F	33	+ 21.1	
Höygaad (1941)		East Greenland		39	+ 13	
Levine & Wilber (1949)		Point Barrow		23	normal	
Bollerud et al. (1950)		Gambell, Alaska			+ 17.1, + 14.0	
Rodahl (1952)		Barter Island			+ 2	39.7
		Anaktuvuk Pass			+ 16	46.3
		Kotzeblue			+ 3	40.2
		Gamble, St Lawrence Island			+ 10	42.9
Brown et al. (1954)		Southampton Island	M	7	+ 29.6 ~ + 21.4	51.7 ~ 47.5
			F	6	+ 33.7 ~ + 27.0	48.9 ~ 46.3
Adams & Covino (1958)		Anaktuvuk Pass	M	6		50
McHattie et al. (1960)		Anaktuvuk Pass	M	5	+ 40	52
Hart et al. (1962)		Pangnirtung, Baffin Island	M	10	+ 22	49
		Canadian Eskimo on white man's diet	M	8	normal	39.7
Milan et al. (1963)		Anaktuvuk Pass	M	6		47.6
Milan & Evonuk (1966)		Wainwright	M	6		38.9
Itoh (1974)	Ainu	Hokkaido	M	14		38.8 (winter) 37.6 (summer)

whites in the United States (range, 3.6 to 7.0 μg/100 ml; mean, 5.1 μg/100 ml). Thus, they suggested that the thyroid gland in the Eskimos is comparatively hyperactive. However, Rodahl (1958) indicated that there was no significant difference between whites, coastal Eskimos and Fort Yukon Indians in thyroid uptake, urinary elimination and blood levels of [131]I. In the case of inland groups abnormally high and very rapid ^{131}I uptakes, abnormally low urinary excretion and low salivary iodine concentrations were observed. This was attributed to exceedingly low iodine intakes in their diet and associated with a high incidence of thyroid enlargement.

Total and free thyroxine concentrations in the plasma of Ainu were compared with the levels of control non-Ainu Japanese subjects (Itoh, 1974). In summer, total thyroxine levels, as determined by the method of competitive protein binding analysis, were 8.6 μg/100 ml $\pm$ 0.35 (S.E.M.) in the controls and 8.2 μg/100 ml. $\pm$ 0.69 in the Ainu. In winter, the levels increased to 10.5 μg/100 ml $\pm$ 0.30 in the controls, while no significant increase was observed in the Ainu (9.6 μg/100 ml $\pm$ 0.44). Plasma free thyroxine levels, estimated by the method based on measuring the rate of dialysis across a semipermeable membrane, were in summer 2.75 mμg/100 ml $\pm$ 0.36 in the controls and 2.33 mμg/100 ml $\pm$ 0.18 in the Ainu; in winter the levels were not significantly different (3.13 mμg/100 ml $\pm$ 0.57 for the controls and 2.69 mμg/100 ml $\pm$ 0.16 for the Ainu).

Thermal and metabolic responses to cold exposure

In the early study of Scholander *et al.* (1957) the critical temperature of nomadic Lapps was shown to be about 27 °C, which is the same as for naked white men living in a temperate climate. They noted that the Lapps are not normally subjected to cold stress, and that they do not have any greater physiological insulation than men living in a temperate climate.

Andersen *et al.* (1960) conducted a field investigation to determine the tolerance of the Lapps to cold. Metabolism and skin and rectal temperatures were determined while the subjects rested and slept naked for a night in a single-blanket sleeping bag, exposed to a moderate cold stress. As compared with unacclimatized white men, the Lapps showed a greater ability to endure the cold night. In general, the Lapps slept well without visible shivering, whereas the controls were prevented from sleeping by the sensations of cold and vigorous shivering. The metabolic rates of the Lapps were close to their basal level, in contrast to the raised metabolism of the unacclimatized control subjects. Some of the Lapps maintained slightly higher peripheral skin temperatures than the control subjects, but had a much greater fall in their rectal temperatures, indicating a greater loss of stored body heat. The nomadic Lapps' response to a moderate cold stress, therefore, resembled that found in the Australian aborigines.

A similar study was carried out by Hart *et al.* (1962) on a group of Eskimo hunters from Cumberland Sound, Baffin Island, and on white controls. The Cumberland Sound Eskimos maintained a resting metabolism that was elevated according to DuBois standard during sleep on warm nights. This elevation was not found in hospitalized men who had been living for an average of six months in Edmonton, Alberta. During exposure to moderate cold, the Cumberland Sound Eskimos and white controls had an elevated metabolism, shivering and a disturbed sleep. Peripheral temperatures were maintained at a higher level in Eskimos than in whites. Because of the absence of marked physiological differences between Eskimos and whites, it was concluded that the principal adaptation of these Eskimos to their climate is technological.

Adams & Covino (1958) observed racial variations to a standardized cold stress of male Eskimos, Negroes and whites. The subjects were subjected nude to an air temperature of 17 °C. After 55 min of exposure, both Eskimos and whites demonstrated an average rise in metabolism of 22 kcal/m²/h above control levels of 55 and 40 kcal/m²/h, respectively. Although the Eskimos possessed a higher metabolic rate and elevated shell and core temperatures than the whites, the pattern of response of these two groups to a standardized cold stress was the same as indicated by the shivering and metabolic response. There apparently exists no true physiological difference between the basic responses of the Eskimos and whites to a whole-body cold exposure.

Peripheral circulation

The hand blood flow of the Eskimos was first studied by Brown & Page (1952) and Brown *et al.* (1954*a*). In a group of Eskimos in the Canadian eastern Arctic it was found that the forearm skin temperature and the forearm and hand blood flow were greater at rest in a room at 20 °C than they were in a control group of white subjects. During immersion of the hand and forearm in water baths at temperatures of 5–42.5 °C, the Eskimos always maintained higher blood flow in hand and forearm than did the controls. In the colder baths the forearm muscle temperature of the Eskimos fell more than it did in the controls, but this was attributed to greater blood flow through the exposed hand and the resulting increased cooling of deep tissues by the venous return.

Hildes, Irving & Hart (1961) found that with a room temperature of 25 °C Eskimo men had a higher circulatory heat loss from their hands than white men during a 30-min immersion in 4 °C water. However, at a room temperature of 15 °C the greater circulatory heat loss from the Eskimos was not seen. Since an inverse relationship was found between hand volume and heat flow per 100 ml hand, when considered on this basis no difference could be made out between the two groups. Thus, they pointed out the possibility that

Table 9.2. *Hand blood flow rates (ml/100 ml hand volume/min ± s.d.) at different temperatures*

Race	Sex	Age (yr)	Number tested	Cold	Warm		Reactive hyper-emia
				(25 °C)	(35 °C)	(35 °C)	
Igloolik	F	20–50	9	36 ± 20	39 ± 19	38 ± 20	63 ± 38
Eskimos		50–70	9	4 ± 2	7 ± 4	7 ± 4	24 ± 11
	M	20–50	38	9 ± 5	14 ± 8	15 ± 9	39 ± 24
		50–80	8	10 ± 6	12 ± 5	14 ± 8	32 ± 16
				(20° C)			(40 °C)
Lapps	M	18–45	13	15 ± 6			55 ± 9
Norwegian fishermen	M	16–52	11	12 ± 7			63 ± 6
Controls	M	21–45	12	11 ± 8			62 ± 6

differences reported in the literature between groups of subjects might be related to differences in hand size.

Krog *et al.* (1960) measured the hand blood flow with local temperatures of 40 °C, 20 °C, 10 °C and a few degrees above zero in Norwegian Lapps, north Norwegian fishermen and a group of control subjects. The mean values of maximum hand blood flow for the three groups of subjects were in close agreement. They suggested that habituation to cold had not produced any detectable change in the morphologically determined maximum dimensions of the vascular bed. Moreover, the mean values of hand blood flow of the three groups did not differ significantly at local temperatures of 40 °C, 20 °C and 10 °C, although the Lapps showed somewhat higher mean blood flow values than the other two groups at the lower temperature (Table 9.2).

The peripheral circulation in the hand of Igloolik Eskimos was studied by Krog & Wika (personal communication). The hand blood flow of the Eskimos did not differ significantly from the values reported in other ethnic groups (Table 9.2). It was also found that women who seemed to be less exposed to cold had a higher maximal blood flow in their hands than men. Especially in the young age groups the values for females were about 150% over the males in the same age group. They concluded that the peripheral vascular bed does not increase in size as a reaction to cold.

The skin circulation of the cheek was investigated by Wika (1971), employing the Xenon-133 clearance method. Igloolik Eskimos, Finnish Skolt Lapps, Norwegian lumberjacks and Norwegian indoor workers were the subjects (Table 9.3). Eskimo men, who were assumed to be the most cold-exposed group, had the least mean skin blood flow. A lower circulation was also found

Table 9.3. Blood flow in the skin of the cheek. Values are expressed in terms of circulatory index as measured by xenon clearance

Sex	Age (yr)	Eskimo		Skolt		Norwegian lumberjack		Indoor worker	
		Number tested	Blood flow	Number tested	Blood flow	Number tested	Blood flow	Number tested	Blood flow
Male	15–19			9	16.3 ± 2.8				
	20–29	14	14.7 ± 2.4	5	18.0 ± 4.2	6	18.3 ± 2.9	9	16.4 ± 2.2
	30–39	14	15.4 ± 3.0	11	14.5 ± 2.9	8	17.3 ± 2.2	8	16.8 ± 2.1
	40–49	8	13.7 ± 1.9	10	14.3 ± 1.7	9	18.7 ± 3.2	4	16.8 ± 3.1
	50–59	4	11.7 ± 1.7	5	13.0 ± 2.3	16	16.0 ± 2.9	5	13.8 ± 1.3
	60–69	3	13.0 ± 1.1	11	12.7 ± 1.4	6	12.8 ± 2.0		
Female	20–29	5	12.5 ± 2.0					7	14.8 ± 5.5
	30–39	4	12.7 ± 0.8						
	40–49 } 50–59	8	12.0 ± 2.4						
	60–69	2	13.2						

in the Skolt Lapps, aged 30–50 yr, than in Norwegian indoor workers belonging to the same age group. The younger Skolt Lapps, however, had as high a skin blood flow as the Norwegian lumberjacks. Women seemed to have a lower skin blood flow than men. Skin blood flow decreases with age. Wika stated that in the Eskimos heat dissipating apparatus may have adjusted to the lower demand for heat dissipation in their cold environment.

Cold-induced vasodilatation reaction

Krog *et al.* (1960) studied the cold-induced vasodilatation reaction (CIVD) on Lapps, Norwegian fishermen and a control group of white subjects. The subjects immersed one hand in stirred ice-water for 30 min and changes in skin temperature of the immersed hand were recorded. The mean maximum skin temperatures reached during the periods of cold vasodilatation were very similar in all three groups of subjects. However, there was a tendency for the onset of the cold vasodilatation to be later in the control subjects than in the other two groups. Krog *et al.* (1969) further studied the reactions on Finnish Lapps. No differences were found between male and female subjects or between age groups in the response pattern to cold stimulation. When the results were compared with those obtained in the earlier experiments, the duration of the initial vasoconstriction was on the same level as in the Norwegian Lapps, and slightly shorter than in the control group. Maximal as well as minimal skin temperature was significantly higher in the Skolt Lapps than in the Norwegian Lapps and in the control group, but the magnitude of the CIVD was not significantly different in these groups (Table 9.4).

Eagan (1963, 1966) reported that the Eskimos did maintain higher finger temperature during ice-water immersion. He also used an air-cooling test. The index and middle fingers of a hand of each subject were exposed bare in a refrigerated box, in which the temperature varied between −20 °C and −25 °C, for 30 min. All of the Eskimos sustained 30 min of finger cooling, but only five of 20 white controls and two of five mountaineers were able to do so. Pain sensations were almost negligible for the Eskimos in both the water- and air-cooling tests. From the results Eagan stated that some sort of genetic cold adaptation may indeed exist in the Eskimos.

Bare hand and facial temperatures during cooling in open air at temperatures between −3 °C and −7 °C were studied by Miller & Irving (1962) in well-clothed Eskimo men, women and children. Two groups of white men served for comparison. In contrast to the groups of whites, Eskimo men showed significantly higher minimum finger temperatures, smaller temperature fluctuations during cooling and higher finger temperatures at the termination of exposure. White men less accustomed to cold cooled more rapidly than Eskimo men during initial exposure. White men accustomed to cold cooled at an intermediate rate. Eskimo women reacted in the same

291

Table 9.4. *Comparisons of cold-induced vasodilatation reactions*

Investigator	Material	Time to first rewarming (min)[a]	Temperature at which first rewarming began (°C)	Average skin temperature during immersion (°C)	Maximum skin temperature (°C)
Krog et al. (1969)	Skolts	5.0	3.6		12.4
Krog et al. (1960)	Lapps	5.4	1.6		8.1
	Norwegian fishermen	6.9	1.7		7.5
	Controls (white)	9.1	1.2		7.7
Eagan (1966)	Eskimos		0.8	4.1	
	Arctic Indians		2.2	5.6	
	Navajo Indians		0.5	2.8	
	Controls (white)		0.2	2.2 (10 min)	
Itoh et al. (1970)	Ainu	5.5	3.3	5.8	8.7
	Controls (Japanese)	6.5	2.2	5.1 (4–20 min)	7.7
Yoshimura & Iida (1950)	Japanese	8.7	2.5	5.4 (5–30 min)	
Nelms & Soper (1962)	Fish filleters	4.5	2.0	4.7	5.7
	Controls (white)	9.9	0.4	1.8 (10 min)	2.9
Iampietro et al. (1959)	Negroes	15.9	1.0	2.7	
	Controls (white)	9.2	2.9	7.2 (5–30 min)	

[a] In these experiments fingers were immersed in ice water at 0 °C.

manner as Eskimo men. Hand discomfort during cooling was reported absent or minor in adult Eskimos, but was marked in most whites. Eskimo and white adults responded similarly to facial cooling, but cheek temperatures of Eskimo children fell more rapidly.

The CIVD of the Ainu was compared with that of several groups of control non-Ainu Japanese subjects (Itoh, Kuroshima, Hiroshige & Doi, 1970). In the summer the response of the Ainu was similar to that of male farmers and students, but slightly higher than that of female students and nurses. In the winter the response of the Ainu was similar to that of male farmers and fish-factory workers, but stronger than that of male fishermen and female students, as evidenced by shorter duration of the initial vasoconstriction, higher skin temperature at which first rewarming began, higher mean skin temperature during immersion and higher maximal skin temperature during immersion. Summarized results indicated that the response of subjects in Hokkaido, the northern island of Japan, particularly that of the Ainu, was considerably higher than the average values reported for Japanese on the main and southern islands of Japan (Table 9.4).

These studies indicate that the CIVD of circumpolar people is evenly increased. In these groups the response is particularly characterized by the earlier onset of cold vasodilatation. The maximal skin temperature reached during the period of cold vasodilatation was very similar in various groups of subjects, including controls. This fact may suggest that the magnitude of the fully established cold vasodilatation in response to immersion of the finger in ice-cold water is very similar in both cold-habituated and control subjects (Krog *et al.*, 1960).

Cold pressor response

The circulation of Eskimos on Southampton Island, Northwest Territories, was studied by Brown *et al.* (1954*b*). They reported that exposure of the hand and forearm to cold water caused a greater elevation of the blood pressure in Eskimos than in control white men. Krog *et al.* (1969) also found considerably greater increase in mean blood pressure in Finnish Lapps when a hand was immersed in circulated ice-water for 15 min. However, results of Lund-Larsen, Wika & Krog (1970), who observed cold pressor response on Greenlanders, were inconclusive because of great individual variations.

On the other hand, a high degree of habituation to cold in extremities of Eskimos was shown by LeBlanc (personal communication). His study was made in the summer at Igloolik. Three tests were used: test I consisted in immersing the hand to the wrist into ice-cold water, in test II the face was immersed into cold water while the subjects breathed freely through a mouthpiece, and test III was a combination of test I and II made simultaneously. The tests lasted 3 min. Every 30 s during the tests, and before and after them, the following parameters were measured: systolic and diastolic blood pressure, heart rate, ECG and ventilation.

When one hand was immersed into ice-cold water, marked increases in blood pressure and heart rate were observed in control subjects, while the pressor response of the Eskimos was less in extent and there was no alteration in heart rate. When the face was immersed, a similar difference in response was found between control and Eskimo groups. This test caused a bradycardia in both white and Eskimo subjects. When the face and hand were immersed simultaneously in cold water, the responses were an approximate cumulation of both tests when given individually. These results indicated that in Eskimos, whether they were hunters, children or women, the cold pressor response was significantly less compared with that in control whites. LeBlanc assumed the possibility of a racial characteristic in this regard.

In Hokkaido, Japan, the cold pressor response was examined in 231 subjects, including 26 Ainu (Itoh *et al.*, 1969*a*). To measure the response, one hand was immersed in a well-stirred water bath at 10 °C for 10 min. Both in summer and in winter Ainu subjects showed considerably high response, while fish-factory workers who work often in fish store-houses at temperatures below −30 °C had low response. Similarly low responses were observed in female fish-factory workers who used to immerse their hands in cold water for the preparation of fish meat for canning. These observations indicate that the cold pressor response is dependent on the living habits of subjects, at least in the same ethnic population, as previously demonstrated by LeBlanc, Hildes & Heroux (1960) and LeBlanc (1962, 1966) on Gaspé fishermen, and the decline of the response is explained by the adaptive process of habituation (Glaser & Whittow, 1957; Glaser, Hall & Whittow, 1959). In the Ainu a significant sex difference was observed in that the response was less in females. A part of this difference might be attributed to the fact that females immerse their hands in cold water for washing, when doing their housework.

It is interesting that the cold pressor response was unexpectedly high in the Ainu compared with that of the non-Ainu Japanese subjects, although the Ainu are self-confident of their strong resistance to cold. The cold pressor response is thought to be concerned with the irritability of the sympathetic nervous system. A highly significant correlation was demonstrated between the cold pressor response and the pressor effect of norepinephrine, suggesting that the cold pressor response is closely related to the liberation of catecholamine. It was also shown that treatment with hexamethonium bromide, a ganglion-blocking agent, caused a marked decline in the cold pressor response. From the results it was inferred that the response is concerned with the extent of norepinephrine release and/or the susceptibility of the sympathetic vasomotor system to catecholamine. Markedly enhanced sensitivity to norepinephrine was found in the Ainu.

Sensitivity to norepinephrine

It has been demonstrated that man's acclimatization to cold is accompanied by the development of nonshivering thermogenesis (Davis, 1961) and that norepinephrine produced a significant increase in the oxygen consumption in cold-exposed man (Joy, 1963; Budd & Warhaft, 1966), while there is at most a minor increase in unacclimatized man. It was proposed that in man, as in various species of mammals, norepinephrine may be the direct mediator of nonshivering thermogenesis occurring with cold acclimatization. Consequently, it could be presumed that the sensitivity to exogenous norepinephrine may be enhanced in circumpolar people. This was clearly demonstrated in the Ainu (Itoh *et al.*, 1970*a*).

In the Ainu, subcutaneous injection of norepinephrine in a dose of 0.025 mg/10 kg caused a marked elevation of the energy metabolism from 40.6 kcal/m²/h $\pm$ 1.86 (S.E.M.) to 46.8 kcal/m²/h $\pm$ 1.77 (+ 15.3%) after 15 min and to 47.6 kcal/m²/h $\pm$ 1.47 (+ 17.2%) after 30 min, while in control non-Ainu Japanese subjects no change was observed following this dose of norepinephrine. When 0.05 mg/10 kg of norepinephrine was injected into controls, the metabolic rate increased moderately. Plasma levels of free fatty acids (FFA) also increased markedly following the injection of norepinephrine in the Ainu. The average values after 0.025 mg/10 kg (602 μEq/l $\pm$ 53.7) was just double the initial level (301 μEq/l $\pm$ 23.5). Although the controls showed no increase in the plasma FFA levels by this dose of norepinephrine, a higher dose of 0.05 mg/10 kg caused a significant elevation. Plasma levels of total ketone bodies at fasting state were significantly higher in the Ainu (1107 μM/l $\pm$ 86.0) in comparison with the levels in the controls. Injection of norepinephrine in a dose of 0.025 mg/10 kg caused a marked increase in the plasma ketone bodies to 1526 μM/l $\pm$ 64.8 after 15 min and to 1370 μM/l $\pm$ 52.1 after 30 min in the Ainu. On the other hand, in the controls the plasma concentration showed rather a slight increase even after 0.05 mg/10 kg of norepinephrine.

The effect of different doses of norepinephrine was also examined on the Ainu. However, injection of 0.01 mg/10 kg caused no significant changes in the above parameters and 0.05 mg/10 kg produced marked ill effects, for instance feelings of tightness in the chest and apprehension, weakness, numbness and/or heaviness in the extremities, pallor of the face, and so on. Increased rate and depth of the respiration and extra-systole were also noticed. Therefore, the study could not be carried through. In the non-Ainu Japanese subjects such ill effects were slight and scarce after the same dose of norepinephrine. The findings of markedly enhanced sensitivity to norepinephrine in the Ainu appear to suggest that this ethnic group is equipped with adaptive mechanism characterized by norepinephrine-sensitive thermogenesis.

Sweat gland activity

The total number of the active sweat glands of 12 Ainu ranged from 1960 $\times$ 10^3 to 1991 $\times$ 10^3 with an average of 1443 $\times$ 10^3 (Kawahata & Sakamoto, 1951). This figure was considerably less compared with those found in other races (Table 9.5). Accordingly, it was thought that the total number of the active glands is smaller in people living in cold climates than in people living in the tropics (Kuno, 1956).

Recently O. Schaefer, J. A. Hildes, P. Greidanus & D. Leung (personal communication) found very significant differences in numbers and degree of response to chemical stimulation of sweat glands when they compared various body surface areas in Eskimos and Caucasians. Eskimos showed greater numbers and greater activity of functioning sweat glands in the face especially around the nose and mouth, which sweat very profusely on slight stimulation. On the other hand, all body surfaces normally covered by fur in winter-time had diminished responses by a factor of approximately 1:2 on the trunk, 1:3 upper extremities, 1:4 lower extremities and 1:5 for feet, apparently in the order of distance from the body core and danger to freezing. It was reasonably assumed that the observed peculiarities of sweating in Eskimos serve very well their environmental climatic and clothing conditions and may therefore constitute true morphological and functional adaptation.

The pattern of sweating response evoked by a standard sweating stimulus was classified by Ohara (1966) into the following four types: type 1, high sweat volume and high salt concentration; type 2, high sweat volume and low salt concentration; type 3, low sweat volume and high salt concentration; type 4, low sweat volume and low salt concentration. According to this author (1972), for non-Ainu Japanese males types 2 and 4, especially type 4, were frequently observed in a group living in Nagoya district (mild climate), while types 1 and 3 were frequent in a group born and living in Hokkaido (colder climate). The observation suggests that types 2 and 4 occurred more frequently in subjects more adapted to heat and, to the contrary, types 1 and 3 were more frequent in subjects less adapted to heat.

Table 9.5. *Total number of active sweat glands in different races (Kuno, 1956)*

Race	No. of subjects	Age (yr)	Number of glands ($\times$ 10^3)		
			Minimum	Maximum	Mean
Ainu	12	13–63	1069	1991	1443
Russians in Manchuria	6	38–58	1636	2137	1886
Caucasians	10	16–26	1800	2894	2469
Eskimos	10	14–44	1727	3068	2836
Japanese	11	2.5–35	1781	2756	2282
Negroes in USA	9	19–36	2024	2371	2179
Chinese in Formosa	11	17–55	1783	3445	2415
Siamese	9	18–36	1742	3121	2422
Filipinos	10	17–42	2642	3062	2800

Ohara (1972) found a marked racial difference between Ainu males and non-Ainu Japanese males born and living in the same district in Hokkaido. Although no large differences were observed between the two groups in living habits and diets, Ainu males showed a high percentage of type 1 when compared with non-Ainu Japanese males, especially in summer. This indicates that Ainu males discharge large amounts of sweat of high salt concentration in summer. From the results he inferred that the Ainu is less adaptable to heat than the non-Ainu Japanese, presumably in part due to their genetic make-up. Ohara also demonstrated a definite sex difference in the pattern of sweating response in both Ainu and non-Ainu Japanese. Females differ distinctly from males in the absence of types 1 and 2 in any season.

Fatty acid metabolism

The serum FFA concentrations of the Point Hope Eskimos were quite different from those of the US white population. The average Eskimo fasting FFA were 1.33 mEq/l for males and 1.94 mEq/l for females, which were markedly higher than the mean values of 0.4 mEq/l for the US whites (Feldman *et al.*, 1972). It was inferred from the high serum FFA and low glucose levels in the Eskimos that FFA play a major role in body energy production.

Marked differences in the plasma FFA levels were found between different groups of male Japanese subjects. The levels were the highest in subjects born on the main island (598 μEq/l $\pm$ 23.0 (S.E.M.)), followed by those born on Hokkaido, the northern island (407 μEq/l $\pm$ 14.8) and the lowest values were obtained in the Ainu (306 μEq/l $\pm$ 22.2). The differences between these groups were highly significant. It is particularly interesting that the plasma levels of FFA showed such stepwise changes according to the experiences in cold exposure. The lowest levels were found in subjects who are considered to be well adapted to cold, while the highest levels occurred in subjects who are not exposed to cold. A similar difference was also observed in females between the Ainu (234 μEq/l $\pm$ 34.2) and non-Ainu Japanese (394 μEq/l $\pm$ 18.4) in the same districts. Moreover, considerably higher levels of plasma FFA, amounting to 634 μEq/l $\pm$ 33.0, were observed in male subjects born on Okinawa which is located between Japan and Formosa. The relation might be offset by a higher turnover of the plasma FFA in the cold-adapted subjects (Itoh *et al.*, 1969*c*).

Under basal conditions, a significant positive correlation was found between energy metabolism and plasma FFA concentration in Japanese groups, while the BMR of the Ainu was negatively correlated to the plasma levels of FFA. Administration of nicotinic acid which inhibits the mobilization of FFA from fat stores caused a decrease in the plasma FFA with concomitant decrease in BMR and increase in the respiratory quotient. In the Japanese, the values of BMR after nicotinic acid in relation to plasma FFA fit on the same regression line as obtained with values at rest. On the other hand, in the Ainu, values after nicotinic acid showed quite different patterns from those observed in the resting state. Here again a significant negative correlation was found when values of BMR after nicotinic acid were plotted against plasma FFA determined simultaneously.

Immersion of the whole body in a cold-water bath at 23 °C for 30 min produced an increase in energy metabolism in Ainu similar to or slightly less than the increase in Japanese subjects immersed at 25 °C. In the water bath Ainu showed a small elevation of plasma FFA and a marked increase in plasma ketone bodies, while control Japanese showed a pronounced increment in plasma FFA and a small rise in plasma ketones.

Under basal conditions plasma ketone-body levels were significantly higher in the Ainu than in the non-Ainu Japanese, in contrast with the low levels of plasma FFA in the former group. Moreover, in the Ainu plasma ketone bodies showed a pronounced elevation after norepinephrine and acute cold exposure, and a highly significant correlation was found between energy metabolism and plasma ketone bodies. Probably in this ethnic group plasma FFA is rapidly converted to ketone bodies in the liver and the ketone bodies may be effectively utilized as metabolic fuel. Thus, Ainu are likely to possess a specific trait in the fat metabolism (Itoh, 1974).

Sugar metabolism

Extremely low incidence of clinically manifest diabetes mellitus has been reported in Eskimos in Alaska, Canada and Greenland. Schaefer (1968, 1969) and Schaefer, Crockford & Romanowski (1972) found that more than 50% of Canadian Eskimos resemble mild diabetics in their delayed insulin response to orally administered glucose but they showed, in contrast to most diabetics, a normal insulin release and normal glucose disappearance when tested for tolerance to intravenously administered glucose. They found 'normalization' of glucose tolerance after protein ingestion as the consequence of earlier insulin release, possibly by 'priming' the acutely-releasable insulin pool by amino acid-responsible insulinotropic hormones such as pancreozymin. It was suggested that the regulatory weakness of glucose metabolism in Eskimos and improvement with antecedent meals may involve factors not only at the gut mucosa but also in the pancreas and possibly elsewhere. Since the traditional diet of Eskimos contained large amounts of protein, lesser amounts of fat and only small amounts of carbohydrate, glucose and other rapidly absorbable sugars were for Eskimos less important factors than amino acids for the stimulation of insulin release.

A diminished sensitivity of the plasma glucose to insulin in the glucose tolerance test was reported by Feldman (personal communication) in Eskimos at Point Hope in Alaska. However, the tolbutamide tolerance test showed adequate amounts of insulin, averaging 140 units at 2 min with an appropriate decline in plasma glucose to about 33% of fasting levels at 10 min.

Fasting levels of blood glucose were 86 and 76 mg% in Point Hope and Sitka Eskimos, respectively (Feldman, personal communication). The levels of the Ainu were 79 mg% (Itoh, 1974) or 86 mg% (Namiki, personal communication).

Very low incidence of diabetes mellitus in the Ainu has long been known. Glucose tolerance tests demonstrated no difference in the changes in blood glucose, plasma immuno-reactive insulin and human growth hormone levels after glucose administration between Ainu and control Japanese subjects (Kuroshima *et al.*, 1972). The only significant difference was a lower level of plasma FFA after a glucose load in the Ainu. Japanese subjects showed a marked rebound of the plasma FFA.

The lactic acid concentration in the blood under basal conditions was 14.5 mg/100 ml in the Ainu, which was not different from the control values.

A lactose intolerance has been found in over 80% of adult Alaskan Eskimos and Indians (Duncan & Scott, 1972). Since these populations traditionally did not drink milk, they are considered to be deficient in intestinal lactase.

Circadian rhythms

A vast body of literature has appeared on human circadian rhythms, but relatively few investigations have been made on indigenous Arctic subjects. Lobban (1967) studied the daily rhythms of renal excretion for water, potassium, sodium and chloride on Indians and Eskimos under natural conditions during the continuous midsummer daylight and in the continuous darkness of midwinter. She reported that the excretory patterns of the Arctic subjects contain a high proportion of abnormalities, such that the averaged patterns for the indigenous groups are less well defined than are those for a control group of subjects from a temperate zone. The loss of definition of the rhythms was most marked in the Eskimo subjects, where differences between day and night excretory rates had virtually disappeared. However, with increasing urbanization and consequent changes in their way of life, it seemed that the circadian rhythms of renal excretion of the Eskimos are also changing (Lobban, 1971). Rhythms were well defined, even in children, but their temporal synchronization differed markedly from that of Arctic incomers and of temperate zone and equatorial subjects. Lobban (1974) further studied the urinary excretion of water and electrolytes in Eskimo subjects in the North Baffin settlement of Pond Inlet in spring (light/dark), summer (light/light), autumn (light/dark) and winter (dark/dark). Seasonal variations in the daily patterns of urinary excretion were observed, but were more closely related to the life-style of the subjects than to the seasonal light/dark variations in the natural environment. Lobban inferred that to some extent the recent and

The human biology of circumpolar populations

marked changes in the social environment of the Eskimo have been accompanied by biological changes in the people themselves.

According to Bohlen (1970, 1971), in Wainwright Eskimos circadian rhythms were demonstrable for oral temperature, urinary potassium and chloride excretion. The rhythms were synchronized irrespective of subtle changes in light intensity, perhaps as a result of social factors. He also found that displacement of Caucasians from temperate to polar latitudes did not dampen the circadian amplitude, expressed as percent of average, nor was there an increase in amplitude of circannual rhythms corresponding to the greater prominence of annual changes in daily photofraction at high latitude. Thus, he suggested the relative importance of social over photic circadian synchronization for man in the Arctic.

References

Adams, T. & Covino, R. G. (1958). Racial variations to a standard cold stress. *J. appl. Physiol.*, **12**, 9–12.

Andersen, K. L., Lønying, Y., Nelms, J. D., Wilson, O., Fox, R. H. & Bolstad, A. (1960). Metabolic and thermal response to a moderate cold exposure in nomadic Lapps. *J. appl. Physiol.*, **15**, 649–53.

Bohlen, J. G. (1970). Circadian and circannual rhythms in Wainwright Eskimos. *Arctic Anthrop.*, **7**, 95–100.

Bohlen, J. G. (1971). Physiological rhythms in man and the polar light regimen. *Abst. 2nd intern. Symp. circumpolar Health*, p. 11.

Bollerud, J., Edwards, J. & Blakely, R. A. (1950). Survey of the basal metabolism of Eskimos. *Arctic Aeromed. Lab. Project Rep.*, 2101–20.

Brown, G. M., Bird, G. S., Boag, T. J., Boag, L. M., Delahaye, J. D., Green, J. E., Hatcher, J. D. & Page, J. (1954a). The circulation in cold acclimatization. *Circulation*, **9**, 813–22.

Brown, G. M., Bird, G. S., Boag, L. M., Delahaye, D. J., Green, J. E., Hatcher, J. D. & Page, J. (1954b). Blood volume and basal metabolic rate of Eskimos. *Metabolism*, **3**, 247–54.

Brown, G. M. & Page, J. (1952). The effect of chronic exposure to cold on temperature and blood flow of the hand. *J. appl. Physiol.*, **5**, 221–7.

Budd, G. M. & Warhaft, N. (1966). Cardiovascular and metabolic responses to noradrenaline in man, before and after acclimatization to cold in Antarctica. *J. Physiol.*, **186**, 233–42.

Crill, G. W. & Quiring, D. P. (1939). Indian and Eskimo metabolism. *J. Nutr.*, **18**, 361–8.

Davis, T. R. A. (1961). Chamber cold acclimatization in man. *J. appl. Physiol.*, **16**, 1011–5.

Duncan, I. W. & Scott, E. M. (1972). Lactose intolerance in Alaskan Indians and Eskimos. *Am. J. clin. Nutr.*, **25**, 867–8.

Eagan, C. J. (1963). Local vascular adaptation to cold in man. *Fed. Proc.*, **22**, 947–52.

Eagan, C. J. (1966). Biometeorological aspects in the ecology of man at high latitudes. *Int. J. Biometeor.*, **10**, 293–304.

Feldman, S. A., Ho, K. J., Lewis, L. A.. Mikkelson, B. & Taylor, C. B. (1972).

Lipid and cholesterol metabolism in Alaskan Arctic Eskimos. *Arch. Path.*, **94**, 42–58.

Glaser, E. M., Hall, M. S. & Whittow, G. C. (1959). Habituation to heating and cooling of the same hand. *J. Physiol.*, **146**, 152–64.

Glaser, E. M. & Whittow, G. C. (1957). Retention in a warm environment of adaptation to localized cooling. *J. Physiol.*, **136**, 98–111.

Gottschalk, C. W. & Riggs, D. S. (1952). Protein-bound iodine in the serum of soldiers and Eskimos in the Arctic. *J. Clin. Endocr.*, **12**, 235–43.

Hart, J. S., Sabean, H. B., Hildes, J. A., Depocas, F., Hammel, H. T., Andersen, K. L., Irving, L. & Foy, G. (1962). Thermal and metabolic responses of coastal Eskimos during a cold night. *J. appl. Physiol.*, **17**, 953–60.

Heinbecker, P. (1928). Studies on the metabolism of Eskimos. *J. biol. Chem.*, **80**, 461–75.

Heinbecker, P. (1931). Further studies on the metabolism of Eskimos. *J. biol. Chem.*, **93**, 327–36.

Hildes, J. A., Irving, L. & Hart, J. S. (1961). Estimation of heat flow from hands of Eskimos by calorimetry. *J. appl. Physiol.*, **16**, 617–23.

Höygaad, A. (1941). Studies on the nutrition and physio-pathology of Eskimos. *Det norske videnskaps-akademi's skrifter, Mat.-Naturv. Klasse.* 1940, No. 9.

Iampietro, P. F., Goldman, R. F., Buskirk, E. R. & Bass, D. E. (1959). Response of Negro and White males to cold. *J. appl. Physiol.*, **14**, 798–800.

Itoh, S. (1974). *Physiology of Cold-Adapted Man.* Sapporo: Hokkaido University Medical Library.

Itoh, S. & Kuroshima, A. (1972). Lipid metabolism of cold-adapted man. In *Advances in Climatic Physiology*, ed. S. Itoh, K. Ogata & H. Yoshimura, pp. 260–77. Tokyo: Igaku-Shoin.

Itoh, S., Kuroshima, A., Doi, K., Moriya, K., Shirato, H. & Yoshimura, K. (1969a). Lipid metabolism in relation to cold adaptability in man. *Fed. Proc.*, **28**, 960–4.

Itoh, S., Kuroshima, A., Doi, K., Yoshimura, K. & Moriya, K. (1969b). Plasma lipid profiles in the winter of Hokkaido inhabitants. *Japan. J. Physiol.*, **19**, 233–42.

Itoh, S., Doi, K. & Kuroshima, A. (1970a). Enhanced sensitivity to noradrenaline of the Ainu. *Int. J. Biometeor.*, **14**, 195–200.

Itoh, S., Kuroshima, A., Hiroshige, T. & Doi, K. (1970b). Finger temperature responses to local cooling in several groups of subjects in Hokkaido. *Japan. J. Physiol.*, **20**, 370–80.

Itoh, S., Shirato, H., Hiroshige, T., Kuroshima, A. & Doi, K. (1969c). Cold pressor response of Hokkaido inhabitants. *Japan. J. Physiol.*, **19**, 198–211.

Joy, R. J. T. (1963). Responses of cold-acclimated men to infused norepinephrine. *J. appl. Physiol.*, **18**, 1209 12.

Kawahata, A. & Sakamoto, H. (1951). Some observations on sweating of the Ainu. *Japan. J. Physiol.*, **2**, 166–9.

Krog, J., Alvik, M. & Lund-Larsen, K. (1969). Investigations of the circulatory effects of submersion of the hand in ice water in the Finnish Lapps, the 'Skolts'. *Fed. Proc.*, **28**, 1135–7.

Krog, J., Folkow, B., Fox, R. H. & Andersen, K. L. (1960). Hand circulation in the cold of Lapps and North Norwegian fishermen. *J. appl. Physiol.*, **15**, 654–8.

Kuno, Y. (1956). *Human Perspiration.* Springfield: C. C. Thomas.

Kuroshima, A., Itoh, S., Azuma, T. & Agishi, Y. (1972). Glucose tolerance test in the Ainu. *Int. J. Biometeor.*, **16**, 193–7.

LeBlanc, J. (1962). Local adaptation to cold of Gaspé fishermen. *J. appl. Physiol.*, **17**, 950–2.

LeBlanc, J. (1966). Adaptive mechanism in human. *Ann. N.Y. Acad. Sci.*, **34**, 721–32.

LeBlanc, J., Hildes, J. A. & Heroux, O. (1960). Tolerabce of Gaspe fishermen to cold water. *J. appl. Physiol.*, **15**, 1031–4.

Levine, V. E. & Wilber, C. G. (1949). Fat metabolism in Alaskan Eskimos. *Fed. Proc.*, **8**, 95–6.

Lobban, M. C. (1967). Daily rhythms of renal excretion in Arctic-dwelling Indians and Eskimos. *Quart. J. exp. Physiol.*, **52**, 401–10.

Lobban, M. C. (1971), Human daily rhythms of physiological function and activity in the Arctic – a changing scene. *Abst. 2nd Intern. Symp. Circumpolar Health*, *p. 8.*

Lobban, M. C. (1974). Seasonal variations in daily patterns of urinary excretion in Eskimo subjects. *Abst. 3rd Intern. Symp. Circumpolar Health, p. 14.*

Lund-Larsen, K., Wika, M. & Krog, J. (1970). Circulation responses of the hand of Greenlanders to local cold stimulation. *Arctic Anthrop.*, **7**, 21–5.

McHattie, L., Haab, P. & Rennie, D. W. (1960). Eskimo metabolism as measured by the technique of 24-hour indirect calorimetry and graphic analysis. *Arctic Aeromed. Lab. Alaska Technical Report AAL–TR–60–43.* Alaska: Fort Wainwright.

Milan, F. A. & Evonuk, E. (1966). Oxygen consumption and body temperatures of Eskimos during sleep. *J. appl. Physiol.*, **22**, 565–7.

Milan, F. A., Hannon, J. P. & Evonuk, E. (1963). Temperature regulation of Eskimos, Indians and Caucasians in a bath calorimeter. *J. appl. Physiol.*, **18**, 378–82.

Miller, L. K. & Irving, L. (1962). Local reactions to air cooling in an Eskimo population. *J. appl. Physiol.*, **17**, 449–55.

Nelms, J. D. & Soper, D. J. (1962). Cold vasodilation and cold acclimatization in the hands of British fish filleters. *J. appl. Physiol.*, **17**, 444–8.

Ohara, K. (1966). Chloride concentration in sweat: Its individual, regional, seasonal and some other variations, and interrelationships between them. *Japan. J. Physiol.*, **17**, 274–90.

Ohara, K. (1972). Salt concentration in sweat and heat adaptability. In *Advances in Climatic Physiology*, ed. S. Itoh, K. Ogata & H. Yoshimura, pp. 122–33. Tokyo: Igaku-Shoin.

Rabinowitch, I. M. & Smith, F. C. (1936). Metabolic studies of Eskimos in the Canadian Eastern Arctic. *J. Nutr.*, **12**, 337–56.

Rodahl, K. (1952). Basal metabolism of the Eskimo. *J. Nutrition*, **48**, 359–68.

Rodahl, K. (1958). In *Cold Injury, Trans. Fifth Conf.*, ed. M. I. Ferrer. New York: Josiah Macy, Jr. Foundation.

Schaefer, O. (1968). Glucose tolerance testing in Canadian Eskimos: A preliminary report and a hypothesis. *C.M.A.J.*, **99**, 252–62.

Schaefer, O. (1969). Carbohydrate metabolism in Eskimos. *Arch. Environ. Health*, **18**, 142–7.

Schaefer, O., Crockford, P. M. & Romanowski, B. (1972). Normalization effect of preceding protein meals on 'diabetic' oral glucose tolerance in Eskimos. *C.M.A.J.*, **107**, 733–8.

Scholander, P. F., Andersen, K. L., Krog, J., Lorentzen, F. V. & Steen, J. (1957). Critical temperature in Lapps. *J. appl. Physiol.*, **10**, 231–4.

Wika, M. (1971). Skin circulation in the cheeks of Arctic populations and adaptation to cold. *Acta physiol. Scand.*, **82**, 39–40A.

Yoshimura, H. & Iida, T. (1950). Studies on the reactivity of skin vessels to extreme cold. I. A point test on the resistance against frostbite. *Japan J. Physiol.*, **1**, 147–59.

10. Work physiology and activity patterns

ROY J. SHEPHARD

Aspects of physiology relating specifically to cold adaptation, including circulatory, hormonal and metabolic adjustments, have been considered in Chapter 9 by Dr Itoh, This chapter, on the other hand, examines the working capacity and activity patterns of individual circumpolar peoples, seeking to draw together material from several published studies (Andersen et al., 1962; Andersen, 1969; Ikai et al., 1971; Lammert, 1972; Rennie et al., 1970; Shephard, Rode & Godin, 1972; Shephard & Rode, 1973; Shephard, 1974) and unpublished data generously made available by my colleagues Dr A. Rode and Mr G. Godin, by other members of the Canadian IBP team, by Dr D. Rennie and Dr E. R. Buskirk from the USA, and by Dr M. Ikai and Dr K. Ishii from Japan. An attempt will be made to seek features common to the circumpolar populations, and to explore the reasons for regional differences.

The Arctic habitat

General features of the Arctic habitat are discussed elsewhere (Chapter 1), but we may note specific features that could modify working capacity. Where reliance is placed upon hunting as the source of food, clothing and fuel, intense physical activity may be a requirement for survival. This could have a significant immediate training effect, improving muscular strength and cardio-respiratory fitness and reducing body fat; in the longer term, it may also exert a selective pressure in that those who lack the constitutional basis for such adaptations of working capacity are more likely to die from starvation or exposure.

Reliance upon food from the hunt could lead to a high fat intake. Apart from the implications for blood lipids (p. 270), this encourages glycogen sparing during effort, with a prolongation of muscular endurance. It may also help to preserve the compact body form, lost in 'white' sugar-eaters. On the other hand, episodic shortages of game could retard growth during adolescence, and in adult life could restrict the capacity for sustained work by giving rise to anemia and limitations of depot fat.

Climate interacts significantly with working capacity. While the environment is very cold for a sedentary person, a hard-working hunter may be over-clothed to the point of vigorous sweating. There is thus a need for highly variable thermal insulation. Water loss in the dry Arctic air may also restrict working capacity.

Genetic isolation often leads to the appearance of disadvantageous genotypes. It could contribute to a restriction in the variance of data, a diminution of secular trends in body size, and (theoretically) the emergence of a people with outstanding working capacity.

General aspects of disease are discussed elsewhere (p. 14). However, in our present context, chronic diseases such as tuberculosis could (i) retard growth, (ii) lead to a permanent impairment of oxygen transport, and (iii) selectively eliminate those members of the community with a poor working capacity.

Population sampling

Sampling is a crucial problem in any population survey (Rose & Blackburn, 1968; Shephard, 1971). Ethical considerations dictate that we should work with volunteers and often Arctic samples have been less than representative of their village or settlement, with this in turn bearing a doubtful relationship to larger circumpolar populations.

The problem can be illustrated by the Canadian data. Our physiology team examined 130 males and 94 females over the age of nine yr living in the village of Igloolik, near the tip of the Melville peninsula. We saw a relatively large percentage of the villagers, 77% of the boys, 65% of the girls, 67% of men aged 20–39 yr, 56% of women aged 20–39 yr, 41% of the older men and 48% of the older women. The influence of sampling procedures upon averaged results was checked by comparing our data with those collected by the medical survey team (Drs Hildes & Schaefer). We obtained nursing station records on 335 villagers over the age of nine yr, 224 of whom were seen by the physiology team (Table 10.1). Ignoring the patients with congenital dislocation of the

Table 10.1. *A comparison of nursing station records for Igloolik villagers tested and not tested by the physiology team*

	Tested	Not tested
Normal health	176 (78.6%)	49 (44.1%)
Primary tuberculosis	9 (4.0%)	12 (10.8%)
Hilar calcification	11 (4.9%)	22 (19.8%)
Secondary or advanced tuberculosis	17 (7.6%)	7 (6.3%)
Fibrosis	9 (4.0%)	13 (11.7%)
Emphysema	2 (0.9%)	2 (1.8%)
Crippled	–	5 (4.5%)
Recent 'coronary' attack	–	1 (0.9%)
Total	224	111

hip, poliomyelitis and a recent coronary attack, there yet remained a much higher proportion of villagers with pulmonary disease among those not seen (56/105) than among those seen (48/176), the difference being highly significant ($x^2 = 33.62$, $P < 0.001$); interestingly enough, it was particularly Eskimos with minor manifestations of tuberculosis (Ghon focus or hilar calcification) who remained untested, suggesting a more cautious attitude to life in those having abnormalities of their chest radiographs. The tendency of villagers with chest disease not to volunteer for physiological evaluation is confirmed (Table 10.2) by the medical history (hospital admissions for tuberculosis, other chest diseases, and chronic cough), examination (clubbing and adventitious sounds), and overall impression of health. Both medical and physiological teams had collected information on static and dynamic lung volumes. Nevertheless, at the time of preparing this report, the medical team had information on only 49 of our sample of 223 and had sought out 79 people whom we had not seen. Mean results for the 49 subjects common to the two groups of investigators were remarkably similar (Table 10.3), and coefficients of correlation between the two sets of data were also satisfactory (for vital capacity (FVC), $r = 0.92$, and for one-second forced expiratory volume ($FEV_{1.0}$), $r = 0.93$). The additional men seen by the physiology group were apparently similar to the jointly studied population, but as might be predicted from Tables 10.1 and 10.2 the medical investigators sought out men with an equal restriction of both FVC and $FEV_{1.0}$. Few women were common to the two series, but again those seen by the medical team alone had poorer results than those seen by the physiology team alone. Discussion with the medical team supports the hypothesis that they tended to select patients with respiratory symptoms for lung-function testing. To a lesser extent, the very healthy Eskimos may have gravitated to the physiology laboratory; however, testing of the best hunters was restricted by the fact that they were seldom present in the village.

No details of sampling procedures are known for other IBP investigations. Ikai *et al.* (1971) tested 21 Ainu males from a population of approximately 275 living at Hokkaido; the likely bias was towards healthy but sedentary subjects. Rennie *et al.* (1970) saw some 40% of the Alaskan Eskimos living at Wainwright, although his sample (57 males, 30 females) was deficient in adult women. He later examined some 10% of the Point Hope and Point Barrow Eskimos (sample 121 males, 110 females); the last group are highly acculturated and by no means a genetic isolate. Lammert (1972) examined 12 men and three women; this cannot have been more than 20% of the Eskimos living at Upernavik in Greenland. The men were said to be seal-hunters, with their principal activity (Kayak paddling) involving mainly arm work. The IBP studies of the Lapps (Ove Wilson, to be published) are unfortunately incomplete as yet; data on this population are thus restricted to the sample of nomadic Lapps tested by Andersen (1969) and Andersen *et al.* (1962) and

307

Table 10.2. *Influence of medical history and examination[a] upon the probability of Igloolik villagers volunteering for physiological testing*

Severity of symptom or sign[b]	Hospital admission for TB		Other chest disease		Chronic cough		Clubbing		Adventitious sounds		Overall impression of health	
	Seen	Not seen	Seen	Not seen	Seen	Not seen	Seen	Not seen	Seen	Not seen	Seen	Not seen
0	74	66	–	17	34	15	150	218	1	3	6	10
1	14	24	18	19	20	13	0	10	137	198	9	142
2	4	21	1	4	17	31	1	3	26	46	4	12
3	2	2	6	17	–	–	0	3	7	24	0	1
x^2	12.96		13.24		11.85		9.06				19.74	
P	$0.01 > P > 0.001$		$0.01 > P > 0.001$		$0.01 > P > 0.001$		$0.05 > P > 0.02$		not significant		< 0.001	

[a] Data kindly provided by Dr O. Schaefer and Dr J. Hildes.
[b] The scales used by Dr Schaefer and Dr Hildes have been compressed to give a reasonable sample size in each category.

Table 10.3. *Canadian data for forced vital capacity (FVC as l BTPS) and one-second forced expiratory volume (FEV$_{1.0}$ as l BTPS) as reported by the physiology team and by the medical team (medical team data by courtesy of Dr Hildes and Dr Schaefer). Data are expressed as means $\pm$ S.D. with the number of observations in parentheses*

		Seen by both teams		Seen by physiology team alone	Seen by medical team alone
Age (yr)	Sex	Physiology team	Medical team		
FVC					
9–19	M	4.91 (2)	4.99 (2)	4.08 (44)	4.23 (2)[a]
9–19	F	–	–	3.59 (32)	–
20–59	M	4.87 $\pm$ 0.73 (41)	4.94 $\pm$ 0.62 (41)	4.92 (28)	4.41 $\pm$ 0.88 (41)
20–59	F	2.82 $\pm$ 0.20 (5)	2.80 $\pm$ 0.20 (5)	3.59 (44)	2.66 $\pm$ 0.89 (24)
FEV$_{1.0}$					
9–19	M	4.06 (2)	4.15 (2)	3.38 (44)	3.56 (2)[a]
9–19	F	–	–	3.02 (32)	–
20–59	M	3.69 $\pm$ 0.70 (41)	3.64 $\pm$ 0.66 (41)	3.86 (28)	3.14 $\pm$ 1.04 (41)
20–59	F	1.99 $\pm$ 0.26 (5)	1.98 $\pm$ 0.21 (5)	2.83 (44)	1.88 $\pm$ 0.78 (24)

[a] Average age higher than 9, 19 yr old children seen by the physiology team alone.

other data on Fenno–Scandia available in the world literature. Andersen (1961) selected 49 men and 21 girls from a population of some 1000 Lapps living in the Kautokeino region; his sample was thus about 10% of the older children and adults, but included no girls over the age of 21 yr.

Methodology

Details of IBP methodology for the measurement of working capacity were standardized by an international working party (Shephard *et al.*, 1968) and summarized by Weiner & Lourie (1969). The Canadian physiology team adhered scrupulously to this recommended protocol. Other investigators introduced minor variations of technique; these will be noted when they appear relevant to the interpretation of disparate results. Procedures used at both Hokkaido (Ainu) and Igloolik were checked biologically by making parallel observations on normal 'white' subjects. Details are given elsewhere (Shephard, 1974), but it may be stressed that results for the 'white' subjects were at the anticipated level, thus ruling out any possibility that results for the circumpolar peoples were distorted by technical artefacts.

The prime season of study was summer in all investigations. However, seasonal comparisons of working capacity were completed on the Ainu (Ikai *et al.*, 1971) and the Igloolik Eskimos (Rode & Shephard, 1973a); little difference was seen during the winter months other than a small increase of sub-cutaneous fat (Table 10.13). The Ainu and the Igloolik Eskimos were screened for disease, and the results reported are for health individuals. In Upernavik, Lappland and Alaska the elimination of diseased subjects (possibly 20% of the sample, *see* Table 10.1) apparently has yet to be carried out.

Details of smoking habits were known only for the Canadian Eskimos. Almost all the adults in this group smoked some cigarettes. However, in many instances, the duration of the habit was 10 years or less, and probably because of financial constraints, the average daily consumption was lower than in most white communities (14.6/day in the men, 8.3/day in the women).

Physical characteristics

Differences of body size and form are important in interpreting many facets of working capacity such as aerobic power, vital capacity, body water and percentage body fat. Body weight alters the energy cost of many activities, while differences of leg length – by altering leverage for both active muscles and recording tensiometer cables – have a more direct influence upon reported leg-strength data.

Physical characteristics are summarized in Table 10.4. While not conforming to the popular stereotype of extreme shortness of stature, many of the older circumpolar peoples are still shorter than the average 'white' person, particularly in Hokkaido and among the Kautokeino, Pasvik and Skolt Lapps (Andersen *et al.*, 1962; Lewin, Jürgens & Louekari, 1970). Regional

Table 10.4. *Physical characteristics of adult Eskimos (20–59 yr)*

Region	Height (cm)	Weight (kg)	Relative weight (kg/cm)	Relative weight (Excess, kg)
Males				
Igloolik	165.6	68.0	0.411	+ 5.7
Hokkaido	161.7	57.1	0.353	+ 2.4
Wainwright	168.6	69.6	0.413	+ 5.1
Point Hope	–	69.2	–	–
Point Barrow	–	77.3	–	–
Upernavik	167.0	65.2	0.390	+ 1.9
Nellim (Pasvik Lapps)[a]	163.4	–	–	–
Suengel (Skolt Lapps)[a]	159.2	–	–	–
Kautokeino (Nomadic Lapps)	~160	~61	~0.38	~+ 2
Females				
Igloolik	155.4	58.0	0.373	+ 5.2
Wainwright	159.4	63.9	0.401	+ 8.4
Point Hope	–	60.6	–	–
Point Barrow	–	65.1	–	–
Nellim (Pasvik Lapps)[a]	152.0	–	–	–
Suengel (Skolt Lapps)[a]	147.8	–	–	–

[a] Data of Lewin *et al.* (1970).

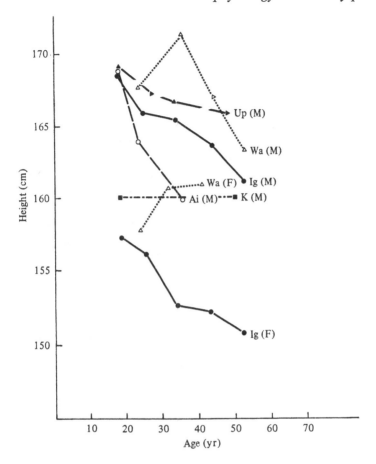

Fig. 10.1. The apparent 'secular trend' in the standing height of circumpolar populations. Ai, Ainu; Ig, Igloolik; K, Kautokeino; Up, Upernavik; Wa, Wainwright. M, males; F, females. Lewin *et al.* (1970) have reported a true secular trend of 1.4 cm/decade for male Pasvik Skolts, 1.3 cm/decade for female Pasvik Skolts, 0.8 cm/decade for male Suengel Skolts and 0.6 cm/decade for female Suengel Skolts.

anthropometry (Lewin & Hedegård, 1971) shows a normal trunk length, but legs that are up to 10 cm shorter than in a 'white' person. The Igloolik Eskimos were also heavy relative to 'white' actuarial standards (Society of Actuaries, 1959), without evidence of excess subcutaneous fat. The Alaskan Eskimos were equally heavy, but in view of the higher skinfold readings (Table 10.9) much, if not all, of their excess weight may be attributable to body fat.

In some communities, a plot of standing height against age (Fig. 10.1) suggests a rapid 'secular trend' to increase of stature; this is greatest for the Ainu, and least consistent for the Nomadic Lapps and the already tall

Eskimos of Wainwright; as early as 1956–7, a survey of Alaskan Eskimos showed little difference of standing height between young and older men (D. W. Rennie, personal communication). Where a decrease is seen, it begins in early adult life and this is difficult to attribute to normal ageing processes such as kyphosis or compression of intervertebral discs. However, before accepting the apparent secular trend at face value, it will be necessary to check the percentage of hybrids in the different age groups. Genetic admixture consequent upon groupings in larger settlements can play an important role in increasing body size. Jamison (1970) found that hybrid Alaskan Eskimos were 4 cm taller than those who were relatively pure Eskimo stock. It will also be necessary to exclude diseased subjects. Thus, the medical team at Igloolik reported a smaller and more variable secular trend than the physiologists; heights measured by both the physiology and medical groups were generally in good agreement, and the discrepancy seemed due to the diseased patients seen by the medical but not by the physiology team – tuberculosis can have an appreciable systemic effect in reducing standing height, particularly if contracted during the growing years (*see* p. 19). While many of the older adults give a history of tuberculosis (p. 14), the disease is now much less prevalent than two or three decades ago. A second possible extrinsic factor is an alteration of diet. Lewin *et al.* (1970) attribute the short stature of the Suengel Lapps to poor nutrition, and the rapid secular trend of the Nellim Lapps to improved nutrition. It would be difficult in these and other Arctic communities to establish by retrospective questioning the precise quantities of food available to adults during their period of growth. In Igloolik, at least, we have the impression that there was episodic starvation before the coming of the 'white' man. At the present time (p. 262), 'native' foods are being increasingly replaced by refined carbohydrates, a change that again seems likely to alter the course of growth.

Maximal oxygen intake

The maximal oxygen intake is widely accepted as the best single measure of capacity to perform work of moderate duration (Shephard *et al.*, 1968; Shephard, 1969). All IBP investigators have made some direct measurements of this variable (Table 10.5), but unfortunately test methods have differed and the total number of subjects has been rather small.

A maximal value should be defined by rigid criteria, including an acceptable plateau of oxygen consumption (increment < 2 ml/kg min STPD for a 5–10% increase of work load), a respiratory gas exchange ratio of approximately 1.15, a heart rate close to the theoretical age-related maximal value, and a blood lactate of 100 mg/100 ml (in young adults). Unfortunately, no investigators except Andersen *et al.* (1962) and Karlsson (1970) have found it practicable to measure blood lactates on the circumpolar peoples, and oxygen

Table 10.5. *Direct measurements of maximal oxygen intake on circumpolar groups*

Population	Age (yr)	Sex and number tested	Heart rate (per min)	Gas exchange ratio	Ventilation (l/min BTPS)	Maximal oxygen intake (l/min STPD)	(ml/kg min STPD)
Hokkaido: Ainu							
bicycle ergometer							
	16–19	M (5)	196 ± 5	1.18 ± 0.10	95.8 ± 6.6	2.72 ± 0.39	46.5 ± 5.3
	20–29	M (3)	189 ± 8	1.23 ± 0.10	85.6 ± 23.1	2.32 ± 0.40	44.4 ± 2.3
	30–39	M (6)	189 ± 11	1.20 ± 0.13	81.3 ± 10.2	2.34 ± 0.34	39.5 ± 7.4
Igloolik Eskimos:							
step good	23 ± 8	M (12)	185 ± 13	1.13 ± 0.03	–	3.52 ± 0.56	52.3 ± 3.6
fair	27 ± 13	M (14)	178 ± 15	1.04 ± 0.03	–	3.03 ± 0.83	48.9 ± 5.3
poor	24 ± 8	M (7)	176 ± 3	0.97 ± 0.03	–	3.27 ± 0.72	54.1 ± 3.0
fair	14 ± 2	F (5)	182 ± 16	1.05 ± 0.03	–	1.95 ± 0.30	39.2 ± 7.0
Nellim Lapps							
?bicycle ergometer							
	15–19	M	–	–	–	~2.8	52
	20–29	M	–	–	–	2.88	~49
	10–14	F	–	–	–	~1.4	35
	20–29	F	–	–	–	1.76	~33
Kautokeino Lapps							
bicycle ergometer							
	10	M } (49)	197	–	–	~1.4	~53
	18–30	M }	197	–	102.8	~3.5	~53
	50	M }	181	–	–	~2.7	~45
skiing	15–38	M (5)	194	–	101.3	2.68	57.4
Point Hope Eskimos[a]							
treadmill							
	11–18	M (14)	180	–	94.8	2.41	44.9
	24	M (2)	186	–	99.8	2.75	42.3
	35	M (2)	181	–	102.1	3.07	41.6
	13–18	F (10)	189	–	69.3	1.86	33.5
	24.7	F (3)	182	–	73.5	1.95	32.0
Upernavik Eskimos							
bicycle ergometer							
	14–19	M (4)	186	0.96	–	2.36	40.1
	20–29	M (3)	191	1.04	–	2.75	40.7
	30–45	M (4)	174	1.04	–	2.35	37.2

[a] Data selected from larger sample on basis of terminal heart rate.

313

plateaux also have been defined rather less precisely than would be acceptable in a base laboratory. Our method of assessing maxima in Igloolik was to subdivide subjects on the basis of the quality of effort as judged by an experienced observer. Substantially higher terminal pulse rates and gas exchange ratios were found in those making a 'good' effort than in those making a 'fair' or 'poor' effort; the 'good' subjects at Igloolik and the Ainu developed final pulse rates and exchange ratios comparable with those of the 'white' population.

There are many difficulties in motivating primitive and non-competitive

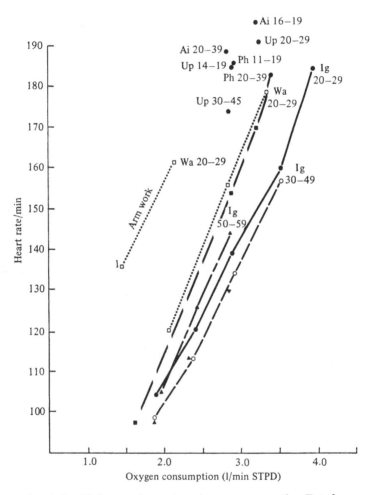

Fig. 10.2. The relationship between heart rate and oxygen consumption. Date for age groups shown. Ig, Igloolik, step test; Wa, Wainwright, step test; Ph, Point Hope, treadmill; Up, Upernavik, bicycle ergometer; Ai, Ainu, bicycle ergometer (Shephard, 1974).

societies to perform all-out effort, and it might seem from Table 10.5 that the Ainu were exercised a little more vigorously than the other circumpolar populations. Nevertheless, the Igloolik Eskimos attained an appreciably higher maximum oxygen intake. Expressing data in ml/kg. min, the Igloolik males had an 18% advantage over the Ainu, a 24% advantage over the Point Hope group, and a 29% advantage over the Upernavik Eskimos. The other

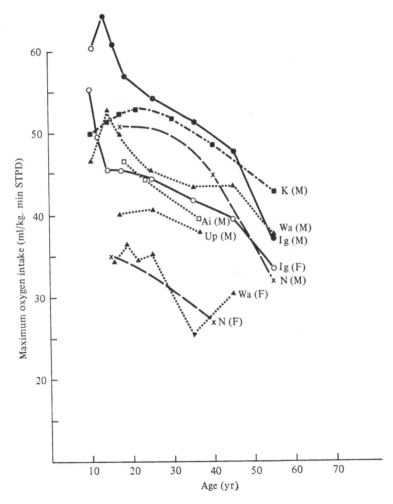

Fig. 10.3. The relationships between age and maximum oxygen intake (ml/kg. min STPD). Direct measurements on Ainu (Ai) and Upernavik (Up) Eskimos and on Kautokeino (K) and Nellim (N) Lapps using a bicycle ergometer. Wainwright Eskimos (Wa), linear extrapolation of step test data (Wa). Igloolik Eskimos (Ig), step test data applied to Astrand nomogram and corrected downwards by 8%. M, males; F, females.

circumpolar peoples with the possible exception of the nomadic Lapps would fit conveniently into the distribution curve for a sedentary 'white' population, but the Igloolik group definitely had an above-average working capacity. The general superiority of the Igloolik male was further demonstrated by the larger body of data on pulse responses to submaximum effort (Fig. 10.2). As in 'white' populations, the plot of oxygen consumption against heart rate was almost linear over the range 50–90% of $\dot{V}_{O2(max)}$. However, the directly measured maximum oxygen intakes of Table 10.5 were some 8% less than would be predicted by extrapolation of the data to a theoretical maximum heart rate. Assuming this to reflect a physiological phenomenon rather than a lack of motivation to maximal effort, prediction procedures based on sub-maximum exercise (Åstrand, 1960; Maritz et al., 1961) would over-estimate the true maximal oxygen intake. However, even if an 8% correction is applied to the submaximum data (Fig. 10.3), the Igloolik Eskimos remain much superior, both to other circumpolar peoples, and also to 'white' populations.

One possible technical problem relates to differences in the mode of exercise (Table 10.5). In young 'white' men, the largest values of $\dot{V}_{O2(max)}$ are obtained by uphill treadmill running, with step-test data 3–4% smaller, and bicycle ergometer results 6–7% smaller (Shephard et al., 1968). The male Eskimos apparently behave similarly (Shephard, 1974), with an average discrepancy of 3.8% between step and bicycle ergometer data; however, in the female the discrepancy is larger, averaging 13.8%. A fair proportion of traditional Eskimo activity involves arm work such as kayak paddling, and one might thus wonder whether a larger aerobic power could be developed by an arm-work ergometer. In one group where comparisons were made (Wainwright), arm work led to a much larger increment of pulse rate than a corresponding amount of leg work (Fig. 10.2); however, in the Kautokeino Lapps, a slightly higher aerobic power was measured during uphill skiing (57.4 ml/kg min) than with bicycle ergometry (55.4 ml/kg min).

Efficiency of effort

Work tolerance depends not only upon maximum aerobic power and body weight, but also upon the efficiency of effort. Ergonomic data (Godin & Shephard, 1973a) suggest that with their current diet, the basal metabolism of the Eskimo is no longer different from that of the 'white' population. Accepting this finding, the net mechanical efficiency of stepping in the Igloolik Eskimos was close to generally accepted values for a double nine-inch step; at 10, 15, 20 and 25 ascents per minute respective efficiency figures were 14.7, 15.4, 15.2 and 13.9% for 130 males, and 13.7, 15.4, 15.9 and 16.2% for 93 females.

The Igloolik Eskimos had not seen a bicycle prior to the arrival of the

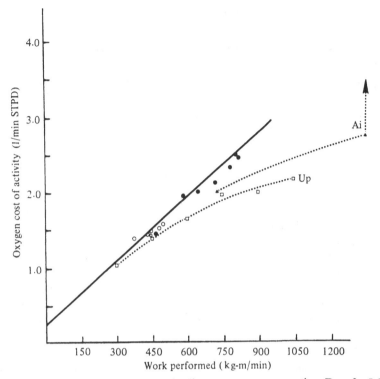

Fig. 10.4. Relation of bicycle ergometer loading to oxygen consumption. Data for Igloolik males, ●—● Igloolik females, ○—○; Ainu males, ▲- - -▲ Upernavik males, □- - -□. The second data point for the Ainu males refers to maximal effort with the arrow indicating an allowance made for anaerobic work (assumed metabolic respiratory quotient of 0.935).

physiology team. We were thus not surprised that their mechanical efficiency on the bicycle ergometer was both low (average 18.4% in the males, 18.1% in the females) and more variable than in the 'white' community (Rode & Shephard, 1973*b*). When the Ainu performed moderate work (720 kg-m/ min), they consumed 5–6% less oxygen than the Igloolik Eskimos (i.e. mechanical efficiency 19–20%). At the lowest work-loads, oxygen consumptions for the Upernavik group were also much as in Toronto. However, at higher loads there was a surprising curvilinearity (Fig. 10.4) and an apparent plateauing of oxygen costs; one must suspect that technical problems (slowing of pedalling rate, mouthpiece leakage, and anaerobic work) developed at this stage. The mechanical efficiency for the Kautokeino Lapps was better than in the Eskimos, being only marginally below that for a 'white' community (Andersen *et al.*, 1962).

The oxygen conductance line

Space does not permit detailed discussion of the behavior of individual links in the oxygen transport line (Rode & Shephard, 1973c, d). Contrary to earlier reports, the $FEV_{1.0}$ of the healthy male Eskimo living in Igloolik was at the anticipated level for 'white' subjects of comparable age, height and smoking history, while values for the female Eskimos were on average 12% greater than predictions. The FVC was 10% above predicted values in the men, and 18% above predicted in the females. Rennie *et al.* (1970) have reported rather similar findings from Wainwright, the male Eskimos having a 10% augmentation of FVC, $FEV_{1.0}$ and maximum breathing capacity.

The residual volume (RV) and total lung capacity (TLC) were estimated from the chest radiographs provided by Drs Hildes and Schaefer. Both were apparently somewhat increased, but this is probably a sampling artefact, in that many of the radiographs were taken to evaluate subjects with respiratory symptoms (Table 10.6). The overall RV/TLC ratio was 39.9% for the men, and 45.1% for the women, but selecting subjects with no history of respiratory disease much more normal ratios were obtained (27.1% for men with an average age of 30.6 yr, and 30.5% for women with an average age of 31.6 yr).

The steady-state carbon-monoxide diffusing capacity was essentially normal. Resting values (ml STPD/mm Hg/min) declined from the young (20–9 yr) to the elderly (50–59 yr), from 21.1 ± 7.0 to 10.9 ± 4.1 in the men, and from 15.3 ± 4.4 to 9.3 ± 3.1 in the women. Readings at 70–75% of $\dot{V}_{O2(max)}$ also showed a normal pattern of decline, from 43.2 ± 10.1 to 30.5 ± 10.5 in the men, and from 29.0 ± 8.0 to 21.1 ± 2.2 in the women. Blood carboxyhemoglobin levels, dropping from 5.4 ± 2.0 to $4.0 \pm 0.8\%$ in the men, and from 2.7 ± 0.6 to $2.1 \pm 1.1\%$ in the women were compatible with the reported cigarette consumption.

The maximum cardiac output and arterio–venous oxygen differences were estimated by a CO_2 rebreathing method. In the adult, results were not remarkable. At 70–75% of $\dot{V}_{O2(max)}$ the stroke volume declined from 126 ± 24 ml in the young (20–29 yr, to 114 ± 20 ml in the elderly men (50–59 yr); corresponding values for the women were 78 ± 16 ml and 70 ± 6 ml respectively. Arterio-venous oxygen differences (133–51 ml/l) showed no significant age or sex trends over the adult span. In children, there was a hypokinetic response to exercise, but this was no greater than would be anticipated in a 'white' community (Table 10.7).

Hemoglobin levels were normal, with few examples of clinical anemia. Although many of the subjects were quite thin, we would conclude that present nutrition is adequate to sustain hemoglobin levels. Whether this will remain true with further expansion of the village population is doubtful. The current yearly energy expenditure is about 367×10^6 kcal (Table 10.11), and about a third of the required calories are met from the hunt (Godin &

Table 10.6. *Total lung capacity (TLC) and residual volume/total lung capacity ratio (RV/TLC). Data from 98 medical radiographs obtained in Igloolik (radiographs by courtesy of Dr Schaefer and Dr Hildes). Data are expressed as means ± s.D. with the number of observations in parentheses*

	Age (yr)				
	20–29	30–39	40–49	50–59	All subjects
TLC (ml BTPS)					
Men	7874 ± 1038(18)	7823 ± 1334(29)	7601 ± 958(9)	8187 ± 441(5)	7802 ± 1170(63)
Women	6372 ± 1456(15)	6424 ± 781(12)	6574 ± 544(5)	6068 ± 733(3)	6392 ± 1073(35)
RV/TLC (%)					
Men	34.9	37.9	41.8	45.0	39.9
Women	40.8	45.9	53.9	48.4	45.1

Table 10.7. *Cardiac volumes (ml and ml/kg body weight) estimated from medical radiographs (radiographs by courtesy of Dr Schaefer and Dr Hildes). Data are expressed as means ± S.D. with the number of observations in parentheses*

| | Age (yr) | | | | |
	20–29	30–39	40–49	50–59	All subjects
Absolute volume (ml)					
Men	739 ± 85(18)	764 ± 111(28)	620 ± 137(9)	726 ± 91(5)	732 ± 115(60)
Women	596 ± 72(15)	556 ± 85(13)	579 ± 43(5)	590 ± 41(3)	578 ± 72(35)
Relative volume (ml/kg)					
Men	10.8 ± 1.3(17)	11.4 ± 1.8(25)	8.6 ± 1.7(8)	11.2 ± 1.3(5)	10.8 ± 1.8(55)
Women	10.7 ± 1.3(8)	8.0 ± 0.3(4)	9.8 ± 1.3(3)	10.2 ± 1.9(3)	9.8 ± 1.6(18)

Shephard, 1973*a*), ringed seal, walrus, caribou and fish providing much of the dietary protein. Further exploitation of game is scarcely practicable, and the needs of a larger community will probably be met by increased store purchases of such items as flour, sugar, tinned milk and soft drinks.

Blood volumes for those men working within the village of Igloolik (94.6 $\pm$ 12.6 ml/kg) were high normal values. Results for the hunters (83.6 $\pm$ 17.7 ml/kg) were somewhat lower, possibly because they were still dehydrated at the time of testing. Cardiac volumes were measured by both the physiology and the medical teams, using the same films. Inevitably, there were small differences of interpretation, particularly in the lateral plates. The coefficient of correlation between the two sets of readings was 0.82 $\pm$ 0.05, with a systematic discrepancy (24 $\pm$ 9 ml). Both teams were in agreement that cardiac volumes were essentially normal, although the results may have been biased downwards by the use of medical radiographs.

We may conclude that the oxygen transport line for the Igloolik Eskimos shows normal or high normal values throughout. This is in keeping with the oxygen consumption data, which although above expectation for a sedentary population was less than would be found in top athletes.

Muscle power, strength and endurance

Data on muscle strength and anaerobic power are collected in Table 10.8. Differences of body weight and leverage hamper direct comparisons with 'white' people. The grip strength of the Eskimos was not outstanding. This may reflect the need to wear clumsy mittens for much of the day. On the other hand, the leg strength of the male villagers was somewhat greater than that of 'white' residents of Igloolik who were tested on the same apparatus, while our figures for the Igloolik women would be quite unusual in a 'white' community. The tensiometric evidence of good muscular development is confirmed by estimates of lean body mass (p. 323). A need to walk through deep snow for much of the year undoubtedly contributes to the maintenance

Table 10.8. *Strength and anaerobic power of circumpolar peoples* (*Shephard, 1974*)

Age (yr)	Igloolik				Wainwright Anaerobic power	
	Grip strength (kg)		Leg strength (kg)		m/s	m/kg s
	M	F	M	F	M	M
20–29	52.0 $\pm$ 5.3	29.8 $\pm$ 4.0	88.3 $\pm$ 17.4	68.9 $\pm$ 17.8	1.67 $\pm$ 0.20	24.2 $\pm$ 2.9
30–39	46.1 $\pm$ 8.1	28.4 $\pm$ 4.9	88.9 $\pm$ 24.1	66.9 $\pm$ 16.5	1.49 $\pm$ 0.13	21.3 $\pm$ 2.6
40–49	42.4 $\pm$ 5.0	29.0	76.5 $\pm$ 15.6	68.1	1.22 $\pm$ 0.31	18.6 $\pm$ 5.4
50–59	38.5	27.3 $\pm$ 2.0	76.9	69.5 $\pm$ 2.7	1.17	17.0
60–69	–	–	–	–	0.99	14.8

of leg strength in Eskimos and 'whites' alike. In the summer, the Eskimo men also carry caribou carcasses on their backs for long distances, while many women carry 2–3 yr-old children in the traditional 'yappa' for much of the working day (*see also* Table 10.16).

Studies at Wainwright show that the anaerobic power of the leg muscles, as measured by the speed of sprinting up a short staircase, is also 10–20% greater than in the average 'white' community. A statistical analysis by Rennie *et al.* (1970) demonstrates an advantage of some 0.2 m/s over much of the adult life span.

Ikai *et al.* (1971) measured the oxygen consumption of the Ainu in the first five minutes following exercise at 75% of aerobic power. Results were some 75% of those found in the 'white' population. However, the difference seems related more to the light body weight of the Ainu (average 57.1 kg) than to any fundamental improvement of the recovery process. Indeed, rather similar figures were established for other lightweight Japanese residents of Hokkaido.

Body composition

Differences in the working capacity of the several circumpolar groups could reflect in part varying body weights and body compositions. The Ainu, Igloolik and Kautokeino males carried very little subcutaneous fat (Table 10.9). However, in the more acculturated Wainwright group, readings were appreciably higher. Igloolik and Kautokeino females carried much more fat than the males, and particularly among the older women readings were at least as large as reported for Wainwright.

None of the many possible formulae for converting skinfold readings to percentage body fat is particularly successful in the Eskimo (Shephard, Hatcher & Rode, 1973). More direct measurements of body composition have been made by the deuterium oxide method (Table 10.10). Absolute volumes of body water were much as previously reported for 'white' subjects. However, when account was taken of stature, readings were higher in the Eskimos. Thus Norris, Lundy & Shock (1963) showed a total body water of 0.225 l/cm in 'white' men, compared with our figure of 0.267 l/cm for the Igloolik male. Again, Young *et al.* (1963) quoted a figure of 0.179 l/cm for 'white' women, compared with our figures of 0.207 l/cm for the Igloolik female. The percentages of body fat derived from the deuterium results are much larger than would be anticipated from the skinfold readings, particularly in the male. While it is possible that the hydration of Eskimo lean tissue differs from the 73% figure accepted for the white population, the main explanation is probably that the Eskimo stores fat elsewhere than in the skin (omentum, mesentery, and retroperineum). This can be considered a positive adaptation to several environmental demands – variable body insulation, a high fat diet, and episodic starvation.

Table 10.9. *Skinfold data for circumpolar populations*

Region and ethnic group	Age (yr)	Sex	Number tested	Triceps (mm)	Sub-scapular (mm)	Supra-iliac (mm)	Total (mm)
Ainu, Hokkaido							
	16–19	M	5	5.6	7.9	5.6	19.1
	20–29	M	3	4.8	6.6	4.4	15.8
	30–39	M	6	5.8	9.9	8.6	24.3
Eskimo, Igloolik							
	20–29	M	30	4.4	6.6	7.2	18.1
	30–39	M	25	4.7	7.3	7.5	19.4
	40–49	M	9	4.1	5.7	5.9	15.7
	50–59	M	5	4.4	6.8	6.0	17.2
	20–29	F	21	10.2	9.0	8.7	28.0
	30–39	F	15	14.0	12.3	10.6	36.9
	40–49	F	6	11.7	10.0	9.0	30.7
	50–59	F	5	28.0	23.8	22.6	74.4
Nomadic Lapps (Kautokeino, 1960)	15–55	M	31	7.0	8.2	6.1	21.3
	15–21	F	11	15.3?[a]	15.5	11.4	42.2
Eskimo, Wainwright							
	25–65 (48)	M	42	9.6	12.2	–	–
	25–65 (42)	F	37	12.2	14.9	–	–

[a] Shown as 25.3 in report of Andersen *et al.*, 1962.

Table 10.10. *Body composition of the Eskimo (Shephard, Hatcher & Rode, 1973)[a]*

Age (yr)	Sex	Number tested	Total body water (l ± 4.7)	(% ± 5.3)	Body fat (kg ± 5.2)	(% ± 7.3)	Lean tissue (kg ± 1.7)	(% ± 2.0)
15–19	M	11	39.3	63.9	7.9	12.7	14.4	23.4
	F	9	31.8	59.0	10.7	19.4	11.6	21.6
20–29	M	12	43.0	62.0	10.6	15.3	15.7	22.7
	F	12	33.5	58.7	12.0	19.8	12.3	21.5
30–39	M	10	46.0	64.5	8.5	11.8	16.8	23.6
	F	10	31.1	54.2	14.3	26.0	11.0	19.8
40–59	F	10	31.8	54.6	15.3	25.5	11.6	20.0

[a] For clarity and economy of space, the averaged standard deviation is shown for each column of data.

The human biology of circumpolar populations

The Igloolik Eskimo thus carries little subcutaneous fat, a rather normal proportion of deep-body fat, and a mass of lean tissue that is large in proportion to standing height. The added lean mass explains why he is overweight relative to actuarial standards (Table 10.4). However, his large aerobic power cannot be explained simply in terms of body weight. Indeed, if data are reported in absolute units (l/min STPD), the advantage of the Igloolik Eskimo is increased to 22% relative to the Upernavik and Point Hope groups, and 34% relative to the Ainu.

Growth, development and ageing of working capacity

General aspects of growth and development are discussed elsewhere (p. 213). This section concentrates on the development of those physical variables concerned with working capacity (Rode & Shephard, 1973e, 1974). Cross-sectional information is available on a number of circumpolar groups (Fig. 10.5).

Skinfold thicknesses of young Igloolik children are generally slight and on the boys there is an almost total absence of subcutaneous fat (Fig. 10.5a). The girls commence to become fatter in their twelfth year, and the thickness of the skinfolds reaches a peak in their sixteenth year. The boys show a small increase in subcutaneous fat just prior to puberty, but with this exception remain thin throughout adolescence. Percentages of body fat among the Wainwright Eskimos show a similar trend (D. W. Rennie *et al.*, personal communication), the boys remaining relatively thin throughout adolescence and the girls showing a progressive increment of adipose tissue.

Data for leg and grip strength indicate that during the pubertal period (11–14 yr), the Eskimo girls have a somewhat higher strength than the boys; this is contrary to the experience of 'white' communities, where as a probable

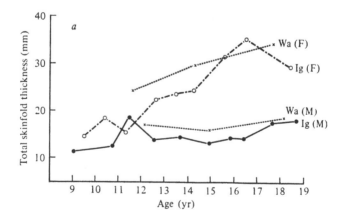

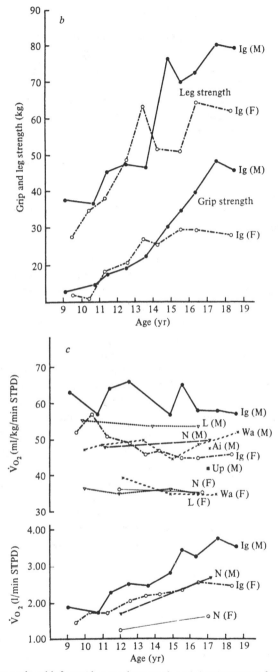

Fig. 10.5. Cross-sectional information on the growth and development of working capacity. Ig, Igloolik Eskimos; Wa, Wainwright Eskimos; Up, Upernavik Eskimos; Ai, Ainu; L, Nomadic Lapps; N, Nellim Lapps. M, males; F, females. *a*, Skinfold thicknesses; *b*, Grip and leg strength; *c*, Aerobic power.

consequence of an earlier menarche and social conditioning girls show little improvement of muscular strength beyond the age of 12 yr (Fig. 10.5*b*). The anaerobic power (vertical velocity, m/s) of the Wainwright boys increases from 1.36 ± 0.16 at the age of 11–13 yr to 1.72 ± 0.28 at 14–16 yr and 1.70 at 17–19 yr. The girls start from a lower figure (1.12 ± 0.30 at 11–12 yr) and show only a small increase to 1.26 at 13–15 yr.

All samples show a progressive increase of absolute aerobic power over the adolescent years (Fig. 10.5*c*). When data are expressed relative to body weight, the boys are seen to have a relatively constant level of cardio-respiratory fitness from 9–19 yr. In many communities, even the young Eskimo girls have poor relative scores, but in Igloolik the 9–10 yr old girls approach the performance of the boys. There is a rapid deterioration with the onset of puberty, but the Igloolik group remains superior to most other circumpolar communities.

Rode & Shephard (1974) discuss the problems of defining growth spurts in populations that are undergoing rapid social change. In addition to the well-recognized problem of averaging, cross-sectional studies underestimate the velocity of growth since older children have been exposed to a less acculturated environment. On the other hand, the usually accepted longitudinal survey reflects not only the normal growth process but the impact of acculturation upon such development. The best information is thus provided by a combination of cross-sectional and semi-longitudinal surveys. Such an approach (Rode & Shephard, 1973*e*, 1974) shows that in the Eskimo boys the peak rate of increase in height precedes spurts in weight and cardio-respiratory power, but is generally coincident with the development of grip and leg-extension strength; in the girls, however, potential growth processes are obscured by a socially-determined decline of intense activity, particularly if education is completed in a larger settlement.

Muscular strength is relatively well maintained with ageing (Table 10.8), but there is a rapid loss of aerobic power beyond the age of 40 years, seen in Igloolik, Wainwright and Nellim but less obvious in the nomadic Lapps (Fig. 10.3). One possible factor contributing to the rapid decline of maximum oxygen intake in the later years of life is the Eskimo tradition that the young hunter gives his parents first choice of such game as he has killed and older men retire early from active life.

Activity patterns and working capacity

How active are the circumpolar peoples? Ikai *et al.* (1971) described the Ainu as a relatively sedentary group, whose main activities were wood-carving and merchandising. At Upernavik, the main activity was kayak fishing, based upon a motor launch; when on board the launch, pulse rates approximated normal resting levels, but when tracking seals in the kayaks, pulse rates of

120–140/min were recorded. Assuming a metabolic rather than an emotional basis for this tachycardia, the cost of kayak paddling would amount to 5–6 kcal/min, a figure that agrees well with oxygen consumption measurements made on Igloolik seal hunters who were using arm-powered canoes. Lammert (1972) estimated that the average hunter spent 25% of the total hunting time in his kayak, work at 50% of aerobic power coming in bursts of 2–3 min duration.

The Norwegian Lapps studied by Andersen *et al.* (1962) were nomadic in life-style, supplementing their reindeer herding by ptarmigan hunting, trapping, and fishing (Andersen *et al.*, 1962; Lewin, 1971); much of their work was done on skis or on foot, with occasional use of a reindeer sleigh. The Nellim Lapps still engage in some fishing and potato growing, but many are now employed wage-earners, particularly in forest industries.

Many of the coastal Alaskan Eskimos (Wainwright, Point Hope and Point Barrow) have continued interest in hunting, although generally power boats are now used. Point Barrow has become a substantial town, and many Eskimos living there have wage-earning employment with various government agencies. Small groups who are still active hunters can be found inland,

Table 10.11. *Overall energy expenditure of Igloolik community* (*Godin & Shephard, 1973*)

Group	Number tested	Days	Yearly energy expenditure (kcal × 10¹¹)
Males			
Hunters	29	161/365 (Hunting)	17.1
		204/365 (Resting)	14.8
Laborers	25	204/365 (Working)	17.1
		161/365 (Resting)	10.1
Sedentary	86	365	78.5
Elderly	7	365	5.9
Children			
0–5 yr	51	365	11.8
5–10 yr	47	365	23.8
10–15 yr	34	365	28.5
Females			
Married	81	365	71.0
Single	37	365	31.1
Elderly	3	365	2.2
Children			
0–5 yr	62	365	14.4
5–10 yr	45	365	22.8
10–15 yr	27	365	18.1
Total			367.2

particularly at Anatulik Pass; here, Rennie has estimated daily energy expenditures at 3600 kcal (personal communication).

The majority of the Igloolik Eskimos like to describe themselves as hunters, but the majority leave the village only rarely, using borrowed equipment. Detailed ergonomic study of the community (Table 10.11) has included a review of available dog-teams, skiddoos and hunting equipment, reported hunting excursions and records of furs sold (Godin & Shephard, 1973a). Only 29 of the 147 men and boys over the age of 15 yr can still be regarded as serious hunters; this group make arduous field trips varying from three days to several weeks in length, and are absent from the village about 40% of the year. A further 20 to 30 men are employed on almost a rotational basis by government and commercial agencies. The remaining 90–100 men are largely unemployed, eking out welfare payments by soap-stone carving and an occasional hunting trip while hoping for an opportunity of government work. Energy costs incurred by the active hunters are summarized in Table 10.12. The demand for short bursts of intensive work is not fully reflected in the 24-h calorie counts. Nevertheless, many forms of hunting demand quite high daily energy expenditures, particularly when related to the small size of the Eskimo. Parallel observations on Eskimos employed within the settlement show a more steady rhythm of moderate work, with appreciably lower total expenditures of energy (average 3280 kcal per day).

The impact of activity patterns on working capacity can be seen in Table 10.13. Three groups of Igloolik males have been distinguished – the traditional Eskimo, still engaged in serious hunting, the transitional Eskimo, largely resident in the village but making occasional hunting trips, and the acculturated Eskimo with regular settlement employment and a 'white' life style. Both in summer and winter, we found a consistent gradient of aerobic power from the true hunters through the transitional Eskimos to the accul-

Table 10.12. *Twenty-four-hour energy costs of different types of hunt. Data for Igloolik Eskimos, based upon 24-h diary records and measured values for oxygen cost of individual activities (Godin & Shephard, 1973a)*

Type of hunt	24-h energy expenditure (kcal)
Fishing summer	4440
ice	4040
Caribou summer	3900
winter	3840
Walrus	3670
Seal ice-hole	3490
boat	3440
floe edge	2530
Average, 8 hunts	3670

Table 10.13. *The influence of seasons and life-style upon skinfold thicknesses and predicted aerobic power (Rode & Shephard, 1973a). Data for Igloolik Eskimos collected during summer (May–August 1970) and winter (January–March 1971)*

| Life-style | Number tested | Average skinfold thickness (mm) | | Predicted aerobic power[a] | | | |
| | | | | l/min STPD | | ml/kg. min | |
		Summer	Winter	Summer	Winter	Summer	Winter
Traditional hunter	20	5.0 ± 0.8	6.4 ± 1.2	3.72 ± 0.52	3.75 ± 0.74	56.6 ± 5.1	56.2 ± 10.1
Transitional	22	6.1 ± 2.7	6.7 ± 1.8	3.64 ± 0.76	3.63 ± 0.71	54.9 ± 10.9	54.9 ± 9.2
Acculturated	18	6.7 ± 2.8	7.9 ± 4.4	3.43 ± 0.64	3.38 ± 0.49	51.2 ± 9.6	50.1 ± 7.8

[a] Values adjusted downwards by 8% to allow for the maximum possible over-estimate of maximum oxygen intake by the prediction method.

Table 10.14. *Coefficients of correlation between an arbitrary index of acculturation (R. MacArthur) and physiological variables measured on Igloolik Eskimos*

	Males (N = 132)		Females (N = 93)	
	r	P	r	P
Age	− 0.02	n.s.	− 0.02	n.s.
Triceps skinfold	0.26	0.004	0.34	< 0.001
Subscapular skinfold	0.20	0.021	0.26	0.011
Suprailiac	0.31	< 0.001	0.36	< 0.001
Σ 3	0.29	0.001	0.34	0.001
Height	− 0.02	n.s.	− 0.02	n.s.
Weight	− 0.02	n.s.	0.12	n.s.
Grip strength	− 0.05	n.s.	− 0.01	n.s.
Leg strength	− 0.15	0.068	0.00	n.s.
$\dot{V}_{O_2}$ (max) l/min	− 0.23	0.009	− 0.06	n.s.

turated wage-earning group. Skinfold thicknesses showed an equally consistent gradient in the opposite direction. The association between hunting and cardio-respiratory fitness was even more pronounced when classification of the hunters was made on the frequency of long (2–3 week) hunting trips. Those who made such trips only 3 or 4 times per year had an aerobic power of 54.5 ± 4.8 ml/kg. min, but those who reported ten or more major trips had an average aerobic power of 62.0 ± 7.5 ml/kg. min.

Dr Ross MacArthur has developed an index of acculturation based on such items as time in school, English vocabulary, type of housing, geographic mobility and wage-income. This, also, is significantly correlated with skinfold thicknesses (in both men and women) and with aerobic power (absolute or relative) in the men (Table 10.14). Such observations do not necessarily prove that the traditional hunting life-style conserves a good working capacity and prevents obesity – it is equally possible that the more fit members of the community are better able to preserve their traditional culture. Nevertheless, if working capacity data vary with acculturation (as shown here), the average results for a given community will be in error unless an appropriate proportion of the often absent hunters is tested.

An unbiased sample of Igloolik males would include 6% very frequent hunters, 11% less frequent hunters, 64% transitional Eskimos and 17% acculturated settlement workers. The average aerobic power calculated for such a sample is 54 ml/kg. min, very close to our experimental value for the young villagers. Thus, the high aerobic power of the Igloolik Eskimos cannot be attributed to the testing of an excessive number of hunters.

Work in a cold climate

The extreme climate of the circumpolar region presents many challenges to a working man. While resting, there is difficulty in conserving body temperature, but during hard work a heavily insulated person may be unable to eliminate sufficient body heat. Reliance upon the variable protection of several layers of caribou clothing rather than an irremovable layer of subcutaneous fat can be considered a positive adaptation to the Arctic habitat. Excessive sweating rapidly degrades the insulating properties of clothing, and it is thus most interesting to observe that, during vigorous work, sweat accumulates on the face rather than on the trunk as in a 'white' person (Shephard & Rode, 1973); this seems to be due to an atypical distribution of sweat glands over the skin surface (*see* p. 296). Shivering, clumsiness, muscular viscosity and the weight of clothing should all increase the energy cost of work in the cold. In practice, this is hard to demonstrate (Godin & Shephard, 1973*a*), perhaps because a Q_{10} effect leads to a diminution of energy expenditures in inactive regions of the body.

Dr O. Schaefer has suggested that the inhalation of Arctic air during work might have long-lasting effects upon respiratory and circulatory function. Our data for healthy Igloolik Eskimos certainly show relatively low $FEV_{1.0}/$ FVC ratios, but as in some classes of athlete (Shephard *et al.*, 1973*b*), this is attributable to an increase of FVC rather than a diminution of FEV. Dr Schaefer has also commented on clinical impressions of a high incidence of pulmonary hypertension and electrocardiographic right-branch bundle block among Canadian Eskimos, and has speculated that this might be a by-product of exposure to intense cold. If there are no associated clinical abnormalities, an alternative possibility is to accept a partial resting right-branch bundle block as a normal expression of well-developed cardio-respiratory fitness (Mitrevski, 1969). We have examined exercise electrocardiograms from 13 Igloolik Eskimos where the medical team found a resting right-branch bundle block (complete in six cases, and incomplete in seven). In most subjects, the anomaly was particularly apparent in leads AVR, V4R, V1 and V2. Exercise electrocardiograms were uniformly taken in the CM5 position. Two showed definite evidence of bundle block, in three there was a suspicion of QRS broadening, and the remaining eight appeared normal. Most of the 13 had a good effort tolerance, and the two cases illustrated (Fig. 10.6) reached stepping rates of 20 and 25 ascents/min respectively when performing a progressive 18-inch step test.

In 'white' Arctic workers, prolonged cold exposure can lead to a persistent dehydration, with a poor working capacity until fluid volumes are restored (Allen, O'Hara & Shephard, 1976). We suspect this factor may also be important in the native circumpolar populations. It was necessary to test some hunters within a few hours of their return to camp, and in these circumstances blood volumes were appreciably lower for the nomads (84 ml/kg) than for Eskimos living within the Igloolik settlement (95 ml/kg). Mechanisms of potential dehydration include an intense respiratory water loss, the wearing

331

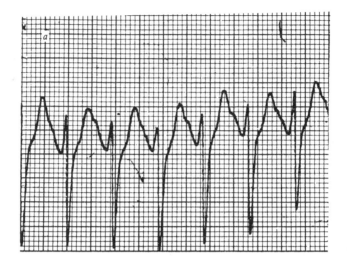

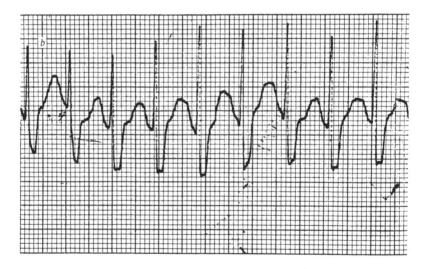

Fig. 10.6. Exercise electrocardiograms (CM5 lead) from two Eskimos showing complete right bundle branch block in Dr Schaefer's resting records. *a*, Subject Ap, 35 yr, stepping 25 ascents/min. Incomplete block in resting record, RR^1 at V4R and V1. *b*, Subject Ag, 55 yr, stepping 20 ascents/min. Complete block in resting record. R^1 at AVF, S at V6, R^1 at V4R.

of an excess of clothing while physically active, and an unwillingness to drink sufficient fluid when travelling. We reasoned that if the difference between nomadic and settlement workers was an expression of cold dehydration, there should be a correlation between cold adaptation and low values for blood volume and total body water. Dr J. LeBlanc measured systolic and diastolic blood pressures and pulse rates during hand immersion, face immersion, and combined hand and face immersion. We developed a cold-tolerance index from a summation of the peak cold-induced changes in these nine measurements. Unfortunately, only four men had undergone both cold acclimatization and body-water studies; nevertheless there were very suggestive correlations between lack of cold acclimatization, a large blood volume (ml/cm, $r = 0.86$, $P \sim 0.07$) and a large body water volume (ml/cm, $r = 0.91$, $P \sim 0.05$).

Seasonal changes in working capacity are shown in Table 10.13. There was no difference of predicted aerobic power between summer and winter seasons, but there was a small increase ($P < 0.001$) in skinfold thicknesses in the winter. While hunting is somewhat less frequent in the coldest months of the year, certain forms of game are most easily found in the winter; expeditions thus continue, and because many individual activities have a higher energy cost when the ground is snow-covered, working capacity tends to be well-maintained.

Working capacity and constitution

A rather small coefficient of variation in working capacity might be anticipated for the circumpolar data, partly because of the small 'genetic pool', and partly because the majority of the population are sufficiently active to realize their genetic potential. Nevertheless, coefficients of variation for the Igloolik and Ainu males are much as in the 'white' population of Toronto (Table 10.15), and only among the older women of Igloolik is there any suggestion of a reduction in the variance of the data (Shephard, 1974).

Table 10.15. *Coefficients of variation for measurements of maximum oxygen intake (ml/kg. min STPD)*

Age (yr)	Igloolik Eskimos M (%)	Igloolik Eskimos F (%)	Ainu M (%)	Toronto 'whites' (Shephard, 1969) M (%)	Toronto 'whites' (Shephard, 1969) F (%)
20–29	18	13	13	16	10
30–39	14	15	18	17	12
40–49	10	14	16	24	30
50–59	31	5	–	19	28

Results for standing height and maximal oxygen intake have been classified with respect to a substantial number of genetic markers (Shephard & Rode, 1973); however, there is no evidence that such classification contributes to the description of either height or maximal oxygen intake.

Effects of child-rearing and disease upon working capacity

Igloolik women aged 15–29 yr were classified according to the size of their families (Table 10.16). The relative aerobic power was similar in all four groups, but the nulliparous had a lower absolute aerobic power and a lower leg strength than those with babies; this supports our thesis that carriage of a baby in a 'yappa' contributes to the development of leg strength. Skinfold thicknesses were larger in those with no children or with only one child; this is probably a reflection of the added energy expenditure involved in raising a young family, although the women with two or more children were naturally somewhat older than those with none or one.

Disease can have a general inhibiting effect upon growth and development. We found that Igloolik Eskimos with a previous history of pulmonary disease were on average 0.7% shorter and weighed 8.5% less than villagers with a history of normal health. In the men, the weight loss was partly fat, but the younger women had greater skinfold thicknesses than those who were healthy, perhaps as a consequence of inactivity during the treatment of their disease. Both sexes showed an association between a history of tuberculosis and poor muscular development, the average deficit of strength amounting to 2.6% for handgrip, and 5.8% for leg strength. There was also a substantial deficit of aerobic power, averaging 12.1% in absolute units (l/min) and 9.0% when related to body weight.

28 of the 224 Igloolik villagers seen by the physiology team had a history of primary tuberculosis and/or hilar calcification with minimal fibrosis. Cardio-respiratory function in this group was close to 100% of the age-related values for normal healthy Eskimos. A further 17 of our subjects had a history of secondary or advanced tuberculosis, and three had emphysema and/or chronic bronchitis with extensive fibrosis. These last 20 cases, 9% of our sample, accounted for most of the functional loss associated with respiratory disease (Shephard & Rode, 1973; Shephard, 1974). Their $FEV_{1.0}$ averaged only 86.2% of the age-adjusted population average, while the FVC was 90.1% of the population average; such figures are reminiscent of results for villagers tested by the medical team but not the physiology team ($FEV_{1.0}$ 85.8% of jointly studied group, FVC 88.0% of jointly studied group). It is clear from Table 10.1 that the medical team saw almost all of the Eskimos with advanced chest disease, whereas the physiology team had the opportunity to test only about 40% of those with pulmonary disease. Such considerations of sampling largely reconcile previous impressions of a high

Table 10.16. *The influence of child-rearing upon working capacity. Data for Igloolik Eskimo women aged 15–29 yr. Data are expressed as means* ± S.D.

Number of children	Number in sample	Predicted aerobic power		Skinfolds (Σ 3, mm)	Leg strength (kg)	Handgrip (kg)	Age (yr)
		l/min STPD	ml/kg min				
0	11	2.49 ± 0.38	47.4 ± 7.3	32.5 ± 10.4	56.3 ± 20.6	28.1 ± 3.0	16.5 ± 1.5
1	5	2.99 ± 0.29	49.7 ± 4.0	32.6 ± 8.3	65.4 ± 18.3	30.4 ± 2.9	20.4 ± 3.6
2	1	2.56	47.9	20.0	60.0	24.0	25.0
3 or more	11	2.80 ± 0.50	48.0 ± 8.3	23.9 ± 7.6	67.6 ± 18.6	29.4 ± 48	25.7 ± 2.3

incidence of respiratory disease and the excellent lung-function results reported by the physiology team (Rode & Shephard, 1973c). It is debatable whether the diseased should be included in population statistics, particularly when a substantial segment of the total population is affected. However, it is vital to note their inclusion or exclusion when evaluating average results for a given community.

Disease can lead to a progressive increase in the working capacity of a community if weaker individuals are eliminated before they reach reproductive age. That such a mechanism may still be operating in the Arctic is suggested by the high incidence of diseases such as tuberculosis, and the very large infant mortality rates for many circumpolar communities. Over the period 1959–68, the average infant mortality rate for Igloolik was 232 per 1000 live births. Two-thirds of the infant deaths occurred at camp (although camp life accounts for but a small percentage of village family-years). Nomadic life was and still is more marked at Igloolik than among other circumpolar populations, and for this reason it could be postulated that the selective effect of disease has had a larger effect in augmenting working capacity in the Canadian Arctic than in the other groups discussed here.

Overall adaptations of working capacity

How well have the circumpolar peoples adapted to the physiological demands of Arctic life? One difficulty encountered by most IBP teams has been the rapid pace of acculturation. The majority of most circumpolar populations no longer find physical activity essential for survival. Even minor journeys are made by skiddoo or power boat, and labour-saving appliances are rapidly becoming available in the home. Much of the day may be spent in well-heated permanent buildings. 'Native' foods meet an ever-decreasing proportion of the total caloric requirements, air travel is increasingly breaking down genetic isolation, and improved medical and nursing care is rapidly controlling the incidence of major diseases.

It is thus hardly surprising that few of the circumpolar peoples show any marked physiological adaptations. Nevertheless, in Igloolik at least, the hunters still show a superior working capacity to the acculturated villagers of the settlement. Despite the introduction of skiddoos, power boats, and automatic weapons, hunting is still hard physical work, not least because a concentration of the population in larger settlements has increased the distance that must be covered in the search for game. While those who choose to hunt show a well-developed aerobic power, it remains unclear how far this is a product of genetic isolation and the pressures of disease and starvation in previous decades, and how far it may be considered a more immediate adaptation to the demands of continuous vigorous activity. Low skinfold readings, the internal storage of body fat, and an unusual distribution of sweat glands serve to meet the demands of arduous journeys with a need for very variable body insulation.

Work physiology and activity patterns

It will be of great interest to repeat the physiological studies in a few years time, when the circumpolar communities have made further 'progress' towards the urban life-style of the 'white' man. In particular, it will be fascinating to watch whether the 'white' stature is finally attained, and whether those of the Igloolik and Kautokeino populations who are currently nomads lose the bonus of aerobic power that currently distinguishes them from both other circumpolar peoples and the sedentary 'white' citizens of more temperate regions.

This study was supported in part by a research grant from the Canadian National Research Council. Specific acknowledgement is made to the journal *Human Biology* for permission to reproduce the material used in Figs. 10.2 and 10.4 and Tables 10.3, 10.8, and 10.15 (previously published in Shephard, 1974).

References

Allen, C., O'Hara, W. & Shephard, R. J. (1976). Changes in body composition during an arctic winter exercise. In *Circumpolar Health*, ed. R. J. Shephard & S. Itoh, pp. 113–18. Toronto: University of Toronto Press.

Andersen, K. L., Elsner, R. E., Saltin, B. & Hermansen, L. (1962). Physical fitness in terms of maximal oxygen intake of nomadic Lapps. *USAF Arctic Aeromed. Lab. Alaska Technical Report AAL TDR–61–53*. Alaska: Fort Wainwright.

Andersen, K. L. (1969). Racial and inter-racial differences in work capacity. *J. biosoc. Sci.*, Suppl. 1, 69–80.

Åstrand, I. (1960). Aerobic work capacity in men and women with special reference to age. *Acta physiol. Scand.*, 49, Suppl. 169.

Godin, G. & Shephard, R. J. (1973a). Activity patterns in the Canadian Eskimo. In *Polar Human Biology*, ed. O. Edholm & E. K. Gunderson. London: Heinemann.

Godin, G. & Shephard, R. J. (1973b). Body weight and the energy cost of activity. *AMA Arch. Env. Health*, 27, 289–93.

Ikai, M., Ishii, K., Miyamura, M., Kusano, K., Bar-Or, O., Kollias, J. & Buskirk, E. R. (1971). Aerobic capacity of Ainu and other Japanese on Hokkaido. *Med. Sci. Sports*, 3, 6–11.

Jamison, P. L. (1970). Growth of Wainwright Eskimos: Stature and weight. *Arctic Anthropol.*, 7, 86–94.

Karlsson, J. (1970). Maximum oxygen uptake in Skolt Lapps. *Arctic Anthropol.*, 7, 19–20.

Lammert, O. (1972). Maximal aerobic power and energy expenditure of Eskimo hunters in Greenland. *J. appl. Physiol.*, 33, 184–8.

Lewin, T. (1971). History of the Skolt Lapps. A literary study. *Suomen Hamm. Toimituksia*, 67, Suppl. 1, 13–23.

Lewin, T. & Hedegård, B. (1971). Anthropometry among Skolts, other Lapps, and other ethnic groups in Northern Fennoscandia. Comparative studies based on earlier investigations. *Suomen Hamm. Tomituksia*, 67, Suppl. 1, 71–98.

Lewin, T., Jürgens, H. W. & Louekari, L. (1970). Secular trend in the adult height of Skolt Lapps. *Arctic Anthropol.*, 7, 53–62.

Maritz, J. S., Morrison, J. F., Peter, J., Strydom, N. B. & Wyndham, C. H. (1961). A practical method of estimating an individual's maximum oxygen intake. *Ergonomics*, 4, 97–122.

337

The human biology of circumpolar populations

Mitrevski, P. J. (1969). Incomplete right bundle branch block (Lead V₁) in athletes. *Med. Sci. Sports*, **1**, 152–5.

Norris, A. H., Lundy, T. & Shock, N. W. (1963). Trends in selected indices of body composition in men between the ages 30 and 80 years. *Ann. NY Acad. Sci.*, **110**, 623–39.

Rennie, D. W., di Prampero, P., Fitts, R. W. & Sinclair, L. (1970). Physical fitness and respiratory function of Eskimos of Wainwright, Alaska. *Arctic Anthropol.*, **7**, 73–82.

Rode, A. & Shephard, R. J. (1973a). Fitness and Season. An arctic study. *Med. Sci. Sports*, **5**, 170–3.

Rode, A. & Shephard, R. J. (1973b). On the mode of exercise appropriate to an arctic community. *Int. Z. Angew. Physiol.*, **31**, 187–96.

Rode, A. & Shephard, R. J. (1973c). Pulmonary function of Canadian Eskimos. *Scand. J. Resp. Dis.*, **54**, 191–205.

Rode, A. & Shephard, R. J. (1973d). The cardiac output, blood volume, and total haemoglobin of the Canadian Eskimo. *J. appl. Physiol.*, **34**, 91–6.

Rode, A. & Shephard, R. J. (1973e). Growth, development and fitness of the Canadian Eskimo. *Med. Sci. Sports*, **5**, 161–9.

Rode, A. & Shephard, R. J. (1974). Growth and development in the Eskimo. *Proceedings of 2nd Canadian Workshop on Child Growth and Development, Saskatoon, November 1972.*

Rose, G. A. & Blackburn, H. (1968). *Cardiovascular survey methods.* Geneva: World Health Organization.

Shephard, R. J., Allen, C., Benade, A. J. S., Davies, C. T. M., di Prampero, P. E., Hedman, R., Merriman, J. E., Myhre, K. & Simmons, R. (1968). The maximum oxygen intake – an international reference standard of cardio-respiratory fitness. *Bull. WHO*, **38**, 757–64.

Shephard, R. J. (1969). *Endurance Fitness.* Toronto: Toronto University Press.

Shephard, R. J. (1971). IBP workshop on fitness of traditional communities. Prague: World Congress of Physiological Sciences, Satellite Symposium on Exercise Physiology.

Shephard, R. J., Rode, A. & Godin, G. (1972). International Biological Programme, Igloolik HA study. Physiology Section report.

Shephard, R. J. & Rode, A. (1973). Fitness for arctic life. The cardio-respiratory status of the Canadian Eskimo. In *Polar Human Biology*, ed. O. Edholm & E. K. Gunderson. London: Heinemann.

Shephard, R. J., Hatcher, J. & Rode, A. (1973). On the body composition of the Eskimo. *Europ. J. appl. Physiol.*, **30**, 1–13.

Shephard, R. J., Godin, G. & Campbell, R. (1973). Characteristics of sprint, medium and long-distance swimmers. *Europ. J. appl. Physiol.*, **32**, 1–19.

Shephard, R. J. (1974). Work physiology and activity patterns of circumpolar peoples. A synthesis of IBP data. *Human Biol.* **46**, 263–294.

Society of Actuaries (1959). *Build and Blood Pressure Study.* Chicago, Illinois.

Weiner, J. S. & Lourie, J. A. (1969). *Human Biology. A guide to field methods.* Oxford: Blackwell.

Young, C. M., Blondin, J., Tensuan, R. & Fryer, J. H. (1963). Body composition of 'older' women, thirty to seventy years of age. *Ann. N.Y. Acad. Sci.*, **110**, 589–607.

11. Behavior

HARRIET FORSIUS

Relatively few studies have been made on behavioral problems among the circumpolar populations within the framework of the International Biological Program Human Adaptability project. The most extensive studies have been made by MacArthur (1974a), among Canadian and Greenland Eskimos and by Carol Feldman (Feldman et al., 1974) among Alaskan Eskimos. Foulks & Katz (1971) and Foulks (1972) have studied Arctic hysteria in Alaska, and social changes in Canada have been studied by Evelyne Latowsky-Kallen (1972, 1974). In connection with the IBP–HA program, studies in child psychiatry and psychology have been carried out by Harriet Forsius (1973) and Leila Seitamo (1972, 2, 6) in North Finland. Asp (1966) and his coworkers have made sociological IBP studies among Finnish Lapps. No IBP work from the Soviet Union is known to the author of this chapter. However, numerous other projects on behavioral problems among Arctic populations have been carried out. To complete the picture of 'Man in the Arctic' and to form a broader-based literature, some of these are referred to here, mainly works which have been presented in connection with IBP workshops or at the International Symposia on Circumpolar Health (Fairbanks, Alaska 1967; Oulu, Finland 1971; Yellowknife, Canada 1974). Reviews of psychological research in the North have been published by Berry (1971, 1974). Most studies deal with cognitive abilities, social change and stress.

Mental abilities

The circumpolar populations who live under extreme external circumstances certainly need substantial mental power in order to survive. Thus anecdotal legends have long praised the mechanical skills of the Eskimos and other northern people, for example, and investigations suggest the existence of a 'northern' cognitive style. Cognitive abilities have also been the focus of research work carried out by behavioral psychologists in the north. However, if we try to assess these mental abilities, it must always be borne in mind that if we compare populations with different cultural backgrounds we are always operating with more or less culturally-bound tests. According to Biesheuvel, (1969) the attempt to construct culture-fair tests only means that these tests succeed in measuring reliably individual differences, though what they measure in different cultures may be far from identical. However, even tests which are of varying significance cross-culturally can be valuable in assessing different

339

ways in which cultural groups have adjusted to their particular environmental requirements.

This point is perhaps most extensively developed by MacArthur who tried to answer two general questions. (a). Which intellectual abilities are least and most related to differences in cultural backgrounds. (b). Which particular cultural influences are related to the development of which intellectual abilities. These questions were approached through field work with 751 subjects in four main ethnic groups: Central Canadian Eskimos (Igloolik and Frobisher Bay, field work in 1969), Nsenga Zambians (Sandwe and Petauke, 1970), Northwest Greenland Eskimos (Aupilagtoq, Kraulshavn and Upernavik, 1971) and a reference group of Alberta Whites (Stratchona County, 1972) (Table 11.1).

The cognitive measures used are described in MacArthur (1973). In examining the patterning of mental abilities in these samples, oblique higher-order methods of factor analysis were used. It was noticed that three important first-order common factors emerge in all the samples, most clearly crystallized in the adolescent Igloolik sample, but recognizable in all the other samples too. These clusters of abilities were called: (a) verbal–educational, for which the marker tests were considered to be arithmetic and oral English (Danish) vocabulary, (b) the spatial-field independence cluster for which the markers were Block Design, Witkin Embedded Figures and Form Assembly, and (c) the inductive-reasoning-from-non-verbal-stimuli cluster for which the Progressive Matrices were considered to be the marker test. These three factors were all highly correlated and showed high loadings on a single third-order factor, so that a general intellectual ability factor may be conceived as running through all three.

It has long since been stated that a test usually favors the members of the culture in which the test is constructed; it would thus be expected that these tests, which are evolved essentially from Euro-American culture, would favor the Alberta White reference group. This was the case especially with the verbal–educational tasks, where all of the corresponding age groups of the indigenous people scored lower than the White group. However, for the Eskimo groups from both the Igloolik and the Upernavik district the spatial tasks showed relatively little difference from the Whites, though the inductive reasoning tasks showed more difference. On the other hand, the Nsenga adolescents were about the same distance below the Whites on all three clusters of abilities. According to MacArthur, this pattern of cognitive abilities reflects the difference in cultural background in the samples: in line with Witkin differentiation theory, as elaborated by Berry (1971), the hunting background of the Eskimo demands minute visual discrimination and spatial awareness for navigation, together with an upbringing encouraging independence and initiative has fostered a broad spatial-field-independence cluster of abilities. An extension of Witkin–Berry theory suggests that the

Table 11.1. *Subsamples and timetable (MacArthur, 1973)*

Age group (yr)	9–12	13–16	17–26	27–36	Total	Year of field work
Number of cases						
Central Eskimos	61	62	29	25	177	1969
Nsenga Africans	65	65	31	31	192	1970
Greenland Eskimos	65	64	23	24	176	1971
Alberta Whites	70	64	36	36	206	1972

Eskimo environment and upbringing have also fostered a cluster of abilities involving inductive reasoning from non-verbal stimuli. This agrees closely with results for two samples of Mackenzie district Eskimos living more than 1000 miles away (MacArthur, 1968, 1969).

On the other hand, for the Nsenga sample, with their agricultural background and upbringing encouraging conformity and obedience, the field-independence factor is not so broadly spatial for the adolescents.

For all Nsenga age-groups the inductive reasoning tasks seem to tap mainly the same abilities as do the verbal and educational tasks. Other workers have also found that perceptual and cognitive abilities which are visually based appear to be relatively well developed in the traditionally hunting societies. Berry & Annis (1974) found generally high levels of performance on Kohs Blocks and Raven's Matrices in a study with three Amerindian samples (Cree, Carrier and Tsimshian). The performances of two Ojibway samples were similarly high, and the ability with embedded figures on the Rod and Frame apparatus exceeded non-native norms (J. W. Berry, J. Kane and T. Mawhinney, personal communication). Bowd (1974) has produced similar evidence.

Carol Feldman *et al.* (1974), made an attempt to test Piaget's general hypothesis that cognitive development follows a logically necessary invariant sequence of stages ability. They used non-verbal tests involving colored blocks as a stimulus for 67 Eskimo children from the North slopes of Alaska and for 59 children from south-western Kentucky. Their results provide evidence for a universal structure in the developmental sequence of cognitive stages.

In an Alaskan Eskimo sample Bock & Feldman (1969) showed evidence of problems of vocabulary growth, which are typical of populations in which schooling is in one language and social communications in another.

The Eskimos generally did well on classification tasks requiring highly abstract concepts, and less well on tasks requiring functional concepts of the sort that vary between different environments. The subjects performed in a similar way in parallel English and Inupiat items on the abstract questions as subjects of European origin. The authors suggest that it can be assumed that the Eskimo would score even higher on less culturally based measures than they do on those used.

As an answer to the question whether the adaptation required by the Eskimo environment is the same as that required in the Southern 48 States (what is the difference, for example, between becoming a good hunter or a good industrial worker?), the authors stress that it seems likely that the ability to form some abstract notions and to use them generatively would be adaptive in any environment.

The abilities of the Eskimo populations to deal with such tasks as block design, form assembly and embedded figures, which are widely established indicators of technological aptitude in Euro-American society, were according to MacArthur scarcely related at all to the degree of transition. Similarly there were no significant differences in the solving of these tasks between the adolescents with an average of 6 yr of schooling and the middle-aged adults with little schooling.

'Thus across ethnic-group, across age-group, and within age-group data confirm in terms of well-known standard psychological tests, and even for child samples, the relatively highly developed spatial and field-independence abilities which the Eskimo has developed in his traditional life.' (MacArthur, 1974*b*.)

In an investigation among a Lapp tribe living in the northeast of Finland, the Svettijärvi Skolt Lapps, Leila Seitamo (1972, 1976) found no correlation between the performance in block design with age and years of schooling. She found that 81 Skolt children aged 6–15 yr showed the same learning ability as 68 Finnish children when the effect of cultural factors was eliminated from the intelligence tests as far as possible (block design). The inferiority of the Skolt group appeared in non-verbal performances typical of an educational society (picture arrangement, coding) and in verbal functions. (The Skolts are bilingual, and have obtained their school education in Finnish; however since 1973 teaching has been partly in their own language.) However, it was found that visual perceptual functions as measured by the Picture Completion test are well developed in the Skolt group. Life in the wilds and a cultural pattern which reinforces exact perceptions were assumed to account for these abilities. In the Skolt group verbal functions improved as a function of length of school career. The verbal–educational factor in this group was very large. No large field-independence factor appeared.

In all the investigations mentioned above we obtain support for the theory that the cognitive abilities of the northern people are a functional adaptation to the demands of their environment, in which perceptual motor skills are highly useful.

Sexual differences in cognitive abilities

In perceptual skills the usual finding is that females perform less well than males. Therefore one striking finding in MacArthur's study of the different ethnic groups (MacArthur, 1971, 1974*a*) was the almost complete lack of significant point-biserial correlation of any of the cognitive ability measures with sex of subject. He suggests that the few sex differences that occurred were probably not biological in origin, but related to idiosyncrasies in the upbringing of a particular age-group in a particular social setting at a particular time.

Bock & Feldman (1969) also found less sex difference in spatial visualizing ability than is typically found in subjects of European origin.

The investigations among the Skolt Lapps support the theories that environmental factors provoke differences among the sexes: Seitamo (1974) found that Skolt Lapp boys tended to score lower than Skolt girls and Finnish boys on several culturally-linked tests. In the minute visual perception test, however, they reached the same standard as the Skolt girls and the Finnish boys.

The over-all poor intellectual readiness for school work of the Skolt boys appears most clearly in non-verbal functions which require visuomotor co-ordination and their readiness for visuomotor pencil–paper work (coding) is especially poor. However, the differences are still clearer in verbal readiness as compared with the Finnish boys. Within the Finnish group the differences between the two sexes tended to be in the opposite direction. The collision of two different cultures with abrupt changes in role playing especially for the males in the Skolt culture and the submissive role of the females in the northern Finnish culture were observed to account for these differences.

Psychomotor performance and sex differences

The same inferiority of performance in the Skolt boys as mentioned above was found by Harriet Forsius (1973) testing the same sample of Skolt Lapp children and Finnish children using the Oseretsky motor performance test. The Skolt boys were poorer in their performance than Skolt girls and Finnish boys and girls. In the Finnish group there was no statistical difference between the sexes. The differences in non-verbal intellectual functions were interpreted to depend both on a slower neurophysiological maturity in boys provoked by environmental circumstances and on cultural factors. The girls receive more support and training in handicrafts, in which finer motor skills are used.

Mental health

The assessment of mental health in a population is a difficult task, since our criteria for mental health are indeterminate and vary within different cultures. Usually we judge the mental health of a population from the incidence of mental disorders, but even here we meet with difficulties concerning terminology, while at the same time little epidemiological information exists on the true incidence and prevalence of mental disorders among the circumpolar peoples.

In different areas we see different patterns of problems depending on the environmental circumstances and the social structure of the populations. Where the traditional means of bonding, enculturation and social control by the family and kin group have remained relatively intact, one finds communities which are cohesive, well integrated, and viable. However, acculturation and the conflicts implied in cultural collisions seem to yield a high incidence of mental and social problems.

344

Behavior

Epidemiology of mental disorders

Foulks & Katz (1971) collected records at the Arctic Health Research Center, Fairbanks, Alaska, of all clinical cases treated in Alaskan Field Hospitals during 1968 and in this way got the prevalence of treated mental disorders. They found that the incidence of treated psychiatric disorder among the native population (783/100 000) was not significantly higher than the figures for the over-all US population. However the pattern of mental disorders for native Alaskans varies from the patterns presented by the over-all US population.

Depressive neurosis, anxiety and alcoholism were more frequently treated in Alaskan natives. Also a difference was noticed between larger villages (900–2500 inhabitants) where control of the inhabitants was achieved through law and governmental rules, inducing low self-esteem, depression and alcoholism. In contrast it is suggested that life in the smaller, more traditional, Eskimo village generates a different type of psychology which results in hysterical conversion as a way for the individual to express his conflicts. It was also stated that the groups who had comparatively longer and more intensive contact with western societies suffered from the highest incidence of psycho-physiological disorders (Table 11.2).

According to Koutsky (1971), among the Alaskan natives, 60% of both Caucasians and natives who were admitted to the Alaska Psychiatric Institute had a positive history of alcohol abuse, but the history of drug abuse was higher among the Caucasians. Different cultures provide different stresses with their characteristic response, no new diagnoses were found and the most usual diagnoses were depression, schizophrenia and alcoholism.

Sampath (1976) graded the population of a settlement in the eastern Canadian Arctic with an Eskimo population of around 500 on a scale from 'sickness' to 'wellness' (based on psychiatric symptomatology) and found that 10% of the respondents were suffering from a severe form of mental disorder, 27% from a moderate, 58% from a mild, and 5% had only minimal symptoms. The females had symptoms of mental disorder to a greater degree than the males. The prevalence rate for schizophrenia was found to be 28/1000, for the affective disorders it was 46/1000, for neuroses 116/1000 and for personality disorders 177/1000. He postulated that the adjustment of the Eskimo population to the social stratification of the 'modern' community is producing various stresses which are then being translated into psychiatric symptoms.

Hellon (1972) discusses the effects of climatic, geographical and cultural factors on mental health and also points out the possibility of disturbed metabolic rhythms depending on these factors as a contributory cause of mental illness. A study in social psychiatry carried out on a random sample of 500 subjects in north Finland, who were compared to 500 subjects from the south of the country, was made by Väisänen (1976). He used psychiatric

345

Table 11.2. Incidence of treated mental disorder according to village size per 100 000 in population specified (Foulkes & Katz, 1968)

	N	Villages < 500	Villages 500–900	Villages > 900	Predominantly Caucasian cities	Total incidence	χ^2	P
		27 876	3176	10 488	13 248			
Depression	53	57	–	335	14	94	70.09	< 0.005
Conversion hysteria	8	26	–	7	0	15	–	
Anxiety neurosis	56	109	–	30	172	103	11.310	< 0.005
Neurosis–other	9	10	–	37	29	15	–	
Psycho-physiological disorders	21	57	–	7	–	38	4.771	$0.1 > P > 0.05$
Alcoholism	165	83	–	283	889	299	182.445	< 0.005
Paranoid personality	23	5	–	0	172	34	–	
Paranoid schizophrenia	31	31	–	15	0	24		
Acute undifferentiated schizophrenia	5	10	–	0	7	11		
Chronic undifferentiated schizophrenia	12	31	–	15	7	24		
Schizophrenia–other	7	10	–	7	14	18		
Schizophrenia–total	34	52	–	52	29	78	–	
Epilepsy	45	62	–	37	172	86	15.826	< 0.05
Total incidence	414	517	–	744	1434	783	107.124	< 0.005

346

interviews, the Cornell medical Index Questionnaire and the psychological tests of Wartegg and Zullinger. A clear difference existed between the two areas in the frequency of clinical neuroses, 20.9% in the north and 14.6% in the south. Psychosomatic disorders and functional disturbances were encountered in 71% and 44% of cases respectively. In the north 1.6% were psychotic and 1.0% borderline cases, in the south 1.1% and 0.7%. A greater number of reactive depressions were located in early autumn (September) than during the other seasons.

Suicidal behaviour

Kraus & Buffler (1976) have gathered data relevant to suicidal behavior in the Alaskan native cultures. They found that the suicide rate increased from 13/100 000 per year for the interval 1961–5 to 25/100 000 per year for the interval 1966–70 with a rate of 33/100 000 for the single year 1970. This doubling was accounted for by an increase in suicide among the young, with the age group 20–5 yr seemingly at the highest risk. Suicide attempts and suicidal gestures, rare in the traditional society, seem to appear among natives living in a transitional native town or a western urban environment, and appear to depend on factors such as disruption of the integrity of the family by alcohol, school and extended hospitalization for tuberculosis. Since 1966 deaths by intoxication have appeared for the first time, and the number of women committing suicide has increased, so that during 1970–3, five female suicides occurred for every nine male suicides (Table 11.3).

The prevalence rate for treatment of attempted suicides was calculated for a rural Alaskan native town, and was found to be 1450/100 000/yr. This rate

Table 11.3. *Average annual age-specific suicide death rates 100 000 for total Alaska native population, 1965–9 and 1970–3 (Kraus & Buffler, 1976)*

Age (yr)	1965–9	1970–3
0–14	0.9	3.2
15–19	39.5	50.0
20–24	47.0	170.6
25–34	51.5	103.7
35–44	16.6	64.4
45–54	27.8	32.2
55–64	30.6	8.9
over 65	42.7	–
Total	21.1	40.0

is higher than any reported for other American native cultures and is approximately ten times the rate of 130/100 000/yr calculated for the city of Los Angeles (Mintz, cited by Kraus & Buffler, 1976).

Arctic hysterias

Common to all circumpolar peoples are those mental disorders generally termed Arctic hysterias, and the seemingly high incidence and dramatic nature of such disorders have given rise to many speculations and much theorizing as to what causal factors common to all Arctic peoples might be involved.

A synthetic approach to this complex question is made by Foulks (1972), who tries to evaluate the problem by studying man in his total ecological context regarding environmental, biological, physiological, physical, psychological and social factors. He has studied the Eskimos in north Alaska and reports ten cases which have been investigated thoroughly. Although Arctic hysteria can be seen as an interaction between the Eskimo personality and the cultural–anthropological situation, both of which are variable, mutually interdependent and subject to alteration by a number of other systems such as environment and demography, different specific factors are involved in individual cases. So, for instance, there are several environmental factors which potentially affect adversely the functioning of the Eskimos' central nervous system. The absence of a regular 24-h alternation of light and dark affects the circadian cycling of the calcium metabolism. Calcium is an essential element in the chemical transmission of the neural impulses, and it has been demonstrated that abnormalities in the physiological functioning of calcium are capable of producing a variety of mental disorders including hysteria-like behavior.

Also dietary habits and circumstances connected with solar ultraviolet radiation influence the intake and absorption of calcium, so that this cumulative chain of events results in serum levels of calcium which fall at the low end of the physiological spectrum. These factors potentiate dysfunction in the Eskimos' central nervous system.

The modern style of living among the Eskimos, in houses heated by coal stoves, causes hot, dry air which in turn dries the mucous membrane of the respiratory tract and furthers infections which often lead to middle-ear disease and meningitis. These diseases in turn can affect the nervous system and lead to different degrees of cerebral damage which are often the cause for hyperactivity and inattention, inability to control the emotions, hypersensitivity to drugs and alcohol and various types of epileptic seizures.

In the traditional Eskimo way of life mutual monitoring of each and every individual in the closely-knit society was both necessary and feasible. Deviant behavior was ridiculed or gossiped about, and individuals were very sensitive to the criticism of others. The child soon learns that his behavior is of intense concern and interest to others. The growing Eskimo must learn to control his aggressive and competitive feelings even under the most trying circumstances lest he be considered an inadequate or foolish person. There are few ways the Eskimo can express such feelings when they are generated, for his life is never private. In such small communities hysterical dissociation through a shamans seance, 'speaking in tongues' or in Arctic hysteria offer the only outlets. As mentioned earlier Foulks found that hysterical neurosis was a phenomenon of the small Alaskan Eskimo villages, while inhabitants in the larger villages demonstrated different patterns of mental disorders manifest in depressions and alcoholism. The behavior manifest by the shaman, by the Eskimo with epilepsy and by the Eskimo with Arctic hysteria is often quite similar. It is impossible to determine which behavior comes first or predominates as a model for the others. Foulks is, however, of the opinion that each obviously contains many causal elements in common with the others, but not all elements interact with the same intensity in individual cases. There were many differences among the ten subjects in terms of which etiological agent acted as the major determinant of the attack of Arctic hysteria. Several subjects had epilepsy, several were diagnosed as schizophrenic, most had low normal serum-calcium levels and one had hypomagnesemia and possible alcoholism. These diagnoses themselves, however, cannot account for the total phenomenon of Arctic hysteria. At best they may play a precipitating role, with the over-all form of the attack being shaped within the context of the other multiple forces mentioned above.

The PhD dissertation of Foulks includes valuable references both to Arctic hysteria and to related mental disorders, and especially to literature concerning the anthropological and historical background.

Problems among children and adolescents

In common with many other workers, Elinor Harvey (1976) has noticed that the type of mental disorder varies according to the size and population composition of the village from which the individual originates. She states that approximately half the student population of a secondary school for Eskimo, Indian and Aleut students in Alaska was referred to the mental health team. Those from the smaller villages with a more homogeneous population were more apt to manifest anxiety and depression, while those from the larger, mixed migrant population villages were more apt to manifest disruptive behavior during intoxication due to alcohol. The latter appeared to have identified with the aggressive behavior of their 'well-settled' large-village neighbors who reveal feelings of disruption and the pain of their resettlement by drinking. It is suggested that the already loosened personal identity ties of adolescents are further exaggerated by the loss of cultural

identity ties caused by the family moving to a larger community. The importance of the work of the mental health team is shown by the fact that among
about 300–400 pupils aged 13–23 yr there were 52 and 40 drop-outs and 35
and 20 suicidal gestures per year respectively in the years 1968–9 and 1969–70,
before the mental health team started its work, while in the years 1971–2 and
1972–3 there were only 12 and 20 drop-outs and four and two suicidal
gestures respectively. The explanation for this positive trend is apparently that
once the attitudes of the staff, particularly the native staff, were modified to
feelings of strength and hopefulness by efforts of the mental health team,
these were likely to become the attitudes of the students, who use the staff as
models for identification.

An attempt made by Forsius (1973) to assess the psychic health of the
Skolt Lapp children in the northern part of Finland showed that the Skolt
Lapp children's mental health was just as stable as that of a control group of
Finnish children living under similar external conditions. Both groups showed
an average of two nervous symptoms per child and the number of 'disturbed
children' was approximately the same in both groups. However the lowest
proportion of disturbed children was found among the Skolt girls (16%) and
the highest among the Skolt boys (34%). The percentages for the Finnish
children were 23% for the girls and 31% for the boys. There was a lower
frequency of recurrent psychogenic abdominal pain among the Skolt children
and a lower frequency of aggressive outbursts, but a higher frequency of
irrational fears. This fact is apparently partly dependent on cultural factors,
as is the fact that there were almost twice as many 'disturbed' Skolt boys as
Skolt girls, for less is expected from the girls, and the attitudes of the parents
differ, as shown by Seitamo (1972), who used Schaefer's scale, 'Children's
Reports of Parental Behavior', on the same sample.

Cultural and social change

Socio-cultural changes are often thought to be induced by the influence of
Western culture. However, it should be realized that changes in a community
can also appear for other reasons. Usually it is assumed that there exists an
association between socio-cultural change and mental health, so that persons
and groups undergoing social and cultural change will experience a certain
amount of psychological discomfort.

Both Vallee (1968) and Chance (1965), as well as Margarete Lantis (1968),
stress that there are numerous environmental, social and cultural factors
involved in influencing personal and social life in the north.

Berry (1976) has tried to assess the effect of cultural change in northern
Canada, and he uses the term 'acculturative stress' to apply to those stresses
that are theoretically or empirically linked to acculturation. It is limited to
those affective states or behaviors which are usually subsumed under the

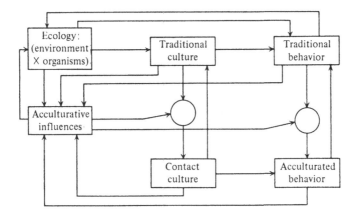

Fig. 11.1. The three-stage ecological-cultural behavioral model of Berry (1976).

term 'mental health'. He gives a basic three-stage model, in which behavior is considered as a function of culture which is in adaptation to ecology, and takes account of acculturative influences (Fig. 11.1).

Berry's studies within three Indian groups, Cree, Tsimshian and Carrier, showed that the greatest acculturative stress was suffered by the Crees, whose traditional migrant life-style differed most from the cultural style of the larger Euro-Canadian society. They have the least interest in assimilating and identify themselves most as 'Indian' people. The opposite was the case for the Tsimshian whose traditional sedentary and stratified culture is least divergent from that of the larger society. The Carrier, who were culturally intermediate, were also psychologically intermediate on these variables.

At the individual level, also, it was stated that acculturative stress was greater for people who were less psychologically differentiated: that is, individuals who were less independent of events in their milieu and were more susceptible to changes due to acculturative influences (Fig. 11.2).

Vallee (1968) points out that socio-cultural change or discontinuity in ways of living does not typically result in breakdown. He refers also to Chance (1968) in stressing that there is an association between emotional instability and the incongruence between cultural goals and the chances of reaching them. It was shown that those Eskimos who identified with the dominant culture were likely to be emotionally stable if they had the opportunity to interact with people from the dominant society. The most unstable, measured on a standardized index, were those who identified with the dominant culture, but did not have any opportunity to interact occupationally and socially with the dominant society. Vallee found an exceptionally high rate of emotional breakdown among Eskimo women of traditionalistic inclination who marry men in close contact with white persons and institutions. These women were

351

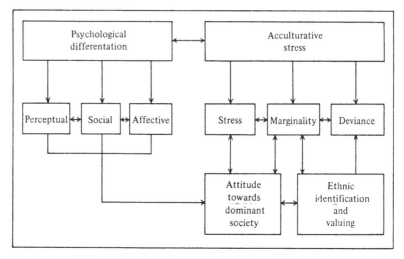

Fig. 11.2. Relationship between acculturation stress and psychological differentiation (Berry, 1976).

enmeshed in two quite different subsystems of subcultures, with the Eskimo one tending to prevail for them, whereas their husbands found the majority culture subsystem most relevant and rewarding.

A corresponding case of incongruence was found by Harriet Forsius (1973) among the Skolt Lapp population in North Finland. Genuine Skolt women who were married to Finnish men, and were apparently under the pressure of the Finnish cultural majority and of two cultures in collision, struggled to fit their children into their surroundings and used significantly more corporal punishment on their children than did the mothers of genuine Skolt families or genuine Finnish families.

Another result of the cultural collision in the same population was shown by Leila Seitamo (1976) who found a lower level in cognitive abilities in the Skolt boys, and a weaker motivation for schoolwork, which resulted in poorer school achievement followed by behavioral disturbances. This is apparently partly depending on the fact that the role of the men has changed more in this transitional Skolt society, and that the men thus have difficulty in fulfilling their role as male objects of identification for the boys.

Patterns in sexual behavior among Eskimo youth on the west coast of Greenland have been investigated by Olsen (1972) who has shown that during the past 250 yr many of the social norms have changed. They must be seen in the light of the cultural and social change in the community and the still more dominating European cultural pattern. In the conflict between the two social norms it seems to have been the European type that has dominated. The sexual behavior pattern in Greenland seems to a great extent to be analogous with that in Scandinavia today. The deviations found seem to be determined by social aspects.

Many authors have speculated on whether there is a special personality connected with the individuals who carry out difficult tasks in lonely situations on long hunting and fishing trips. The Skolt Lapps have been studied from a socio-anthropological view by Pelto (1962) who tries to correlate a personality characteristic which he names 'social uninvolvement' with the Skolt Lapp individualistic social system. He shows, with the help of a modified TAT (picture test) instrument, that the reindeer herders are socially relatively uninvolved compared to other Skolt members of their community. This cognitive characteristic may have an adaptive advantage for herdsmen, hunters and fishermen who must be able to carry out important detailed tasks in solitude. His investigations suggest that the acculturation among the Skolt Lapps, in personality terms, is from a relative social uninvolvement (associated with the traditional culture of reindeer herding and hunting), towards a greater social involvement. The changes appear mainly among the younger men, but in the Skolt Lapp homes the socialization practices may still be oriented to the individualistic, uninvolved configuration of personality. Thus in the Skolt community the direction of acculturation is opposite to the situation in many other pre-industrial peoples.

Anyhow, the Skolt Lapp community is becoming less and less integrated internally as a result of rapid technological and social developments in Finnish Lapland.

One contributing factor is the ever more widespread use of the snowmobile. The effect this has on the life of the Arctic people is evaluated by Pelto in another work (1971).

The snowmobiles have in many indirect ways increased social interdependence but their use demands the management of technical problems. The Lapp 'pre-adaptation' to snowmobile technology appears to be enhanced by the egalitarian and individualistic structure of socio-economic life in the region. The economic independence and physical separation of individual households has encouraged people to seek their own solutions to technological problems through trial and error. In real life and in folk tales, Lapp protagonists are often faced with situations that require the invention of temporary substitutes or other technical solutions to meet material needs. In their idle hours, the Lapps exchange stories about their various solutions to technical problems of reindeer herding, finding their way in unfamiliar territory and maintaining equipment. Nowadays, these exchanges of 'folklore' often revolve around technical features of their snowmobiles.

The technical adaptiveness of the Skolt appears to depend to a great extent on certain attitudinal and psychological features including the patience and attention to physical detail that makes it possible for the Lappish home mechanic to put his machine back together after he has taken it apart.

The human biology of circumpolar populations

Pelto has tried to trace the internal processes through which economic and social differentiation occurs. What are the characteristics of the individuals who succeed as compared with those who are less successful?

At least in the past, the individual Skolts in their struggle for existence have concentrated on the competition between themselves and their non-human environment instead of an aggressive competition with other persons. In the past, a man's adaptive success depended to a very great degree on his effective management of directly perceived information from and interaction with the reindeer, the physical terrain, and his personal equipment (sleds, harness, fishing gear etc.).

Now the adaptive focus is shifting rapidly to much more socially involved sectors of behavior, especially to relationships with the employers of paid labor as well as commercial operators and tourists. But some of the socially uninvolved personal characteristics of these people have continued to be of adaptive significance. The more successful Skolt adaptors appear to exhibit a flexible mixture of individualistic and more socially involved personal style.

Pelto uses the term 'techno-economic' differentiation to show that adaptation is effected by means of material things – technological inventories. He presents a 'Material Style of Life' index to measure the social differentiation, which according to him is not yet very pronounced among the Skolt Lapps.

MacArthur (1974a) has tried to assess status variables which might be considered to indicate degree of transition towards the technological society among the ethnic groups he had investigated, as mentioned earlier. Partly using factor analyses of sixteen variables, he chooses the following: (1) Transitional family ratings by priest for Igloolik and village/town for the other ethnic groups. (2) Wages versus land. (3) Modern appliances in home. (4) Time in school (self). (5) Use of English (or Danish) away from school or work. (6) Occupation of head of family (self for adult).

Asp (1966) and his co-workers have made sociological studies among the Finnish Lapps within the framework of the IBP program, especially as a continuation of Asp's work of the Finnish Lapps which was started in the early sixties.

Psycho-physiological studies

A comparative study of psycho-physiological reactions in three ethnic groups has been conducted by R. Eide, Professor at Bergen. His hypothesis was that a basic psycho-physiological reaction such as the startle pattern could have evolved differently in different groups according to the demands of the environment, and that people living in a rough environment demanding quick and vigorous actions would have a stronger startle than people habitually living under less dramatic conditions. The startle reaction may be considered a basic 'defense reaction'. It is found in animals down to the reptiles, and in children from the age of about three months.

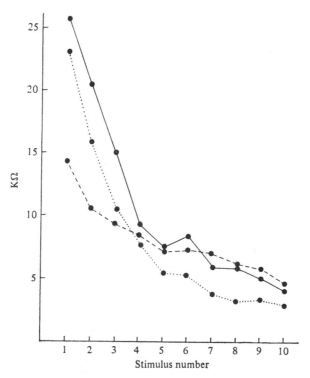

Fig. 11.3. Galvanic skin reactions to a series of auditory stimuli in different ethnic groups.
●——●, Lapps; ●· · ·●, Eskimos; ●- - -●, Finns.

He investigated three groups: A population of Eskimos in Aupilagtoq, Greenland, a group of Skolt Lapps and Inari Lapps in north Finland and Finns of caucasoid stock who had immigrated in the same region as the Lapps. The samples tested were considered roughly representative of the different populations, containing both sexes with age distributions from seven to 70 yr. The test consisted of repeated strong auditory stimulation and galvanic skin reactions and heart-rate responses were measured.

The main findings are shown in Fig. 11.3. It will be seen that Lapps and Eskimos show substantially (and significantly) higher initial galvanic skin reactions than Finns, but that there is little difference in reaction levels after ten stimulations.

These results substantiate the hypothesis that Eskimos and Lapps would need a stronger physical 'defense reaction' pattern than Finns. Eide stresses that this is only a tentative explanation. What has been demonstrated is only that Eskimos and Lapps show stronger galvanic skin reactions (a measure of peripheral sympathetic activity) to auditory stimulation. It cannot be excluded that cultural factors may have contributed to these results, although such factors are less likely to influence physiological reaction patterns than purely psychological traits. When parents' reactions were correlated with those of their children, several significant positive correlations were found: strength of the startle reaction tends to 'run in families'. Both environmental and genetic factors may have contributed to this finding.

Heart-rate reactions were only measured in Finns and Lapps. Lapps had higher reactions but not significantly so. There was a significant tendency for heart-rate reactivity to decrease as a function of age. This might indicate a somewhat reduced sympathetic drive on the heart in the older age groups. The author's conclusions are that racial differences in psycho-physiological reaction patterns possibly exist and that this should be investigated in more ethnic groups with a greater number of stimulus and response variables.

References

Asp, E. (1966). The Finnicization of the Lapps, a case of acculturation. *Annales Universitatis Turkuensis, Turku.* SER. B–TOM. 100.

Berg, G. (ed.). (1973). Circumpolar Problems. Habitat, Economy and Social Relations in the Arctic. In *A Symposium for Anthropological Research in the North, September 1969.* Wenner-Gren Center Intern. Symp. Series, Vol. 21. Oxford: Pergamon Press.

Berry, J. W. (1971). Psychological research in the north. *Anthropologica*, 13, 143–57.

Berry, J. W. (1974). Psychological research in the north: some recent findings. *Canadian Psychological Association Annual Meeting, Windsor, June 1974.*

Berry, J. W. (1976). Acculturative Stress in North Canada–Ecological, Cultural and Psychological Factors. In *Circumpolar Health*, ed. R. J. Shephard & S. Itoh, pp. 490–7. Toronto: University of Toronto Press.

Berry, J. W. & Annis, R. C. (1974). Ecology, Culture and Psychological Differentiation. *Int. J. Psychol.*, No. 3.

Biesheuvel, S. (1969). *Methods for the measurements of psychological performance.* IBP Handbook No. 10.

Boag, T. J. (1966). *Mental Health in the Arctic.* Excerpta Medica International Congress Series, No. 150. Proceedings of the IV World Congress of Psychiatry, Madrid.

Bock, R. D. & Feldman, Carol (1969). Cognitive Studies Among the Residents of Wainwright, Alaska. Paper presented at the convention of the American Association for the Advancement of Science, Boston.

Bowd, A. (1974). Practical abilities of Indians and Eskimos. *Canad. Psychol.*, 15, 281–90.

Briggs, Jean (1970). *Never in Anger.* Cambridge, Mass.: Harvard University Press.

Chance, N. A. (1965). Acculturation, Self-Identification, and Personality Adjustment. *Amer. Anthrop.*, 67, 372–93.

Chance, N. A. (1968). Implications of Environmental Stress. Strategies of Developmental Change in the North. *Arch. environ. Health*, 17, 571–7.

Eide, R. (1971). Group Differences in Physiological Reactions and Habituation to Auditory Stimulation. At the Second International Symposium on Circumpolar Health, 21–7 June, Oulu, Finland.

Feldman, C. F., Lee, B., McLean, J. D., Murray, J. & Pillemer, D. (1974). *The Development of Adaptive Intelligence*. San Francisco: Jossey-Bass.

Forsius, Harriet (1973). The Finnish Skolt Lapp Children. A child psychiatric study. *Acta Paediat. Scand.*, Suppl. 239.

Foulks, E. F. (1972). *The Arctic Hysterias of the North Alaskan Eskimo*. Washington: American Anthropological Association.

Foulks, E. F. & Katz, S. (1968). *The mental health of Alaskan natives.*

Foulks, E. & Katz, S. (1971). Mental Health in Arctic Alaska. Paper at Second International Symposium on Circumpolar Health, 21–4 June, Oulu, Finland.

Harvey, Elinor B. (1976). Five years of psychiatric consultation at a boarding school in Alaska for Eskimo, Aleut and Indian children. In *Circumpolar health*, ed. R. J. Shephard & S. Itoh, pp. 517–25. Toronto: University of Toronto Press.

Hellon, C. P. (1972). Mental Health Problems in the Arctic. *Inter-Nord*, **12**, 333–7.

Kallen-Latowsky, E. (1972). Eskimo youth: the new marginals. At the Scandinavian Meetings, IBP–HA, 6–8 June.

Kallen-Latowsky, E. (1974). *Social change, stress and marginality among Inuit Youth*. York University.

Kraus, R. & Buffler, P. (1976). Suicide in Alaskan Natives: A Preliminary Report. In *Circumpolar health*, ed. R. J. Shephard & S. Itoh, pp. 556–7. Toronto: University of Toronto Press.

Koutsky, C. D. (1971). Alaskan Psychiatry. At the Second International Symposium on Circumpolar Health, 21–4 June, Oulu, Finland.

Lantis, M. (1968). Environmental Stresses on Human Behaviour. *Arch. Environ. health*, **17**, 578–85.

MacArthur, R. S. (1968). Some differential abilities of Canadian native youth. *Int. J. Psychol.*, **3**, 43–51.

MacArthur, R. S. (1969). Some cognitive abilities of Eskimo, White and Indian-Metic pupils aged 9–12 years. *Canad. J. behav. Sci.*, **1**, 50–9.

MacArthur, R. S. (1971). Mental abilities and psycho-social environments: Igloolik Eskimos. IBP–HA Project, Igloolik, Northwest Territories. Annual Report No. 3. Toronto: Department of Anthropology.

MacArthur, R. S. (1973). Some ability patterns: central Eskimos and Nsenga Africans. *Int. J. Psychol.*, **8**, 239–47,

MacArthur, R. S. (1974*a*). Cognitive Abilities of Eskimos of Igloolik, Canada and Upernavik District, Greenland. IBP–HA Project, Igloolik, Northwest Territories. Annual Report No. 6. Toronto: Department of Anthropology.

MacArthur, R. S. (1974*b*). Construct validity of three New Guinea performance scale sub-tests: central Eskimos and Nsenga Africans. In *Readings in cross-cultural psychology*, ed. J. L. M. Dawson & W. J. Lonner. Hong Kong: University of Hong Kong Press.

Olsen, G. A. (1972). Pattern in sexual behaviour in Greenland. *Inter-Nord*, **12**, 312–15.

Pelto, P. J. (1962). *Individualism in Skolt Lapp society*. Helsinki: Finnish Antiquities Society.

Pelto, P. J. (1971). Social Uninvolvement and Psychological Adaptation in the Arctic. At the Second International Symposium on Circumpolar Health, 21–4 June, Oulu, Finland.

Pelto, P. J. (1973). *The Snowmobile Revolution: Technology and Social Change in the Arctic*. Menlo Park, California: Cummings Publishing Company, Inc.

Sampath, H. M. (1976). Modernity Social Structure, and Mental Health in the

Canadian East Arctic. In *Circumpolar Health*, ed. R. J. Shephard & S. Itoh, pp. 479–89. Toronto: University of Toronto Press.

Seitamo, Leila (1971). Das elterliche Verhalten in der Wahrnehmung der Kinder. In *Bericht uber das IV Internat Symp: Der Mensch in der Arktisch*, ed. W. Lehmann. *Anthropol. Anz.*, **33**, 114.

Seitamo, Leila (1972). Intellectual functions in Skolt and Northern Finnish Children with special reference to cultural factors. *Inter-Nord*, **12**, 338–43.

Seitamo, Leila (1976). Psychological adaptation of Skolt Lapp children in culture change. In *Circumpolar health*, ed. R. J. Shephard & S. Itoh, pp. 497–507. Toronto: University of Toronto Press.

Vallee, F. G. (1968). Stresses of Change and Mental Health Among the Canadian Eskimos. *Arch. Environ. Health*, **17**, 578–85.

Väisänen, E. (1976). Mental Health in North Finland. In *Circumpolar health*, ed. R. J. Shephard & S. Itoh, pp. 513–17. Toronto: University of Toronto Press.

Wallace, A. F. C. & Ackerman, R. E. (1960). *An Interdisciplinary Approach to Mental Disorder Among the Polar Eskimo of Northwest Greenland*. Anthropologica N.S.

12. Summary

F. A. MILAN

The outlined objectives of the research reported in this volume were to elucidate the biological and behavioral processes responsible for the successful adaptation and perpetuation of circumpolar human populations to the environment and its resources and, one should add, to each other.

C. Ladd Prosser (1964), in his introduction to the volume *Adaptation to the Environment*, wrote that the concept of biological adaptation involves many generalizations and speculations and that this concept has intrigued scientists since the time of Herodotus and Pliny. John Bligh (1973) stated that 'a living organism cannot be considered in isolation from its environment any more than the flame of a candle can be considered in the absence of the atmosphere', and additionally, 'the structures and functions of each species have been gradually shaped by the natural selection of those chance mutations which afforded an individual organism some advantage in its struggle to survive and procreate in its particular environment'.

In 1963, Milan asserted that 'Remnants of primitive populations with relatively simple technologies living in harsh environments characterized by heat, cold, extreme aridity or high altitude, can provide laboratory subjects for the systematic study of both morphological and physiological adaptations to the environment and its resources. These peoples can provide information about the limits of variability in vital function, which the evidence shows are considerably different from the limits in sedentary urban man of our era.'

Yet, Professor W. S. Weiner (1966), the IBP–HA Convenor, in discussing some of the major problems in human population biology and the reasons for undertaking a world-wide comparative study of human adaptability, wrote prophetically that 'vast changes are affecting the distribution, population density and ways of life of human communities all over the world. The enormous advances in technology make it certain that many communities which have been changing slowly, or not at all, will relatively soon be totally transferred. We are therefore in a period when the biology of the human race is undergoing continuous change measured in terms of health, fitness and genetic constitution.'

In reviewing the previous chapters in this volume, the theme of 'continuity and change' seems to be an all-pervading influence. The circumpolar populations are no longer isolated and the contacts with 'southerners' have already had profound social and genetic effects on northern peoples.

The human biology of circumpolar populations

This volume, in addition to discussing evidence for adaptation in circumpolar populations, also presents a detailed description of many aspects of their human biology. Perhaps, for our purposes, the quotation below applies: 'For the general biologist, the best and perhaps the only measure of adaptive fitness to a particular environment is the extent to which the organisms of the species under consideration can occupy the environment, make effective use of its resources, and therefore multiply abundantly in it.' (Dubos, 1965.)

Demography

Eskimos, earlier isolated and found in relatively small numbers in small reproductively isolated demes across the Arctic from Siberia to Greenland, nearly became extinct due to the effects of communicable diseases, for which they had no inborn resistance, introduced by Europeans. Skolt Lapps were similarly affected. Always having demonstrated good reproductive potentialities, the Eskimo populations increased markedly with the introduction of public health and a reduction of mortality, especially that of tuberculosis, an assured adequate caloric intake and a lag in the introduction of family planning programs. This brought down the median age and increased the dependency ratio. Nevertheless, the 'demographic transition', which required 200 yr in Europe, was accomplished in only 25 yr among the Eskimo populations.

The long-term genetic isolation of Eskimo populations was broken in the 1880s so that at Point Hope, Alaska, some 35% of the population now have non-Eskimo genes. This figure is 27% at Wainwright, Alaska, but only 6% at Igloolik in the Canadian Arctic due to their relative geographical isolation. The concept of 'pure' with regard to race is now fallacious.

Due to small population size at the village or deme level, and the concomitant limited mate choice, consanguinity is still relatively common, especially for cousin marriages, and the calculated inbreeding effect was similar at Wainwright, Igloolik and among Finnish Lapps. Now out-migration of young women from their home villages and marriage with men from the dominant population is increasing in Alaska and Greenland.

In Finland, the Lapps are considered as a minority population and at Inari, the Skolt Lapp gene pool already contains 28% of genes of Finnish origin.

Genetics

In chapter 3, the Soviet scientists, Drs Rychkov and Sheremet'eva, using serological and morphological data mostly available in the Russian scientific literature, have presented a theoretical discussion of the genetics of the circumpolar Eurasian populations. Pointing out the equilibrium between population numbers and ecological resources, they have calculated the breeding population size to be 253 $\pm$ 20 persons in Eurasia and 267 $\pm$ 31 in the Arctic of the western hemisphere. These small populations are affected by a process termed 'random genetic drift', a chance loss or fixation of genes, or, put in another way, the result of the accumulation of sampling fluctuations over generations. But in Eurasia, the authors state that gene migration has over-come 'drift'. And by using an 'island' isolate population model, followed by a linear 'stepping-stone' model of population structure, they have calculated a gene migration coefficient. From the latter model, which reveals that a gene can permeate the populations from the Kola Peninsula to Chukhotka in some 70 generations, or approximately 1750 years, they consider all of the circumpolar populations as a 'totality'. They then calculated gene frequencies for the genetic markers used, weighted according to population size, which they claim characterize the entire area. These frequencies, however, show a disruption of Hardy–Weinberg equilibrium with a marked deficiency of heterozygotes in the ABO and MN system. In using weighted gene frequencies for A, B, N, and P_2 it was found that the circumpolar populations in the USSR were more similar to the Lapps and Eskimos outside Eurasia than to the southern Eurasian populations. This, they state, points out the strong commonalities or genetic specificness of the circumpolar populations of Europe, Asia and America.

Finally, in looking at some loci, viz: ABO, P, transferrin T_f, and group component GC, deviations were found from expected frequencies. This indi-cated some selective pressure at these loci, according to the authors. In the Rhesus system this was true for the RH negative gene, Rh-, (d) and in the Lewis system.

In the ABO frequencies this population is very similar to the 'classical' Mongoloids (Cavalli-Sforza & Bodmer, 1971). Unfortunately, varieties of A (A_1 and A_2) were not determined, for A_2 is supposedly a non-Mongoloid gene.

Readers should refer to Cavalli-Sforza & Bodmer (1971) and the references to chapter 3 for a clarification of the Soviet approach.

In chapter 4, Drs A. Erikkson, W. Lehmann and N. Simpson have ex-amined genetic polymorphs in human blood to reveal genetic affinities and biological distances between and among Eskimos, Lapps, Ainu and some Siberian populations. According to E. B. Ford's (1945) original definition, a genetic polymorphism relates to the occurrence in the same population of two or more alleles at a gene locus, each with an appreciable frequency maintained by a mechanism other than chance. This assumes a selective advantage, or an adaptive advantage of the heterozygote. Since thus far in man the heterozygote conferring resistance against malaria and the Rh in-compatability factor are the only proven examples of selective advantage, the authors of this chapter were also interested to see if a case could be made for adaptive alleles in the circumpolar populations. However, in small Arctic populations, the authors warn that the variation could be due to 'founder effect', 'genetic drift' or 'population mixing'. They then cautiously speculate that the histocompatibility genes on the white blood cells (HL– A^{12}, A^{13} and A^{21}), which are completely absent in circumpolar populations, and the allele Inv^1, with a higher frequency among circumpolar populations than other populations, may have some adaptive significance. Additionally, the ABO and MN systems and their gene frequencies suggest clines around the Pole.

Genetic distance calculations were confounded by admixture with considerable gene flow into the aboriginal populations so that the 'distance networks' represented the present genetic resemblances, according to the authors, rather than a phylogenetic descent tree.

In this chapter are also described the frequencies of other genetic traits among circumpolar populations; such as: PTC, ear wax dimorphism, INH inactivation, pigmentation, middle phalangeal hair, dermatoglyphics, pathological disorders of known genetic inheritance and aspects of the immunogenetic systems. In the latter systems, the authors conclude, are to be found evidences of adaptation.

Craniofacial studies

Dr A. A. Dahlberg, in chapter 5, has presented the findings of studies of the dentition and cranial and facial development among the Ainu, Eskimo, Lapps and populations of the Soviet Union. He stated that there is no evidence in these data of adaptive significance to be found in the morphology of the face or teeth of any of these peoples. The Eskimo dental morphology was found to be quite similar to that of American Indians and Mongoloids but different from that in Caucasians and Negroes. Lapps were found to be quite different from both Caucasians and Mongoloids with a highly individual dental complex. From a public health point of view, the dental health of Arctic populations is very poor with the high caries incidence related to the introduction of refined sugars in recent decades.

The Arctic eye

In chapter 6, Dr Henrik Forsius has summarized studies undertaken on the eyes of Eskimos and Lapps. Considering that the eye is almost continuously exposed to low temperatures and blowing snow in winter and high levels of ultraviolet radiation in the spring, there was very little evidence for adaptation of the eye, or the eye region itself, to these conditions. Although corneal diameters were some 5 to 10% smaller in Eskimos than in Caucasians, there were no differences in the size of the palpebral fissure, the opening between the top and bottom lids. The anterior chamber of the eye, which is bounded anatomically in the front by the cornea and a small part of the sclera and in the back by the iris, and which contains the aqueous fluid, although theoretically it should be deep as a low temperature protecting device, was found to be extremely shallow in Greenlandic, Canadian and Alaskan Eskimos. This anatomical peculiarity predisposes these populations to glaucoma in old age. Glaucoma, or a 'rock hard' eye, in this instance is due to the inability of the aqueous, generated at the ciliary body, to flow out through the *sinus venosus sclerae*, or the Canal of Schlemm, at the corneoscleral junction, thereby increasing the intraocular pressure. Additionally, there has been a tremendous increase in near-sightedness in children and adolescents in Alaska and Canada due, apparently to an increase in the use of the eye for 'near work', as in school. Nutritional changes may also be involved.

The Eskimo eye also suffered 'climatic insults'. Acute snow blindness, or *keratoconjunctivitis*, an inflammation caused by the reflection of ultraviolet rays from the snow surface in spring, was as common in Eskimos as others. Other corneal tissue disorders were also prevalent. Corneal scarring due to an allergic response to the tuberculosis bacterium and called *phlyctenular keratoconjunctivitis*, or PKC, a blister on the eyeball, was also common. In Bethel, Alaska, for example, it was seen in 1587 persons among a sample of 4087 Eskimos.

Additionally, color blindness was found to be almost non-existent in a number of Arctic populations and where found was probably due to racial admixture.

Anthropometry

In chapter 7, Drs Thord Lewin and Paul Jamison, as sub-editors, have reviewed the published anthropometric measurements for stature and weight of Eskimos and Lapps, and some unpublished measurements of Icelanders at Husavik, to discuss adult morphology and growth and development. These populations, except Icelanders living in a community settled by Gardar the Swede in the early 800s, have all experienced considerable gene flow from outside their ethnic enclaves. Nevertheless, Lappish and Eskimo children demonstrated a slower growth rate, although they are the same with respect to both weight and stature, and they lag behind Iceland children in these parameters. All of the Eskimo and Lappish populations demonstrated a secular growth trend, that is, children are now taller than their parents, due presumably to dietary changes. Obesity is now seen among some Eskimos, but it is much less at Igloolik where the traditional life-style has not been changed as much as elsewhere in the Arctic.

The human biology of circumpolar populations

Nutrition

In chapter 8, Dr H. H. Draper has reviewed the results of the nutritional studies conducted on Eskimos, Lapps and the Ainu. The Ainu were found to be similar to low-income Japanese living in the same area. Nomad Lapps were similar to settled Lapps in dietary intakes. Norwegian Lapps, however, consumed more protein than the settled Norwegian population in the same geographical area.

The greatest effects of recent dietary changes were observed among the Eskimos. August Krogh and his wife Marie (1913) investigated the diet of Greenlanders at Ammassalik in 1908 and reported them to be 'the most carnivorous people on the face of the earth'. At that time protein consumption averaged 280 g/day. Nutritionists consider 40 to 50 g of protein per day more than adequate. The diet of Eskimos has always fascinated western scientists because they seemed to survive in the far north mainly on protein and with no obvious source of many of the essential vitamins. Their major nutritional problem was to generate sufficient glucose in the absence of carbohydrate to operate, among other things, the nervous system. Elsewhere, Draper (1978) stated that glucose homeostasis was maintained primarily by synthesis from dietary amino acids. Therefore extra protein was required to furnish the amino acids required for glucose synthesis beyond those normally required for protein synthesis. Additionally, the polyunsaturated fats in Eskimo foods that were derived from seal and whale oils contributed to their low serum cholesterol levels and the concomitant low prevalence of heart disease.

Presently, Eskimos are consuming a diet containing some 30 to 40% carbohydrates derived from purchased, processed foods. The most striking impact of refined sugars has been to increase dental disease. Carbohydrates have also had an impact on growth and are probably implicated in the secular growth trend, in that children are now taller than their parents. Although it could be claimed that the introduction of western foods has led to an assured and adequate caloric intake, has led to better growth rates, an earlier onset of menarche and a greater survival of new-borns, these effects are counterbalanced by the adverse health effects. A common nutritional disorder in Canadian and Alaskan Eskimos now is a moderate iron deficiency anemia.

Heredity disorders – inborn errors of metabolism – have been found in Eskimos and Lapps which are causing difficulties in adapting to new diets. These metabolic defects are traceable to the absence of action of a single specific digestive enzyme, which among the Lapps, at least for lactase, has been shown to be caused by a single autosomal recessive gene. Adult-type lactose malabsorption, caused by a very low lactase activity in the jejunum which results in a marked reduction of the hydrolysis of lactose to glucose and galactose, was seen in Alaska, Greenland and among Finnish Lapps. Unhydrolyzed lactose is not absorbed and passes through the intestines un-

364

changed resulting in, among other symptoms, abdominal cramps and fermentative diarrhea. Evidence for primary deficiency of sucrase, the enzyme that splits polysaccharides into monosaccharides which are then absorbed, was also found in Greenland and Alaska. Dr B. Boettcher (personal communication), who examined electrophoretic patterns of serum and salivary amylase isoenzymes in Wainwright Eskimos, found the patterns different from that in Caucasians. This enzyme, derived from the parotid gland and the pancreas, is involved in starch digestion.

It thus seems that Eskimos, at least, had adapted well to their earlier diets, having had strong, sound teeth, low obesity and high lean-body mass, and that recent dietary changes have had great impacts on their health. Continued research is warranted in this area, especially, as pointed out by Draper, in the metabolism of bone.

Physiology

In chapter 9, Dr Shinji Itoh has reviewed the studies of physiological function in circumpolar populations. In response to a standardized, whole body, cold exposure, he concluded that there were no differences between the responses of Eskimo and white subjects. Although the evidence for a better peripheral circulation in Arctic peoples is inconclusive, the evidence shows a difference in the cold induced vasodilatation response in Eskimos as compared to others. They uniformly show an earlier onset of cold vasodilatation and uniformly warmer fingers. Since a major problem of well-clad Eskimos is to lose heat when exercising, it is not surprising to find that they have more active sweat glands distributed extensively around the face, which is exposed to the environment, than elsewhere. Studies of circadian rhythmicity in physiological functions showed rhythms synchronized to social factors rather than being entrained to photic stimuli, as one would anticipate.

Work physiology

In chapter 10, Dr Roy Shephard has reviewed the studies of physical activity and the measurements of physical activity in the circumpolar populations. Except at Igloolik, where still active hunters show a superior working capacity, few of the populations studied showed any marked physiological adaptations in this parameter. This, Dr Shephard concluded, is due to the reduction in physical exercise now required for living in the Arctic. Concomitantly, Igloolik hunters are leaner than their kinsmen in Alaska.

Behavior

In chapter 11, Dr Harriet Forsius has reviewed the studies conducted on the

365

behavioral aspects of circumpolar populations. Cognitive abilities were examined in a number of these populations. Cognition is here defined as 'the alert qualities of the mind in contact with the environment and able to think and reason' (*Stedman's Medical Dictionary*, 1966). Carol Feldman, in following J. Piaget's scheme for the documentation of the growth of the child's organization of reality and concepts of space, causality and time concluded that the cognitive development of Alaskan Eskimo children was organized in an invariant sequence of stages as among children elsewhere. Evidence showed that the cognitive abilities of the northern peoples examined were a functional adaptation to the demands of their environment, in which perceptual motor skills were useful.

Unfortunately, due to cultural change and 'acculturative stress', earlier adaptive behaviors seen in these kin-bound societies such as non-aggressiveness, non-competitiveness, sharing, etc. seem not always useful in the 'new' societies. Alcoholism, leading to family disintegration, accidents, suicidal behaviors, violence and differential arrest rates for 'natives' versus 'non-natives', has increased markedly in the Arctic in recent years. Alcohol usage of a destructive nature probably implies individual emotional difficulties and is also a social indicator of the health of the society. It would seem that social and behavioral disorders have replaced many of the infectious disorders that were prevalent in the past. Public health authorities are aware of these facts and are presently concentrating much of their energies and resources on the mental health status of Arctic populations.

Conclusions

This volume presents a series of what might be considered base-line studies on the demography, physical and mental health, physical fitness, growth and the genetic constitution of a number of Arctic human populations during the period 1967–74 under the auspices of the International Biological Program. The studies reveal the changes that have occurred in these populations in response to contact with other societies and a changed diet. In view of the increasing economic importance of Arctic resources – petroleum, minerals, hydroelectric power, etc. – and the rapidity of northern development in the USSR, Alaska, Arctic Canada, Greenland and Fenno-Scandia one can assume that these changes will continue. It is the opinion of the Editor that the impacts of these developments on the health and well-being of the circumpolar populations should be monitored on an international scale to ensure their continual survival as Arctic Peoples. In the future, they may have much to offer as models for our own survival in the Arctic regions of an increasingly over-populated world.

References

Bligh, J. (1973). *Temperature regulation in mammals and other vertebrates.* Amsterdam: North Holland Press.

Cavalli-Sforza, L. L. & Bodmer, W. F. (1971). *The genetics of human populations.* San Francisco: W. H. Freeman & Co.

Dubos, Rene (1965). *Man adapting.* New Haven & London: Yale University Press.

Krogh, A. & Krogh, M. (1913). *A study of the diet and metabolism of Eskimos.* Copenhagen: Bianco Lund.

Milan, Frederick A. (1963). An experimental study of thermoregulation in two arctic races. PhD thesis, University Microfilms, Inc., Ann Arbor, Michigan.

Prosser, C. Ladd (1964). Perspectives of adaptation: Theoretical Aspects. In *Handbook of Physiology,* ed. D. B. Dill, E. F. Adolph & G. C. Wilber. Washington, DC: American Physiological Society.

Stedman's Medical Dictionary (1966). Baltimore, Md: Williams & Wilkins Company, Waverly Press, Inc.

Weiner, J. S. (1966). Major problems in human population biology. In *The Biology of Human Adaptability,* ed. P. T. Baker & J. S. Weiner. Oxford: Clarendon Press.

Index

ABO blood groups, genetic markers
in circumpolar peoples, 42–3, 55, and in
those south of Arctic Circle, 61
four clines of, round N Pole, 118–20
gene frequencies of: 64, 67, 91, 92, 94–5;
in 16 communities of Greenland
Eskimos, 102–3; variances of, 70, 71,
72, 75
Hardy-Weinberg equilibrium for, 58, 59
in saliva, 182
acculturation of Eskimos
measure of extent of, 354
rapidity of, 336
stress of, 344, 345, 349, 350–4
and working capacity, 330
acetyl transferase, in inactivation of isoniazid, 134
acid phosphatase of red cells, genetic
marker, 84, 100
gene frequencies of, 91, 92, 93, 96–7
negatively correlated with glutathione
reductase? 93, 100
activity patterns, and working capacity,
326–30
adenosine deaminase of red cells, genetic
marker
gene frequencies of, 96–7, 108–9
genetic distances measured with, 124–7
adenylate kinase, genetic marker
gene frequencies of, 91, 96–7, 100
genetic distances measured with, 124–7
age (adults)
and body measurements, 244–7; stature,
311–12
and maximum oxygen uptake, 315, 326
and muscular strength, 321, 326
Ainu
anthropometry: skinfold thickness, 264,
265, 323; stature, 310, 311
craniofacial studies, 170–1
diet, 264–6
eye color, 136
gene frequencies, 81, 82–3, 94, 96, 98, 108,
110, 113, 118, 129; different from both
Eskimos and Lapps, 112–13; same as
in Eskimos, different from Lapps, 91–3,
100–1, 106–9, 112; same as in Lapps,
different from Eskimos, 115–18

genetic distances from Eskimos and
Lapps, 124–8
Japanese admixture, 82, 83
Mongoloid traits predominate over Caucasian, 170
physiology: basic metabolic rate, 286, and
thyroid activity, 287; cerumen dimorphism, 132–4; free fatty acids in
plasma, 275, 297–8, and effects on, of
glucose, 279, and of norepinephrine,
275, 295; isoniazid inactivation, 135;
ketone bodies in plasma, 275, 295, 298,
and effect on, of norepinephrine, 276,
277, 295, 298; number of sweat glands,
296, and activity, 297; PTC tasting,
131; responses to cold, 292, 293, 294;
sensitivity to norepinephrine, 295
working capacity: development, 325;
efficiency of effort, 317; heart rate and
oxygen uptake, 314; maximum oxygen
uptake, 313, 315, 320
Alaska, Eskimos of, see Eskimos, Alaska
alcoholism, 345, 346, 349, 366
Aleuts
gene migration, 66
genetic markers, 42
language, 38, 39
population, 14; effective reproductive
size of, 48, 49; reproductive age limits
in, 46
Altai highland people, gene migration, 66
American Indians
dental traits, 171
genetic distance between Eurasian Arctic
peoples and, 38
physiology: cerumen dimorphism, 133;
isoniazid inactivation, 135
Amur and Sakhalin peoples, gene migration,
66
amylase, salivary: electrophoretic patterns
of isoenzymes of, in Eskimos and
Caucasians, 365
anaemia, iron-deficiency, 267–8, 364
anthropometry, 363
methods, 217–18
peoples studied, 213–17
results: adults, 242–53; children, 218–
242

369

Index

antibodies
  to different pathogens, 155–6, 158
  γ-globulins and, 157
antigens, histocompatibility, *see* histo-compatibility antigens
antitrypsin alleles (alpha₁), genetic markers: gene frequencies for, 98–9, 112, 114, 121
anxiety neurosis, 346
archaeology, of circumpolar peoples, 10, 38, 44, 77
Arctic
  climate, 7–9; and working capacity, 305–6, 330–3
  geography, 6
  man in, 9–11
  migrations into, 33, 39–40
Arctic Slope Regional Corporation, Alaska, 20
area developed by an elementary population, in Eurasian circumpolar zone, 46
Australoids, possible affinity of Ainu with, 82, 170, 171
auto-antibodies, rare in Lapps, 156

Barrow, N Alaska: climatic data, 8
basal metabolic rate, 285, 286, 316
beards, 171, 195
birth control, among Eskimos, 19, 29–30, 32
birth rates of Eskimos
  Alaska, 16, 21, 22, 23; bottle-feeding of infants and, 17
  Greenland, 27
births: ages of parents at first and last of a family, Eurasian Arctic, 46, 47
blood, peripheral circulation of
  cheek, 289–91
  hand, 288–9
blood groups, as genetic markers, 38, 42–3, 54–6, 58
  see also individual blood groups
blood pressure, Lapps, 150
blood volume
  acclimatization to cold and, 333
  in hunters, 321, 331
bone of forearm: mineral loss from, with age, 282
Buryat people, gene migration, 66

C3 component of serum, genetic marker, 96–7, 123
calcium
  content of, in native and commercial foods, 262, 264
  serum levels of, and transmission of nerve impulses, 348, 349

Canada, Eskimos of, *see* Eskimos, Canada
carbohydrate
  in commercial foods, 262
  percentage of, in diet: Danes, 259; Eskimos, 259, 264; Lapps, 257, 258
cattle herders, cultural-economic type, 40, 41
Caucasian genes, in circumpolar peoples, 91, 93, 100–1, 109, 114, 129
  in Eskimos, Alaska and Canada, 29
Caucasoid and Mongoloid groups, in circumpolar peoples, 169–70
cerumen (ear-wax), gene frequencies for dry and wet types of, 132–4
cholesterol of plasma
  in Alaskan Eskimos, 273, 274, 276, in Finns, 274, and in Lapps, 274, 276
  relation between vitamin E and, 269, 270
cholinesterase of serum, genetic marker
  one gene for, not present in Ainu and Eskimos (except at one place in Alaska), 84, 91, 100
  second gene for, in all original circumpolar peoples? 121, 122
chromosomal aberrations, auto-antibodies and, 156
Chukchi, 'coastal' and 'reindeer', 40
  assimilation of Siberian Eskimos into, 28
  ethnic origins, 50
  gene frequencies, 118, 120
  gene migration, 52, 66
  genetic markers, 42, 43, 89
  language, 38, 39
  move into Yukaghir territory, 40
  population structure (coastal), 47
  reproductive age limits, 46; effective reproductive size of population, 48, 49
cigarette-smoking, by Canadian Eskimos, 310, 318
circadian rhythms, synchronized to social factors rather than to photic stimuli, 299–300, 365
cognitive abilities, 366
  developmental sequence of, 342
  sex differences in, 343–4
  those visually based, well developed in hunting peoples, 340, 342, 343
cold
  acclimatization to, and blood volume and body water content, 333
  pressor response to, 293–4
  thermal and metabolic responses to, 287–8
  vasodilatation induced by, 291–3, 365
color blindness, almost absent in circumpolar peoples, 205–6
  where present, due to racial admixture? 363

Index

congenital dislocation of hip, 152, 153
corneal thickness, 197
dental traits, 183–5, 186
dermatoglyphics, 145, 147–50
eye colour, 137, 138, 200, 201
gene frequencies, 95, 97, 99, 101, 108, 113, 117, 118
genetic distances, 124–8
hair colour, 135, 136
inbreeding, 31
physiology: cerumen dimorphism, 133; intra-ocular pressure, 206, 207; isoniazid inactivation, 135; PTC tasting, 131
stature and weight, 310; secular trend in stature, 311, 312
Lapps: Finland: Skolt: Nellim (from Pasvik), 30, 89, 92
stature, 312
working capacity: activity patterns, 327; development, 325; maximum oxygen uptake, 313, 315
Lapps: Finland: Skolt: Svettijärvi (from Suengel), 30, 89, 92
diet, 257, 258
stature, 312
Lapps: Finland: Utsjoki, 89, 92
cerumen dimorphism, 133
diet, 257, 258
gene frequencies, 95, 97, 99, 107, 108, 110–11, 113, 117, 118
Lapps: Kola peninsula (USSR), 'fisher' and 'mountain', fishermen and reindeer hunters, later reindeer breeders, 40
dental traits, 186, 187
dermatoglyphics, 145
ethnic origins, 50
gene migration, 52, 66
genetic markers, 42, 54
reproductive age limits, 46; effective reproductive size of population, 48, 49
Lapps: nomadic (Kautokeino)
skinfold thicknesses, 322, 323
stature, 310; secular trend in, 311
tolerance to cold, 287
weight, 310
working capacity: development, 325; maximum oxygen uptake, 313, (age and) 315
Lapps: Northern (USSR): dental traits, 186, 187
Lapps: Norway, 87, 92
activity patterns, 327
anthropometry, 216–17; hip width, 235, 236, and stature, 234, 252; shoulder width, 232, 233, and stature, 234, 251; sitting height, 229, and stature, 231, 250; skinfold thicknesses, 240, 241;

stature, 220, 222, and age, 221; upper arm circumference, 238, 239; weight, 224, 226, (and age) 225, (and stature) 228, 249
congenital dislocation of hip, 152
dental traits, 183
diet, 258–9
gene frequencies, 95, 97, 99, 107, 108, 110–11, 113, 117, 118
PTC tasting, 131
skin colour, 138
Lapps: Sweden, 87–8, 92
gene frequencies, 95, 97, 99, 107, 108, 115
hair color, 136
PTC tasting, 131
Lewis blood groups, genetic markers
in circumpolar peoples, 42, 43, 56, 65, and those south of Arctic Circle, 62
frequency variances of genes for, 70, 71, 73, 76
life expectancy of Eskimos: in Alaska, 17, and in Greenland, 26–7
lipid metabolism: in Ainu, 276, in Eskimos, 270–4, 276, and in Lapps, 276–7
lipoprotein of serum, 96–7, 123
lung
forced expiratory volume (FEV), 307, 309, 318; ratio of, to FVC, 331; tuberculosis and, 334
residual volume (RV), 318, 319
total capacity (TLC), 318, 319
vital capacity (FVC), 307, 318; ratio of FEV to, 331; tuberculosis and, 334
Lutheran blood group, genetic marker, 94–5, 116

Mati people, 'meadow' and 'mountain': dental traits, 187
mental abilities, 340–3
cultural background and assessment of, 339–40
sex differences in, 343–4
mental disorders, epidemiology of, 345–7
mental health, 344–5; of children, 349–50
mental health teams, Alaska, 349–50
Metis (French-Canadian-Indian hybrids), compared with Fort Chimo Eskimos for skinfold thicknesses, and weight at different ages, 259, 260
migration
into Arctic, 33, 39–40
to central communities: Canada, 25; Greenland, 27
to coastal settlements, Alaska, 20
to urban areas, Alaska, 31, 32
of women away from small Eskimo communities, 22, 28, 32, 360

Index

MN blood groups, genetic markers, 56, 58
in circumpolar peoples, 42, 43, 102–3
four clines of, round N Pole, 118–20
gene frequencies of, 91, 120; variance of, 51
genetic distances measured with, 127
Hardy-Weinberg equilibrium for, 58, 59
MNS blood groups, genetic markers, 54, 55, 58
in circumpolar peoples, 42, 64, and those south of Arctic Circle, 61
gene frequencies of, 64, 94–5, 119, 120; variance of, 70, 71, 72, 73, 75
Mongolian spot (blue), in sacral region of new-born children: common in Asian peoples, rare in Lapps, 139
Mongoloid and Caucasoid groups, in circumpolar peoples, 169–70
muscle power (grip strength, leg strength, anaerobic power), 321–2
in boys and girls, 324, 325, 326
extent of acculturation and, 330
in women, number of children and, 334, 335
myopia; rare in Greenland Eskimos and NE Siberians, increasing among other circumpolar people, with schooling and 'near work', 199, 263

Nenets, NW Siberia, 'forest' and 'tundra', 40
dental traits, 186, 187, 188
ethnic origins, 50
gene frequencies, 118, 120
gene migration, 52, 66
genetic markers, 42, 43
language, 37, 39
Nganasans, NW Siberia, reindeer breeders, 40
ethnic origins, 50
gene frequencies, 118, 120
gene migration, 52, 66
genetic markers, 43
language, 38, 39
move to Yukaghir territory, 40
norepinephrine
effects in Ainu and Japanese: on plasma free fatty acids, 275, 295, and ketone bodies, 276, 277, 295, 298
mediator of non-shivering thermogenesis? 295
sensitivity to, in Ainu, 295; associated with pressor response to cold, 294
Norway, Lapps of, see Lapps: Norway
Norwegians: gene frequencies, 92, 108, 113
nutrition, see diet

obesity, increasing prevalence of, 254, 363
oestrogens, especially relaxin, and congenital dislocation of hip, 153–4
oxygen
arterio-venous difference in, Igloolik Eskimos, 318
maximum uptake of (measure of working capacity), 312–16, (boys and girls) 325, 326; acculturation and, 330; age and, 315, 326; coefficients of variation for, 333; life style and, 329, 330, 337; season and, 329, 333

P blood group system, genetic marker, 56
in circumpolar peoples, 102, and those south of Arctic circle, 62
gene frequencies of, 65, 94–5, 106, 107; variance of, 70, 71, 73, 76
genetic distances measured with, 127
Palaeo-Asiatic language group, 38
palpebral fissure, 362–3
narrow in Mongoloid peoples, 195
width decreases with age, 197
phenyl thiocarbamide (PTC): ability to taste, as genetic marker, 38, 42–3
gene frequencies of, 65, 67, 130–2; in populations south of Arctic Circle, 62; variances of, 70, 71, 73, 76
phosphoglucomutase, genetic marker
gene frequencies of, 96–7, 101, 106, 107
genetic distances measured with, 124–127
6-phosphogluconate dehydrogenase, genetic marker, 84, 95–7, 115, 116
phosphorus: higher content of, in native than in commercial foods, 262
pinguecula (yellowish spot on cornea), 202, 204
population density
Eurasian circumpolar region, 44
Lapps, 87
population structure
of Eskimos: Alaska, 21, 22; Canada, 23–4, 26
of Eurasian circumpolar peoples, 47–8
in Soviet Arctic, 28
populations
equilibrium between environment and, 41, 44–5
of Eskimos: in Alaska, at different periods, 14–15, 17, 18–19, 20; in Canada, in different areas, 23, 24–5; in different communities, 28; in different countries, 13–14; in Greenland, 26–7
of Eurasian circumpolar peoples: average (optimum) size of, 44–6, and area

Index

smallpox, introduced to Eskimos in Alaska, 14, 15
snow blindness, 203, 363
snowmobiles, 353
social uninvolvement, of fishermen, hunters, and herdsmen, 353
startle reaction, 354-6
stature
  adults, 242-8
  age and, 221, 311-12
  children, 218-23
  hip width and, 234, 252
  measurement of, 217
  secular increase in, 243, 248, 249, 254, 311-12, 363, 364
  shoulder width and, 234, 251
  sitting height and, 231, 250
  weight and, 227-8, 249, 310
strabismus, rarer in circumpolar than in other peoples, 206
sucrose
  consumption of, and dental caries, 364
  intolerance to, from lack of sucrase, 280-282, 365
  taste sensitivity for, 172
suicidal behavior, 347-8, 366
superoxide dismutase, cytoplasmic, as genetic marker: gene frequencies of, 91, 96-7, 101
sweat glands
  numbers of, in different races, 296
  number and activity of, greater on face than on trunk in Eskimos, 296, 331, 336, 365
  patterns of response of, to chemical stimulus, 296-7
Sweden, Lapps of, see Lapps: Sweden
Swedes
  gene frequencies, 92
  isoniazid inactivation, 135

taste sensitivity, see phenyl thiocarbamate, and under sucrose
thiamine, status of Eskimo children with respect to, 267-8
Thule Eskimo culture, 10, 85
thyroid activity, 285-7
timnodonic acid (eicosapentaenoic), in plasma fatty acids of Greenland Eskimos, 270, 271
transferrin alleles, genetic markers, 57, 58
  in circumpolar peoples, and those south of Arctic Circle, 62
  gene frequencies of, 65, 67, 98-9, 112, 114-15, 121; variances, 70, 71, 73, 76
tuberculosis
  corneal scarring associated with, 208, 363

in Eskimos, 14, 15, 16; control of, 14, 16, 17, 19, 25, 360; death rate from, 16, 25
  in populations tested for working capacity, 306, 307, 334, 336
  and stature, 312
Tungus-Evenks, central Siberia: population structure, 47, 48
Tungusic-Manchu languages, 38, 39
Turkic languages, 38, 39
twins, per 1000 births, 29

Ugro-Finnic peoples, migrate to circumpolar zone, 39-40
Upernavik, west Greenland: climatic data, 8
upper arm circumference
  adults, 244, 252
  children, 237, 238-9
  method of measuring, 217-18
Uralic-Altai linguistic community, 38
USSR
  Eskimo and Russian populations of Arctic area of, 28
  genetic markers of circumpolar peoples included in (Behring Strait to Kola peninsula), 54-8

vasodilatation reaction, to cold, 291-3
venereal diseases, introduced to Eskimos in Alaska, 14, 15
vitamin A
  high intake of, by Lapps, 258
  higher content of, in native than in commercial foods, 262
  plasma levels of, Alaskan Eskimos, 268
vitamin B components: higher content of, in native than in commercial foods, 262
  see also riboflavin, thiamine
vitamin C
  lower content of, in native than in commercial foods, 262, 264
  plasma levels of, Alaskan Eskimos, 268
vitamin D, 154
vitamin E, in plasma of Alaskan Eskimos, 268
  relation of cholesterol to, 269, 270

Wainwright, Alaska, 20
water
  acclimatization to cold, and body content of, 333
  body loss of, in dry air: and working capacity, 300, 331, 332; and blood volume, 321
  percentage of body weight in, 322
weight, 217
  and age, 259, 260